ANALYTIC GEOMETRY AND ALGEBRA

1. **Distance** between two points (x_1, y_1) and (x_2, y_2) is given by
$$d = \sqrt{(x_2 - x_1)^2 + (y_2 - y_1)^2}.$$

2. **Slope** of line containing (x_1, y_1) and (x_2, y_2):
$$m = \frac{y_2 - y_1}{x_2 - x_1}.$$

3. **Midpoint** of segment with (x_1, y_1) and (x_2, y_2) as endpoints:
$$\left(\frac{x_1 + x_2}{2}, \frac{y_1 + y_2}{2}\right).$$

4. **Equation of line** containing (x_1, y_1) and (x_2, y_2):
$$y - y_1 = m(x - x_1) \quad \textit{point-slope form.}$$

5. **Slope** of line given by $Ax + By = C$ is $-A/B$ provided $B \neq 0$.

6. The **solution** of the linear equation $ax + b = 0$ is $x = -b/a$.

7. The **solutions** of the quadratic equation $ax^2 + bx + c = 0$ are
$$x = \frac{-b \pm \sqrt{b^2 - 4ac}}{2a}.$$

   If $b^2 - 4ac < 0$, there are no real solutions.

8. The **linear system**
$$Ax + By = E$$
$$Cx + Dy = F$$
has a unique solution provided $AD - BC \neq 0$. It is given by
$$x = \frac{FD - EB}{AD - BC} \quad \text{and} \quad y = \frac{AF - CE}{AD - BC}.$$

9. **Inequalities Properties:**
   (i) If $a < b$, then $a + c < b + c$.
   (ii) If $a < b$ and $c < d$, then $a + c < b + d$.
   (iii) If $a < b$ and $c > 0$, then $ac < bc$.
   (iv) If $a < b$ and $c < 0$, then $ac > bc$.

CALCULUS FOR THE MANAGERIAL, SOCIAL, AND LIFE SCIENCES

CALCULUS FOR THE MANAGERIAL, SOCIAL, AND LIFE SCIENCES

Bevan K Youse
Emory University

with the assistance of
F. Lane Hardy
DeKalb Community College

West Publishing Company

St. Paul   New York   Los Angeles   San Francisco

Copyediting: Beth Watson
Artwork: Bob Gallison
Interior design: Judith Fletcher Getman
Composition: York Graphic Services, Inc.
Production management: York Production Services
Cover art: Based on a design by Primo Angeli

Copyright © 1984 By WEST PUBLISHING CO.
50 West Kellogg Boulevard
P.O. Box 3526
St. Paul, Minnesota 55165

All rights reserved

Printed in the United States of America

Library of Congress Cataloging in Publication Data

Youse, Bevan K
  Calculus for the managerial, social, and life sciences.

  Includes index.
  1. Calculus. I. Title.
QA303.Y73 1984   515   83-21911
ISBN 0-314-77888-8

CONTENTS

PREFACE	ix

1 FUNCTIONS AND LINEAR MODELS — 1

1.1 Introduction	3
1.2 The Real Number Line and Linear Inequalities	4
1.3 The Rectangular Coordinate System and Slope of a Line	12
1.4 Equations of Lines	21
1.5 Linear Systems and Their Applications	28
1.6 Functions and Their Graphs	39
1.7 Applications of Linear Functions	50
1.8 Summary and Review Exercises	64
Biographical Sketch (René Descartes)	69

2 DIFFERENTIAL CALCULUS AND ITS APPLICATIONS — 71

2.1 Introduction	73
2.2 The Limit of a Function and Continuity	74
2.3 The Newton Quotient	82
2.4 The Derivative of a Function	96
2.5 Product, Quotient, and General Power Rules for Differentiation	108
2.6 The Differential and Marginal Analysis	116
2.7 Summary and Review Exercises	128
Biographical Sketch (Isaac Newton)	133

FURTHER APPLICATIONS OF THE DERIVATIVE   **135**

3.1 Introduction   137
3.2 Increasing and Decreasing Functions   138
3.3 Relative Maxima and Relative Minima   150
3.4 More Applied Maximum and Minimum Problems   163
3.5 Related Rates (Optional)   182
3.6 Summary and Review Exercises   187
Biographical Sketch (Gottfried Wilhelm Leibniz)   191

EXPONENTIAL AND LOGARITHMIC FUNCTIONS   **193**

4.1 Introduction   195
4.2 The Chain Rule and Inverse Functions   196
4.3 Exponential and Logarithmic Functions   204
4.4 Differentiation of Exponential and Logarithmic Functions   211
4.5 Exponential Growth and Continuous Interest   217
4.6 Implicit and Logarithmic Differentiation (Optional)   227
4.7 Summary and Review Exercises   236
Biographical Sketch (Carl Friedrich Gauss)   239

THE ANTIDERIVATIVE AND ITS APPLICATIONS   **241**

5.1 Introduction   243
5.2 The Antiderivative of a Function   244
5.3 Antiderivatives Involving the Exponential and Logarithmic Functions   250
5.4 Applications of the Antiderivative   254
5.5 Integration by Substitution   261
5.6 Integration by Parts   264
5.7 Differential Equations (Optional)   270
5.8 Summary and Review Exercises   278
Biographical Sketch (Sonya Kovalevsky)   281

6 THE DEFINITE INTEGRAL AND ITS APPLICATIONS — 283

- 6.1 Introduction — 285
- 6.2 The Definite Integral — 286
- 6.3 More on Area — 301
- 6.4 Other Applications of the Definite Integral (Optional Topics) — 309
  - A. Average Value and Lorentz Curve — 309
  - B. Consumer's Surplus and Producer's Surplus — 314
  - C. Probability Density Function — 319
  - D. Volume by Disk Method — 321
- 6.5 Summary and Review Exercises — 324
- Biographical Sketch (Georg F. B. Riemann) — 327

7 MULTIVARIABLE CALCULUS AND ITS APPLICATIONS — 329

- 7.1 Introduction — 331
- 7.2 Rectangular Coordinate System in Three Dimensions — 332
- 7.3 Functions of Two Variables — 339
- 7.4 Partial Derivatives — 347
- 7.5 Tangent Planes and the Total Differential — 354
- 7.6 Applied Maxima and Minima Problems — 360
- 7.7 Maxima and Minima with Constraints — 371
- 7.8 Method of Least Squares (Optional) — 380
- 7.9 Summary and Review Exercises — 387
- Biographical Sketch (Joseph Louis Lagrange) — 392

8 SEQUENCES AND MATHEMATICS OF FINANCE — 393

- 8.1 Introduction — 395
- 8.2 Arithmetic Sequences — 396
- 8.3 Geometric Sequences — 405
- 8.4 Interest: Simple and Compound — 412
- 8.5 Continuous Interest — 421
- 8.6 Interest on Periodic Savings — 426
- 8.7 Installment Payments and Retirement Income — 433
- 8.8 Summary and Review Exercises — 441
- Biographical Sketch (John von Neumann) — 445

APPENDIX A TABLES A1

Table 1 Compound Interest $(1 + i)^n$ A4
Table 2 Present Value of a Dollar $(1 + i)^{-n}$ A6
Table 3 Exponential Functions e^x A8
Table 4 Four Place Logarithms (Base 10) A11

APPENDIX B ALGEBRA REVIEW B1

B.1 Exponents B3
B.2 Polynomials, Algebraic Expressions, and Factoring B8
B.3 Quadratic Equations B12

APPENDIX C ANSWERS TO ODD-NUMBERED EXERCISES AND SOLUTIONS TO SELECTED EXERCISES C1

INDEX I1

PREFACE

This text is designed for a one-semester (one- or two-quarter) course in calculus for the managerial, social, and life sciences. It discusses the basic concepts and techniques of calculus which are important to these areas. Furthermore, in order to give students a better understanding of not only the scope of the subject but also its usefulness in their field of interest, it discusses many different types of applications.

One major goal of the text is to present a correct, though nonrigorous, development of the basic ideas and techniques of differential and integral calculus. Formal proofs for most theorems are omitted with the conviction that it is easier for the instructor to include proofs of theorems when appropriate than it is for the instructor (or student) to omit proofs which are inappropriate for the goals of the course or the background of the student.

In addition to the inclusion of many examples and a large variety of applications, another important feature of the text is the inclusion of *Practice Problems* at points in the text where new ideas or techniques are introduced. The *Practice Problems*, along with their answers which follow immediately, should not only help but also encourage the student to read the text. Each chapter has a *Summary* and a set of *Review Exercises* covering all the nonoptional topics discussed in the text. Answers to all odd-numbered exercises in the exercise sets as well as complete solutions to selected exercises are given in Appendix C. Furthermore, the answers to *all* review exercises are included in Appendix C.

Most of the review topics needed to read the text are provided in Chapter 1. More important, Chapter 1 provides the student with an opportunity to practice constructing mathematical models for practical problems using concepts and techniques which are reasonably familiar. (For students who need a further review of algebra, an *Algebra Review* is given in Appendix B.)

In Chapters 2 and 3, the derivative is defined, basic differentiation theorems are developed, and many applications of the derivative are discussed. The chain rule for differentiation, a discussion of inverse functions, the derivatives of the exponential and logarithmic functions, and applications of the exponential functions are given in Chapter 4. In Chapter 5, the antiderivative is defined, basic antidifferentiation theorems are developed, and many applications of the antiderivative are discussed. The definite integral and its applications are presented in Chapter 6. Chapter 7 is devoted to multivariable calculus and its applications. In Chapter 8, arithmetic and geometric sequences are discussed and important mathematics of finance applications are investigated.

There are many persons who deserve special thanks for their contribution to this text. First, I thank Professor F. Lane Hardy for his valuable assistance which continued from the writing of the manuscript through the reading of proofs of pages. I wish to thank Peter Marshall, Editor at West Publishing Company, for his many helpful suggestions and tireless efforts to improve the manuscript at each stage of its development; his devotion to and interest in this project were truly exceptional. I also wish to thank Barbara Fuller, Production Editor, for adroitly and cheerfully handling the myriad of details that such a project requires. My grateful appreciation to Lorraine N. Ruff for a superior job in typing the manuscript. Finally, I acknowledge with sincere gratitude the many helpful suggestions given by the reviewers whose names appear under *Acknowledgments*.

Bevan K Youse
Atlanta, Georgia
December 1983

ACKNOWLEDGMENTS

Patricia Bannantine
Marquette University, Wisconsin

Lowell W. Beineke
Indiana University—Purdue Un. at Fort Wayne

Carole Bernett
William Rainey Harper College, Illinois

Robert Bix
University of Michigan, Flint

PREFACE

William D. Blair
Northern Illinois University

John F. Busovicki
Indiana University of Pennsylvania

Frederick J. Carter
St. Mary's University, Texas

Mary Jane Causey
University of Mississippi

Carl Cuneo
Essex Community College, Maryland

Richard M. Davitt
University of Louisville

Garrett Etgen
University of Houston

David Fonken
Austin Community College

William Roger Fuller
University of Portland

John T. Gresser
Bowling Green State University, Ohio

Herbert Gindler
San Diego State University

Joseph Howard
New Mexico Highlands University

D. Frank Hsu
Fordham University, New York

Bruce A. Jensen
Portland State University

Alonzo F. Johnson
West Virginia University

Robert A. Moreland
Texas Tech University

Rose Novey
Saginaw Valley State College, Mich.

Susan S. Nutter
Prestonburg Community College, Kentucky

Walter Roth
University of North Carolina, Charlotte

Wesley Sanders
Sam Houston State University

Bernard L. Schroeder
University of Wisconsin, Platteville

Linda B. Sherrell
Louisiana State University

Clifford Sloyer
University of Delaware

John Spellman
Southwest Texas State University

Dale Thoe
Purdue University

CHAPTER 1

FUNCTIONS AND LINEAR MODELS

1

1.1 INTRODUCTION
1.2 THE REAL NUMBER LINE AND LINEAR INEQUALITIES
1.3 THE RECTANGULAR COORDINATE SYSTEM AND SLOPE OF A LINE
1.4 EQUATIONS OF LINES
1.5 LINEAR SYSTEMS AND THEIR APPLICATIONS
1.6 FUNCTIONS AND THEIR GRAPHS
1.7 APPLICATIONS OF LINEAR FUNCTIONS
1.8 SUMMARY AND REVIEW EXERCISES

1.1 INTRODUCTION

Calculus, a cornerstone of modern mathematics, is unsurpassed in its beauty (mathematics as an art) and in its usefulness (mathematics as a tool). With basic techniques and concepts from algebra and geometry, you should be prepared to embark on the adventure of discovering calculus. You should also find the new ideas and their applications quite worthy of investigation.

This chapter is devoted to some precalculus concepts which are not only useful in solving applied problems but also essential in the development of calculus. (Additional review topics are discussed in Appendix B.) Another important feature of this chapter is that it provides you with an opportunity to practice constructing models for practical problems. Examples of problems you shall learn to solve in this chapter are:

1. Two trucks must make deliveries over a route totaling 500 miles. Truck A gets 15 miles per gallon and Truck B gets 10 miles per gallon. If the total fuel consumption for the deliveries cannot exceed 40 gallons, what is the minimum number of miles that Truck A must travel? (Example 3, Section 1.2.)

2. A woman estimates that 30% of her gross income will go to taxes. If she wants to save each year 10% of her after-tax income and needs $20,000 a year for living expenses, what should her minimum gross income be? (Example 5, Section 1.2.)

3. A political scientist obtained the following information for a mayoral race in which a Democrat named Smith won the election. The total number of Republicans and Democrats voting was 29,900. An exit poll found that 15% of the Republicans and 80% of the Democrats voted for Smith. If these two groups cast a total of 15,600 votes for Smith, how many Republicans voted? How many Democrats voted? (See Exercise 24, Section 1.5.)

4. The mayor of a city determines that three housing subdivisions can be located from City Hall as follows.

   Subdivision A: 3 miles east and 2 miles north
   Subdivision B: 7 miles east and 10 miles north
   Subdivision C: 11 miles east and 3 miles south

   Suppose a fire station is to be built equidistant from each of the three subdivisions. What should be its location in reference to City Hall? (Exercise 13, Section 1.5.)

5. A manufacturer of television sets concludes from previous introductions of new models that 15,000 sets can be sold with no advertising. Also based on past sales records, it is assumed that for each $1000 spent on advertising an additional 50 sets will be sold. If the initial (fixed) costs to produce the television sets were $4,500,000, if each set costs $500 to manufacture, and if each set sells for $650, what would be the profit (loss) with an advertising budget of $1,000,000? (Example 1, Section 1.7.) ∎

1.2 THE REAL NUMBER LINE AND LINEAR INEQUALITIES

Since calculus has at its foundation the set of real numbers, we begin our development of calculus with a brief review of the real number system and the real coordinate line. In this section, we also define *less than* on the set of real numbers and learn how to solve practical problems that can be described by linear inequalities.

real numbers

The set of **real numbers** consists of the set of *rational* and the set of *irrational* numbers. *Rational numbers* are real numbers that can be expressed as the ratio p/q of two integers p and q where $q \neq 0$; they can also be written as finite decimals or repeating infinite decimals. For example, $\frac{1}{2} = 0.5$, $\frac{2}{3} = 0.666\ldots$, $-\frac{3}{2} = -1.5$, and $6 = \frac{6}{1}$ are rational numbers. The set of *irrational numbers*, real numbers that are not rational, contains such numbers as $\sqrt{2}$, $\sqrt[3]{7}$, $\sqrt[5]{14}$, and π; their decimal representations are infinite and nonrepeating.

The real numbers can be paired with the points on a line in the following way. Choose any point on a line, call it the **origin,** and assign 0 to this point; zero is called the *coordinate* of the origin. Next select another point R as in Figure 1.1 and assign 1 as its coordinate. This determines a *unit segment*. A positive number c is assigned to the point c units to the right of the origin (in the direction of R). If a point is d units to the left of the origin, then $-d$ is assigned to the point.

less than

A real number a is said to be **less than** a real number b, de-

FIG. 1.1. Coordinate line.

1.2 THE REAL NUMBER LINE AND LINEAR INEQUALITIES

noted by $a < b$, if and only if there is a *positive* number x such that $a + x = b$. For example, since $2 + 3 = 5$, $2 < 5$. The *less than* concept is used to define *greater than* in the following way: b is greater than a, denoted by $b > a$, if and only if $a < b$. Furthermore, $a \leq b$ if and only if either $a < b$ or $a = b$, and $a \geq b$ if and only if either $a > b$ or $a = b$. The statements $a < b$, $a > b$, $a \leq b$, and $a \geq b$ are called **inequalities**.

inequalities

If P and Q are two points on the number line with coordinates a and b, respectively, then $a < b$ if and only if P is to the left of Q. (See Figure 1.2.) It should be geometrically obvious that if $a < b$ and $b < c$, then $a < c$; this is called the *transitive property* for inequalities. (Recall that there is a similar equality property.)

FIG. 1.2. $a < b$ and $b < c$ imply $a < c$.

Although operations on inequalities and equalities are similar, there are significant differences. For real numbers a, b, and c, the basic **inequality properties** are as follows.

inequality properties

PROPERTY 1 If $a < b$, then

$$a + c < b + c \quad \text{and} \quad a - c < b - c$$

(The same number can be added or subtracted from each side of an inequality without changing the direction of the inequality.) For example, since $4 < 7$, $4 + 8 < 7 + 8$ and $4 - 8 < 7 - 8$.

PROPERTY 2 If $a < b$ and $c < d$, then

$$a + c < b + d$$

(Inequalities can be added provided the direction of the inequality sign for each is the same.) For example, since $-2 < 7$ and $5 < 8$, $-2 + 5 < 7 + 8$.

PROPERTY 3 If $a < b$ and $c > 0$, then

$$ac < bc \quad \text{and} \quad \frac{a}{c} < \frac{b}{c}$$

(Each side of an inequality can be multiplied or divided by the same positive number without changing the direction of the inequality.) For example, since $3 < 8$ and $2 > 0$, $(3)(2) < (8)(2)$.

PROPERTY 4 If $a < b$ and $c < 0$, then

$$ac > bc \quad \text{and} \quad \frac{a}{c} > \frac{b}{c}$$

(The direction of an inequality is reversed if each side is multiplied or divided by a *negative* number.) For example, since $8 < 11$ and $-5 < 0$, $-40 > -55$.

The *solution set* of the linear equation $2x + 6 = 20$ is the set of all real numbers which when substituted for x make the equality a true statement. To solve such an equation we usually subtract 6 from each side of the equality and then multiply each side by $\frac{1}{2}$. To solve the linear inequality $2x + 6 < 20$, we may do the same thing; that is,

$$2x + 6 < 20,$$
$$2x < 14,$$
$$x < 7.$$

The set of real numbers satisfying the original inequality consists of the real numbers less than 7. The graph of this solution set is shown in Figure 1.3. *Note:* The "hole" at 7 indicates that 7 is not in the solution set.

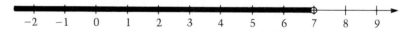

FIG. 1.3. $x < 7$.

Linear inequality

EXAMPLE 1 Find the solution set of $5x + 7 < 8x + 11$.

Solution

$$5x + 7 < 8x + 11$$
$$5x - 8x < 11 - 7 \quad \text{(Property 1)}$$
$$-3x < 4$$
$$x > -4/3. \quad \text{(Property 4)}$$

PRACTICE PROBLEM Solve $17x + 33 > 22x - 7$.

Answer: $x < 8$. ■

1.2 THE REAL NUMBER LINE AND LINEAR INEQUALITIES

Let us now look at some practical problems whose solutions are obtained from linear inequalities.

Maximum unit cost

EXAMPLE 2 Assume that the combined cost of one Domina watch and one Sasco watch is $100. If it is known that 5 Domina watches and 3 Sasco watches can be bought for $424 or less, what is the maximum cost of one Domina watch?

Solution: Let x equal the cost of a Domina watch. Then

$100 - x$ is the cost of one Sasco watch,
$5x$ is the cost of 5 Domina watches, and
$3(100 - x)$ is the cost of 3 Sasco watches.

Since the combined cost cannot exceed $424, it follows that

$$5x + 3(100 - x) \leq 424$$
$$5x + 300 - 3x \leq 424$$
$$2x \leq 124 \quad \text{(Property 1)}$$
$$x \leq 62. \quad \text{(Property 3)}$$

The maximum cost of one Domina watch is $62.

Maximum delivery distance

EXAMPLE 3 Two trucks must make deliveries over a route totaling 500 miles. Truck A gets 15 miles per gallon and Truck B gets 10 miles per gallon. If the total fuel consumption for the deliveries is not to exceed 40 gallons, what is the minimum number of miles that Truck A must travel?

Solution: First, recall that

$$(\text{miles per gallon})(\text{gallons used}) = \text{distance traveled}.$$

Therefore,

$$\text{gallons used} = \frac{\text{distance traveled}}{\text{miles per gallon}}.$$

Let x be the number of miles Truck A travels. Then

$500 - x$ is the number of miles Truck B travels,
$x/15$ is the number of gallons of fuel Truck A uses, and
$(500 - x)/10$ is the number of gallons of fuel Truck B uses.

Since the total fuel consumption cannot exceed 40 gallons,

$$\frac{x}{15} + \frac{500 - x}{10} \leq 40.$$

Multiplying each side of the preceding inequality by the least common denominator 30, we obtain

$$2x + 1500 - 3x \leq 1200.$$

Consequently,

$$-x \leq -300 \quad \text{(Property 1)}$$
$$x \geq 300. \quad \text{(Property 4)}$$

Therefore Truck A must travel at least 300 miles of the delivery route.

PRACTICE PROBLEM Two trucks must make deliveries over a route totaling 600 miles. Truck A gets 15 miles per gallon and Truck B gets 12 miles per gallon. If the total fuel consumption for the deliveries cannot exceed 42 gallons, what is the minimum number of miles that Truck A must travel?

Answer: 480 miles. ■

Minimum unit cost

EXAMPLE 4 Suppose the total cost of one reel mower and one rotary power is $372. If the total cost of 3 reel mowers and 7 rotary mowers is not more than $2204, what is the minimum cost of a reel mower?

Solution: Let x be the cost of one reel mower. Then

$372 - x$ is the cost of 1 rotary mower,
$3x$ is the total cost of 3 reel mowers, and
$7(372 - x)$ is the total cost of 7 rotary mowers.

Since the total cost of the 10 mowers cannot exceed $2204,

$$3x + 7(372 - x) \leq 2204$$
$$3x + 2604 - 7x \leq 2204$$
$$-4x \leq -400$$
$$x \geq 100.$$

Consequently, one reel mower costs at least $100.

PRACTICE PROBLEM In Example 4, what is the maximum cost of one rotary mower?

Answer: $272. ∎

Minimum salary requirement

EXAMPLE 5 A woman estimates that 30% of her gross income will go to taxes. If she wants to save each year 10% of her after-tax income and needs $20,000 a year for living expenses, what should her minimum gross income be?

Solution: Let x represent her gross income in dollars. Since 30% of her gross income goes to taxes,

$$x - 0.30x = 0.70x$$

is her income after taxes. Since $20,000 is needed for living expenses and since

$$0.10(0.70x) = 0.07x$$

is needed for savings, she wants her net income to be at least (greater than or equal to)

$$20,000 + 0.07x.$$

Hence, her salary x must be such that

$$0.70x \geq 20,000 + 0.07x,$$
$$0.63x \geq 20,000,$$
$$x \geq \$31,746.03.$$

PRACTICE PROBLEM A man estimates that 35% of his gross income will go to taxes. If he needs a total of $18,000 a year for living expenses and if he wants to deposit 15% of his after-tax income into a savings account each year, what should his minimum gross income be?

Answer: $32,579.19. ∎

Minimum sales to make a profit

EXAMPLE 6 The *fixed costs* (costs independent of the level of production) for publishing a given textbook are $74,250. If each book costs $18 to produce and sells for $27, how many textbooks must be sold to make a profit?

Solution: Since

total cost

$$\text{total cost} = (\text{variable costs}) + (\text{fixed costs})$$

the total cost in dollars for producing x textbooks is

$$18x + 74{,}250.$$

Since

revenue

$$\text{revenue} = (\text{selling price per unit})(\text{number of units sold}),$$

the revenue from selling x textbooks is $27x$. A profit results when revenue exceeds costs; that is, when

$$27x > 18x + 74{,}250,$$
$$9x > 74{,}250,$$
$$x > 8250.$$

Thus 8251 textbooks must be sold to make a profit.

EXERCISES

Set A—Solutions to Linear Inequalities

In Exercises 1 through 20, find the solution set for the given inequalities.

1. $3x - 5 < 7x + 13$
2. $8x - 23 < 12x + 32$
3. $7 - 8x \geq 5x + 13$
4. $6 - 4x > 7x - 15$
5. $-6x + 6 \leq -1 - 5x$
6. $-8 - 5x \geq -4x + 9$

1.2 THE REAL NUMBER LINE AND LINEAR INEQUALITIES

7. $7(4x + 2) < 2x + 5$
8. $5 - 9x > -6(4 + 9x)$
9. $3(1 + 8x) \leq 3x + 3(3 + 4x)$
10. $3(-3x - 7) \geq 5(-6 - 5x)$
11. $\dfrac{5x - 7}{4} < \dfrac{3x - 2}{5}$
12. $\dfrac{2x + 5}{7} > \dfrac{2 - 3x}{4}$
13. $\tfrac{2}{3}x - \tfrac{4}{7} \leq \tfrac{1}{6}x + \tfrac{5}{42}$
14. $\tfrac{3}{4}x + \tfrac{7}{2} < \tfrac{11}{6}x - \tfrac{5}{3}$
15. $\tfrac{3}{4}x + \tfrac{5}{2} > \tfrac{2}{3}x - \tfrac{5}{6}$
16. $\tfrac{4}{5}x + \tfrac{3}{5} \geq \tfrac{5}{9}x + \tfrac{7}{3}$
17. $0.3x + 1.7 < 0.25x + 3.75$
18. $0.06x + 1.4 > 0.04x - 2.8$
19. $1.25x + 8.1 > 0.75x - 9.3$
20. $-1.5x - 9.3 \leq -1.7x + 5.7$

Set B—Applications Involving Linear Inequalities

Maximum unit cost

21. Assume that the combined cost of one 19-inch television set and one 21-inch television set is $912. Suppose a shipment containing seven 19-inch sets and four 21-inch sets can be purchased for $4620 or less. What is the maximum cost per set of the 19-inch sets?

Minimum unit cost

22. Assume that the combined cost of one 19-inch television set and one 21-inch television set is $858. Suppose a shipment containing nine 21-inch sets and six 19-inch sets can be purchased for $6642 or less. What is the minimum cost per set of the 19-inch sets?

Minimum delivery distance

23. Two trucks must make deliveries over a route totaling 240 miles. The pickup truck gets 12 miles per gallon and the panel truck gets 10 miles per gallon. If the total fuel consumption for the deliveries is not to exceed 22 gallons, what is the minimum number of miles that the pickup truck must travel?

Maximum delivery distance

24. Two trucks must make deliveries over a route totaling 300 miles. The pickup truck gets 14 miles per gallon and the panel truck gets 12 miles per gallon. If the total fuel consumption for the deliveries cannot exceed 23 gallons, what is the maximum number of miles that the panel truck can travel?

Maximum and minimum unit costs

25. (a) Suppose the total cost of one reel mower and one rotary mower is $576. If the total cost of 5 reel mowers and 9 rotary mowers is no more than $4336, what is the minimum cost of one reel mower?
    (b) What is the maximum cost of one rotary mower?

Minimum salary requirement

26. A man estimates that 35% of his gross income will go to taxes. If he needs a total of $30,000 a year for living expenses and if he wants to save each year 6% of his income after taxes, what should his minimum gross income be?

Minimum sales to make a profit

27. A company's fixed costs for producing metal storage containers are $150,000. Each storage unit costs $9 to produce and sells for $12. How many units must be sold so that the company makes a profit?

Minimum sales to make a profit

28. The fixed costs for publishing a certain textbook are $60,000. If each book costs $16 to produce and sells for $21.50, how many books must be sold for the company to make a profit?

1.3 THE RECTANGULAR COORDINATE SYSTEM AND SLOPE OF A LINE

We can use two real coordinate lines to construct *a two-dimensional coordinate system* in a plane. Let us consider one way to do this and then investigate some further ideas which are fundamental in the development of calculus and its applications.

First, in a plane draw two mutually perpendicular number lines intersecting at their origins, one horizontal and the other vertical. (See Figure 1.4.) The usual orientation is to pair the positive real numbers with the points above the origin on the vertical number line and with the points to the right of the origin on the horizontal number line. These two number lines are called the **coordinate axes.** The horizontal coordinate axis is called the *x*-axis (or *first coordinate axis*), and the vertical coordinate axis is called the *y*-axis (or *second coordinate axis*).

The ordered pair $(a, 0)$ is associated with the point on the *x*-axis having a as coordinate. Similarly, the ordered pair $(0, b)$ is associated with the point on the *y*-axis having b as coordinate. Let P be any point in the plane not on either coordinate axis. If the line containing P and drawn parallel to the *y*-axis (perpendicular to the *x*-axis) intersects the *x*-axis at $(a, 0)$ and if the line contain-

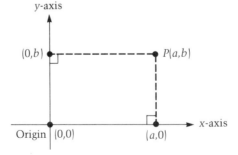

FIG. 1.4. Cartesian coordinate system in the plane.

1.3 THE RECTANGULAR COORDINATE SYSTEM AND SLOPE OF A LINE

ing P and drawn parallel to the x-axis (perpendicular to the y-axis) intersects the y-axis at $(0, b)$, then the ordered pair (a, b) corresponds to the point P. We sometimes write $P(a, b)$ to state this fact. (See Figure 1.4.) This provides a method for associating an ordered pair of real numbers with each point in the plane. The reverse of this technique provides a method for associating exactly one point in the plane with any ordered pair of real numbers. This coordinate system for pairing the points in a plane in a one-to-one fashion with ordered pairs of real numbers is called the **two-dimensional rectangular coordinate system** (or **cartesian coordinate system**).

two-dimensional rectangular coordinate system

If (x_1, y_1) and (x_2, y_2) are two different points in the coordinate plane, then from Figure 1.5 and the Pythagorean theorem we conclude that the **distance** D between the two points is given by

distance

$$D = \sqrt{(x_2 - x_1)^2 + (y_2 - y_1)^2}.$$

Note that if the points are on a line parallel to the x-axis, then $y_1 = y_2$ and a right triangle is not formed. However, if $x_1 < x_2$, as in Figure 1.5, then since $y_2 - y_1 = 0$, the formula yields

$$\sqrt{(x_2 - x_1)^2} = x_2 - x_1,$$

which is the distance between the two points. Since similar remarks hold for points on a line parallel to the y-axis, the formula is valid for any two points in the plane.

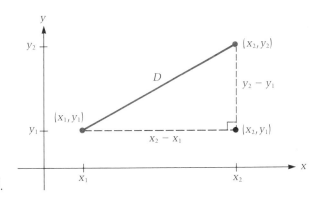

FIG. 1.5. Distance formula: $D = \sqrt{(x_2 - x_1)^2 + (y_2 - y_1)^2}$.

Distance between two points

EXAMPLE 1 Find the distance between the points $(-3, 5)$ and $(3, 13)$.

Solution: $D = \sqrt{(3+3)^2 + (13-5)^2} = \sqrt{36 + 64} = \sqrt{100} = 10.$

PRACTICE PROBLEM Find the distance between the points $(-1, 4)$ and $(4, 16)$.

Answer: 13. ■

Let (x_1, y_1) and (x_2, y_2) be two different points in the coordinate plane and let (x_0, y_0) be the coordinates of the midpoint of the line segment joining the two given points. Although x_0 and y_0 could be found in terms of x_1, y_1, x_2, and y_2 by using the distance formula, it is much easier to use Figure 1.6 and the fact that if parallel lines cut off equal segments on one transversal (a line intersecting the parallel lines), then they cut off equal segments on any other transversal. Since on the real number line the coordinate of the midpoint between any two points is the arithmetic mean (average) of the coordinates of the two points, we conclude that the **midpoint** of the line segment with (x_1, y_1) and (x_2, y_2) is

midpoint

$$\left(\frac{x_1 + x_2}{2}, \frac{y_1 + y_2}{2}\right).$$

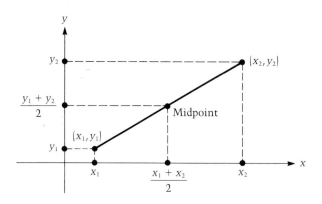

FIG. 1.6. Midpoint: $\left(\dfrac{x_1 + x_2}{2}, \dfrac{y_1 + y_2}{2}\right).$

1.3 THE RECTANGULAR COORDINATE SYSTEM AND SLOPE OF A LINE

Midpoint and distance

EXAMPLE 2

(a) Find the coordinates of the midpoint of the line segment having (2, 5) and (6, 11) as endpoints.

(b) Use the distance formula to show that the distances from the midpoint to each endpoint of the segment are equal.

Solution

(a) The midpoint is

$$\left(\frac{2+6}{2}, \frac{5+11}{2}\right) = (4, 8).$$

(b) The distance between (2, 5) and (4, 8) is

$$\sqrt{(4-2)^2 + (8-5)^2} = \sqrt{13}.$$

The distance between (6, 11) and (4, 8) is

$$\sqrt{(4-6)^2 + (8-11)^2} = \sqrt{13}.$$

Location of postal pickup stations

EXAMPLE 3 In a given city, two postal pickup stations are located from the central post office A as follows.

Pickup station B: 3 miles east and 4 miles north
Pickup station C: 9 miles east and 6 miles north

(a) Suppose a third pickup station P is to be located midway (on a line) between these two stations. What would be its location from the central post office? (b) How far would the third pickup station be from the central post office? (c) How far would it be from each of the other stations?

Solution

(a) Note that if the central post office A is located at the origin, then points B and C in Figure 1.7 can be used to represent the locations of the first two pickup stations. Since

$$\left(\frac{3+9}{2}, \frac{4+6}{2}\right) = (6, 5)$$

is the midpoint of the segment with B and C as endpoints, the

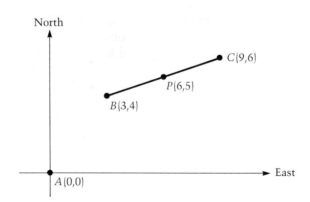

FIG. 1.7.

third pickup station P should be located 6 miles east and 5 miles north of the central post office.

(b) The distance d between $P(6, 5)$ and the central post office $A(0, 0)$ is

$$d = \sqrt{(6-0)^2 + (5-0)^2} = \sqrt{61} \doteq 7.81 \text{ miles.}$$

(c) The distance D between $P(6, 5)$ and $B(3, 4)$ (which is the same as between P and C) is

$$D = \sqrt{(6-3)^2 + (5-4)^2} = \sqrt{10} \doteq 3.16 \text{ miles.}$$

slope

If (x_1, y_1) and (x_2, y_2) are any two points for which $x_1 \neq x_2$ (that is, if the line containing the two points is not parallel to the y-axis), then the **slope** of the line containing (x_1, y_1) and (x_2, y_2), denoted by m, is defined by

$$m = \frac{y_2 - y_1}{x_2 - x_1}.$$

Essentially, the slope is the ratio of the change in y to the change in x between the two points. Lines (L) with positive slope rise to the right, lines with negative slope fall to the right, and lines with

1.3 THE RECTANGULAR COORDINATE SYSTEM AND SLOPE OF A LINE

zero slope are horizontal (parallel to the x-axis). If $x_1 = x_2$, then the two points are on a vertical line (a line parallel to the y-axis); slope is not defined for such lines. (See Figure 1.8.)

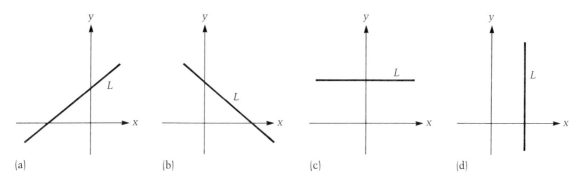

FIG. 1.8. Slope. (a) $m > 0$. (b) $m < 0$. (c) $m = 0$. (d) m undefined.

For two points (x_1, y_1) and (x_2, y_2) with $x_1 \neq x_2$, note that

$$\frac{y_2 - y_1}{x_2 - x_1} = \frac{-(y_2 - y_1)}{-(x_2 - x_1)} = \frac{y_1 - y_2}{x_1 - x_2}.$$

This means, for example, that to find the slope of the line containing $(2, 3)$ and $(7, 5)$ either point can be called (x_1, y_1).

Slope of a line

EXAMPLE 4 Find the slope of the line containing the points $(2, 5)$ and $(7, 9)$.

Solution: $m = \dfrac{9 - 5}{7 - 2} = \dfrac{4}{5}$.

PRACTICE PROBLEM Find the slope of the line containing the points $(2, -7)$ and $(5, 9)$.

Answer: $m = \frac{16}{3}$. ∎

parallel

It is not difficult to prove that two nonvertical lines are **parallel** if and only if they have the same slope. It is also true, though not as easy to prove, that two nonvertical lines are

perpendicular

perpendicular if and only if the slope of one line is the negative reciprocal of the slope of the other. (See Figure 1.9.)

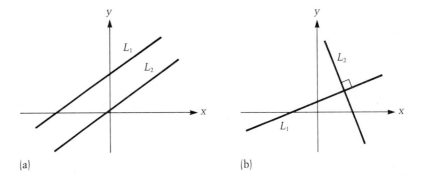

FIG. 1.9. (a) Parallel lines, $m_1 = m_2$. (b) Perpendicular lines, $m_1 = -\dfrac{1}{m_2}$.

Locations of subdivisions

EXAMPLE 5 The mayor of a city determines that three housing subdivisions can be located from City Hall as follows.

Subdivision A: 3 miles east and 1 mile north
Subdivision B: 6 miles east and 8 miles north
Subdivision C: 9 miles east and 3 miles north

(a) Construct a coordinate system and exhibit the locations of the subdivisions. Put City Hall at the origin of the coordinate system.

(b) Determine the locations of the midpoints between A and B, B and C, and A and C.

(c) Construct the perpendicular bisectors of each segment AB, BC, and AC. (The perpendicular bisectors of the sides of a triangle all meet in one point. Later we will determine the coordinates of their point of intersection for triangle ABC.)

Solution

(a) See Figure 1.10.

(b) The midpoint P of AC is $(6, 2)$. The midpoint Q of BC is $\left(\frac{15}{2}, \frac{11}{2}\right)$. The midpoint R of AC is $\left(\frac{9}{2}, \frac{9}{2}\right)$.

(c) Since the slope of the line containing A and C is $\frac{1}{3}$, the slope of the perpendicular bisector is -3. Hence, if we move 1 unit to the left of P and 3 units up, then the resulting point $(5, 5)$ will be another point on the perpendicular bisector of AC. The other perpendicular bisectors can be constructed in a similar manner.

1.3 THE RECTANGULAR COORDINATE SYSTEM AND SLOPE OF A LINE

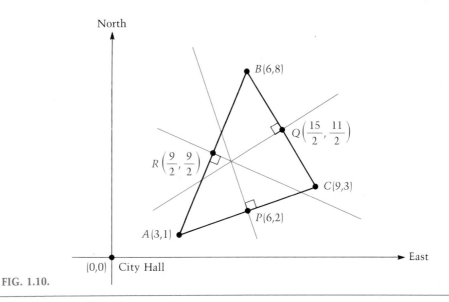

FIG. 1.10.

EXERCISES

Set A—Slope of a Line

Find the slope of each line containing the point P and Q given in Exercises 1 through 10.

1. $P(3, 7)$ and $Q(8, 10)$
2. $P(-3, 6)$ and $Q(7, 12)$
3. $P(-2, 1)$ and $Q(2, 4)$
4. $P(-3, -7)$ and $Q(2, 5)$
5. $P(-7, 3)$ and $Q(5, -4)$
6. $P(-8, 3)$ and $Q(5, -5)$
7. $P(3, -1)$ and $Q(-3, -9)$
8. $P(-4, 6)$ and $Q(5, 6)$
9. $P(-\frac{5}{2}, 7)$ and $Q(\frac{7}{2}, -5)$
10. $P(\frac{3}{4}, -\frac{1}{2})$ and $Q(\frac{19}{4}, \frac{11}{2})$

Set B—Midpoint of a Line Segment

In Exercises 11 through 20, find the midpoint of each line segment with P and Q as endpoints.

11. $P(3, 7)$ and $Q(8, 10)$
12. $P(-3, 6)$ and $Q(7, 12)$
13. $P(-2, 1)$ and $Q(2, 4)$

14. $P(-3, -7)$ and $Q(2, 5)$
15. $P(-7, 3)$ and $Q(5, -4)$
16. $P(-8, 3)$ and $Q(5, -5)$
17. $P(3, -1)$ and $Q(-3, -9)$
18. $P(-4, 6)$ and $Q(5, 6)$
19. $P(-\frac{5}{2}, 7)$ and $Q(\frac{7}{2}, -5)$
20. $P(\frac{3}{4}, -\frac{1}{2})$ and $Q(\frac{19}{4}, \frac{11}{2})$

Set C–Length of Line Segment

In Exercises 21 through 30, find the distance between the points P and Q.

21. $P(3, 7)$ and $Q(8, 10)$
22. $P(-3, 6)$ and $Q(7, 12)$
23. $P(-2, 1)$ and $Q(2, 4)$
24. $P(-3, -7)$ and $Q(2, 5)$
25. $P(-7, 3)$ and $Q(5, -4)$
26. $P(-8, 3)$ and $Q(5, -5)$
27. $P(3, -1)$ and $Q(-3, -9)$
28. $P(-4, 6)$ and $Q(5, 6)$
29. $P(-\frac{5}{2}, 7)$ and $Q(\frac{7}{2}, -5)$
30. $P(\frac{3}{4}, -\frac{1}{2})$ and $Q(\frac{19}{4}, \frac{11}{2})$

Set D–Applications for a Rectangular Coordinate System

Locations of subdivisions

31. The mayor of a city determines that three housing subdivisions can be located from City Hall as follows.
    Subdivision A: 3 miles east and 2 miles north
    Subdivision B: 7 miles east and 10 miles north
    Subdivision C: 11 miles east and 3 miles south
    (a) Construct a coordinate system and exhibit the locations of the subdivisions. Put City Hall at the origin of the coordinate system.
    (b) Determine the locations of the midpoints between A and B, B and C, and A and C.
    (c) Construct the perpendicular bisectors of each segment AB, BC, and AC.

Locations of subdivisions

32. The mayor of a city determines that three housing subdivisions can be located from City Hall as follows.
    Subdivision A: 5 miles west and 4 miles north
    Subdivision B: 3 miles east and 5 miles south
    Subdivision C: 10 miles east and 2 miles north

1.4 EQUATIONS OF LINES

(a) Construct a coordinate system and exhibit the locations of the subdivisions. Put City Hall at the origin of the coordinate system.
(b) Determine the locations of the midpoints between A and B, B and C, and A and C.
(c) Construct the perpendicular bisectors of each segment, AB, BC, and AC.

Distances of warehouses

33. A company has two warehouses, A and B, and plans to open a new warehouse, C. In relation to company headquarters, warehouse A is 40 miles west and 60 miles north, warehouse B is 100 miles east and 40 miles north, and warehouse C will be 20 miles east and 140 miles north. Which of warehouses A and B will be closer to C?

Distances of warehouses

34. Suppose the company described in Exercise 33 decides to open a fourth warehouse, D, midway (on a line) between warehouses A and B.
(a) Give D's location from the company headquarters.
(b) Is warehouse D closer to warehouse A or warehouse C?
(c) How far is warehouse D from company headquarters?

1.4 EQUATIONS OF LINES

In this section we begin to consider the problem of graphing sets of points in the plane whose coordinates satisfy equations in x and y. We also begin the important reverse task of learning how to determine equations in x and y that describe given graphs (sets of ordered pairs of real numbers). In particular, these two problems are investigated for linear equations in x and y and for lines in the coordinate plane. (Other equations and graphs will be investigated in subsequent chapters.)

Let (x_1, y_1) and (x_2, y_2) be any two points in the rectangular coordinate plane such that $x_1 \neq x_2$ and let S be the set of points on the line containing the two points. An equation defining S is called an *equation of the line*; it can be found as follows. A point (x, y) different from (x_1, y_1) is on the line (in the set S) if and only if the slopes

equation of a line

$$\frac{y - y_1}{x - x_1} \quad \text{and} \quad \frac{y_2 - y_1}{x_2 - x_1}$$

are equal. (See Figure 1.11.) Therefore, the point (x, y) is on the line if and only if

$$\frac{y - y_1}{x - x_1} = \frac{y_2 - y_1}{x_2 - x_1}$$

or

$$y - y_1 = \frac{y_2 - y_1}{x_2 - x_1}(x - x_1).$$

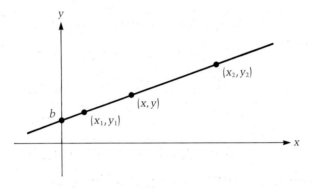

FIG. 1.11.
$y - y_1 = m(x - x_1)$
or $y = mx + b$.

two-point form

The preceding equation is called the *two-point form* for an equation of the line. Since m represents the slope of the line,

$$y - y_1 = m(x - x_1)$$

point-slope form

is also an equation of the line; it is called the *point-slope form*. In the point-slope form, if $x_1 = 0$ and $y_1 = b$, then the equation reduces to $y - b = m(x - 0)$, $y - b = mx$, or

$$y = mx + b.$$

slope-intercept form

Since $y = mx + b$ is an equation of a line with slope m crossing the y-axis at $(0, b)$, it is called the *slope-intercept form*.

Equation of a line

EXAMPLE 1 Find an equation of the line containing $(3, 5)$ and $(7, 8)$.

1.4 EQUATIONS OF LINES

Solution: The slope of the line is

$$\frac{8-5}{7-3} = \frac{3}{4}.$$

Using this slope and the point (3, 5), we obtain

$$y - 5 = \left(\frac{3}{4}\right)(x - 3).$$

Simplifying the preceding equation yields $3x - 4y = -11$. (See Figure 1.12.) *Note:* If the point (7, 8) is used in the formula, we obtain

$$y - 8 = \left(\frac{3}{4}\right)(x - 7)$$

or

$$3x - 4y = -11$$

which, of course, is the same as the previous result.

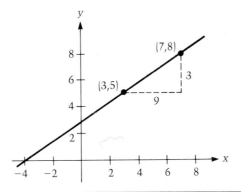

FIG. 1.12. $3x - 4y = -11$.

Equation of a line

EXAMPLE 2 Determine an equation for each of the following lines in the coordinate plane: (a) the line containing (3, −7) having slope $-\frac{8}{5}$; (b) the line having 3 as slope and 9 as y-intercept; and (c) the line parallel to the x-axis containing (5, 3).

Solution

(a) $y - (-7) = -\frac{8}{5}(x - 3)$. Such an equation is often written in the form $Ax + By = C$; it is obtained as follows.

$$5(y + 7) = -8(x - 3)$$
$$5y + 35 = -8x + 24$$
$$8x + 5y = -11$$

(b) $y = 3x + 9$.
(c) Since the line has slope 0 and y-intercept 3, its equation is $y = 3$.

PRACTICE PROBLEM Determine an equation for each of the following lines in the coordinate plane: (a) the line containing $(-4, 2)$ having slope $\frac{3}{7}$; (b) the line having slope -2 and y-intercept 5; and (c) the line parallel to the y-axis containing $(-3, 9)$.

Answer:

(a) $y - 2 = (\frac{3}{7})(x + 4)$, or $3x - 7y = -26$.

(b) $y = -2x + 5$.

(c) $x = -3$. ∎

Any equation that can be written in the form

$$Ax + By = C$$

linear equation in two variables

where A and B are not both zero, is called a **linear equation in two variables**. Any straight line in the coordinate plane can be represented by a linear equation in two variables. Furthermore, any linear equation in two variables has a straight line as its graph. Provided $B \neq 0$, we can solve $Ax + By = C$ for y and obtain

$$y = -\frac{A}{B}x + \frac{C}{B}.$$

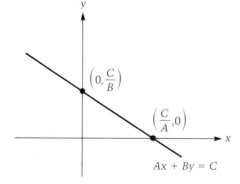

FIG. 1.13.
$y = -\frac{A}{B}x + \frac{C}{B}$;
$m = -\frac{A}{B}$.

1.4 EQUATIONS OF LINES

Since the preceding equation is in the slope-intercept form, we see that $Ax + By = C$ is a line with slope $-A/B$ and y-intercept C/B. (See Figure 1.13.) If $B = 0$ and $A \neq 0$, the line is parallel to the y-axis and intersects the x-axis at $x = C/A$.

Graph of a line

EXAMPLE 3 Graph the equation $3x + 5y = 15$. That is, graph the set of points S defined by

$$S = \{(x, y) \mid 3x + 5y = 15\}$$

(read, "the set of all points (x, y) such that $3x + 5y = 15$").

Solution: (See Figure 1.14.) Since the graph (in two dimensions) of a linear equation in x and y is a straight line, two points whose coordinates satisfy the equation are enough to determine the graph. However, plotting an extra point provides a check on computations. Usually, the x-intercept (where $y = 0$) and the y-intercept (where $x = 0$) are two of the points plotted.

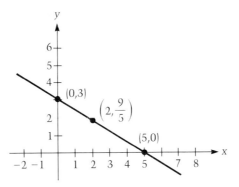

FIG. 1.14. $3x + 5y = 15$.

The fact that the line $Ax + By = C$, $B \neq 0$, has slope $-A/B$ is quite useful in determining an equation of a line. For example, consider the line containing the point $(1, 5)$ with slope $-\frac{2}{3}$. Since $-\frac{2}{3} = -A/B$, we can let $A = 2$ and $B = 3$. Its equation has the form $2x + 3y = C$, where C is a constant to be determined from the fact that $(1, 5)$ must satisfy the equation. Since $2(1) + 3(5) = 17$, $(1, 5)$ satisfies the equation if and only if $2x + 3y = 17$. Consequently, an equation of the line is $2x + 3y = 17$. (It is also true that the equation could have the form $-2x - 3y = C$, but in this case $C = -17$ and the equivalent equation $-2x - 3y = -17$ is obtained. In fact, A and B could be any constants such that $-A/B = -\frac{2}{3}$.)

Equation of a line

EXAMPLE 4 Find an equation of the line with slope $\frac{7}{5}$ containing the point $(2, 3)$.

Solution: An equation of the line has the form $7x - 5y = C$. (*Note:* This is equivalent to $7x + (-5)y = C$, where $A = 7$ and $B = -5$, and $-A/B = \frac{7}{5}$.) For $(2, 3)$ to satisfy the equation, $7(2) - 5(3) = C$ and $C = -1$. Hence, an equation of the line is $7x - 5y = -1$.

PRACTICE PROBLEM Find an equation of the line with slope $\frac{9}{4}$ containing the point $(1, 2)$.

Answer: $9x - 4y = 1$. ∎

Equation of perpendicular bisector

EXAMPLE 5 Find an equation of the perpendicular bisector of the line segment with $P(-1, 4)$ and $Q(7, 10)$ as endpoints.

Solution: The midpoint of the line segment with $(-1, 4)$ and $(7, 10)$ as endpoints is $(3, 7)$. The slope of the line containing $(-1, 4)$ and $(7, 10)$ is $\frac{3}{4}$, and therefore, the slope of the line perpendicular to it is $-\frac{4}{3}$. An equation of the perpendicular bisector of the segment is

$$y - 7 = -\frac{4}{3}(x - 3) \quad \text{or} \quad 4x + 3y = 33.$$

PRACTICE PROBLEM Find an equation of the perpendicular bisector of the line segment with $(-3, 6)$ and $(5, 10)$ as endpoints.

Answer: $2x + y = 10$. ∎

EXERCISES

Set A—Equation of a Line

In Exercises 1 through 10, find an equation of the line containing the given point and having the given slope.

1. $(-7, -7)$, $m = 0$
2. $(-2, -7)$, $m = \frac{5}{7}$
3. $(2, -8)$, $m = $ undefined
4. $(1, 2)$, $m = 4$
5. $(-2, 2)$, $m = -2$
6. $(8, -5)$, $m = -\frac{7}{3}$

1.4 EQUATIONS OF LINES

7. $(7, -2)$, $m = \frac{4}{3}$
8. $(3, -3)$, $m = -\frac{3}{2}$
9. $(2, -6)$, $m = \frac{11}{4}$
10. $(\frac{2}{3}, \frac{3}{5})$, $m = 0$

In Exercises 11 through 20, find an equation of the line containing the two given points.

11. $(2, 6)$ and $(-7, 9)$
12. $(-1, 3)$ and $(2, -9)$
13. $(-3, 5)$ and $(-3, 8)$
14. $(-7, 3)$ and $(8, 1)$
15. $(0, 6)$ and $(7, -4)$
16. $(0, -2)$ and $(6, 0)$
17. $(\frac{2}{3}, 1)$ and $(\frac{1}{2}, \frac{5}{6})$
18. $(-\frac{3}{4}, \frac{1}{2})$ and $(3, \frac{1}{4})$
19. $(1.5, 0.8)$ and $(2, 0.6)$
20. $(1.5, 1)$ and $(-2.5, 0.2)$

Parallel lines
21. Find an equation of the line containing $(-1, 4)$ and parallel to the line $3x - 5y = 11$. Sketch a graph of each line in the same coordinate plane.

Parallel lines
22. Find an equation of the line containing $(-3, 1)$ and parallel to the line $4x + 7y = 15$. Sketch a graph of each line in the same coordinate plane.

Perpendicular bisector
23. Find an equation of the perpendicular bisector of the line segment with $(-4, 2)$ and $(2, 5)$ as endpoints. Sketch a graph of the segment and the line in the same coordinate plane.

Perpendicular bisector
24. Find an equation of the perpendicular bisector of the line segment with $(-2, -4)$ and $(7, 6)$ as endpoints. Sketch a graph of the segment and the line in the same coordinate plane.

Equation of line
25. Find an equation of the line containing the point $(-1, 4)$ and the midpoint of the segment with $(-3, 1)$ and $(5, 3)$ as endpoints.

Equation of line
26. Find an equation of the line containing the point $(2, 5)$ and the midpoint of the segment with $(-2, -1)$ and $(8, -3)$ as endpoints.

Parallel lines
27. Find an equation of the line containing $(\frac{2}{3}, -\frac{3}{5})$ and parallel to the line $6x + 10y = 11$.

Parallel lines
28. Find an equation of the line containing $(-\frac{3}{7}, \frac{3}{4})$ and parallel to the line $7x + 6y = 4$.

Perpendicular bisector
29. Find an equation of the perpendicular bisector of the line segment with $(\frac{5}{6}, -\frac{1}{3})$ and $(-\frac{1}{2}, \frac{3}{4})$ as endpoints.

Perpendicular bisector
30. Find an equation of the perpendicular bisector of the line segment with $(-\frac{1}{4}, \frac{2}{3})$ and $(\frac{5}{6}, -\frac{1}{3})$ as endpoints.

Temperature scales
31. (a) The relationship between the Fahrenheit and Celsius temperature scales is linear. If 0°C corresponds to 32°F and if 100°C corresponds to 212°F, what is a linear equation in F and C that expresses this relationship.

(b) Use part (a) to determine the Celsius temperature that corresponds to 100°F.

(c) Use part (a) to determine the Fahrenheit temperature that corresponds to 40°C.

1.5 LINEAR SYSTEMS AND THEIR APPLICATIONS

A pair of linear equations of the form

$$Ax + By = E \qquad (1)$$
$$Cx + Dy = F \qquad (2)$$

linear system in two variables

is called a **linear system in two variables.** (*Note:* It is assumed for Eq. (1) that not both A and B are zero and for Eq. (2) that not both C and D are zero.) As we know, the solution set for the linear equation $Ax + By = E$ (or $Cx + Dy = F$) contains infinitely many ordered pairs of real numbers, the ordered pairs associated with the points on the graph of the equation. The **solution set** of the given linear system is the set containing exactly those ordered pairs of real numbers in the solution sets of both equations; that is, the solution set is the intersection of the solution sets for Eqs. (1) and (2). In this section we learn how to solve a linear system in two variables by two methods: (1) the method of elimination and (2) the method of substitution. We also learn how linear systems can be used to solve practical problems.

solution set of linear system

Intersection of lines

EXAMPLE 1 Find the point of intersection of the lines $2x - 3y = 3$ and $4x + 5y = 17$. (See Figure 1.15.)

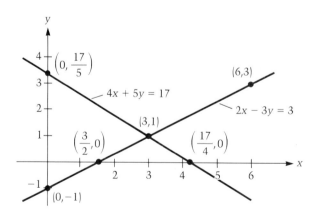

FIG. 1.15.

1.5 LINEAR SYSTEMS AND THEIR APPLICATIONS

Solution: The point of intersection (x, y) is on both lines and hence must be in the solution set for each equation in the following linear system:

$$2x - 3y = 3,$$
$$4x + 5y = 17.$$

If numbers x and y exist which make both equalities true statements, then they also satisfy the system obtained by multiplying the first equation by 2. (This is done to make the coefficients of x the same in each equation.) That is, they satisfy

$$4x - 6y = 6,$$
$$4x + 5y = 17.$$

Subtracting the second equation from the first, we find that

$$-11y = -11,$$
$$y = 1.$$

Substituting $y = 1$ in either of the original equations, we get $x = 3$. The point of intersection of the two lines is $(3, 1)$.

elimination technique

The preceding system of equations was solved by what is called the **elimination technique.** Basically, one or both of the equations are multiplied by constants selected to make either the coefficients of x equal or the coefficients of y equal. Then that variable is eliminated by subtracting one equation from the other, and the resulting equation in one variable can be solved in the usual way.

Ticket sales

EXAMPLE 2 A theater with 500 seats takes in $1347.50 for a given performance that is sold out. If a child's ticket costs $1.75 and an adult's ticket costs $3.00, how many children and how many adults attended the performance?

Solution: Let x be the number of children and let y be the number of adults attending the performance. Then

$$x + y = 500. \tag{1}$$

The children's tickets bring in $1.75x$ and the adults' tickets bring in $3.00y$; hence,

$$1.75x + 3.00y = 1347.50. \qquad (2)$$

Solving Eq. (1) for y and substituting in Eq. (2), we get

$$1.75x + 3.00(500 - x) = 1347.50,$$
$$1.75x + 1500 - 3x = 1347.50,$$
$$-1.25x = -152.50,$$
$$x = 122.$$

Therefore, 122 children and $500 - 122 = 378$ adults attended the performance.

substitution technique

In the preceding example, the solution to the linear system was obtained by what is called the **substitution technique.** In this procedure, one equation is solved for one of the variables, and then this expression is substituted into the other equation resulting in a linear equation in one variable. This equation is solved and its solution is substituted into either of the original equations to obtain the solution to the system.

PRACTICE PROBLEM The total sales from tickets to a concert attended by 12,000 persons came to $98,875. If only two types of tickets were available, one selling for $6.50 and one selling for $9.00, how many tickets of each type were sold?

Answer: 3650 of the $6.50 tickets and 8350 of the $9.00 tickets. ∎

Location of fire station

EXAMPLE 3 Consider the information given in Example 5, Section 1.3, and suppose a fire station is to be built equidistant from each of the three subdivisions. What should be its location in reference to City Hall?

Solution: Notice in Figure 1.16 that the three subdivisions are located at the points $A(3, 1)$, $B(6, 8)$, and $C(9, 3)$. The midpoint P of AC is $(6, 2)$ and the midpoint Q of BC is $(\frac{15}{2}, \frac{11}{2})$. The slope of the line containing A and C is $\frac{1}{3}$, so the slope of the perpendicular bisector of AC is -3. The equation of the perpendicular bisector of AC is

$$3x + y = 20. \qquad (1)$$

Similarly, the slope of the line containing B and C is $-\frac{5}{3}$, so the

1.5 LINEAR SYSTEMS AND THEIR APPLICATIONS

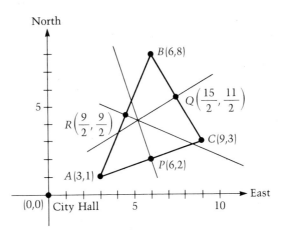

FIG. 1.16.

slope of the perpendicular bisector of BC is $\frac{3}{5}$. The equation of the perpendicular bisector of BC is

$$3x - 5y = -5. \qquad (2)$$

Since the perpendicular bisectors intersect in a point equidistant from A, B, and C, we find the intersection of the lines given by Eqs. (1) and (2). Subtracting Eq. (2) from Eq. (1), we obtain

$$6y = 25,$$

$$y = \frac{25}{6}.$$

Substituting this value for y in Eq. (1), we get

$$3x + \frac{25}{6} = 20,$$

$$3x = \frac{95}{6},$$

$$x = \frac{95}{18}.$$

Therefore, the fire station should be located $\frac{95}{18} \doteq 5.28$ miles east and $\frac{25}{6} \doteq 4.17$ miles north of City Hall.

Population density

EXAMPLE 4 The total combined land area of California and Georgia is approximately 214,000 square miles. In 1970 the population density for California was 128 persons per square mile and for Georgia 79 persons per square mile.

(a) Use the fact that in 1970 the combined populations of the two states was approximately 24,550,000 to approximate the land area of each state.

(b) Find the population of each state in 1970.

Solution

(a) Let x be the land area in square miles of California and let y be the land area in square miles of Georgia. Then

$$x + y = 214,000. \qquad (1)$$

Since the population density for California in 1970 was 128 persons per square mile, the population was $128x$ persons. Similarly, the population of Georgia was $79y$. Hence

$$128x + 79y = 24,550,000. \qquad (2)$$

Solving Eq. (1) for y and substituting into Eq. (2), we get

$$128x + 79(214,000 - x) = 24,550,000,$$
$$128x - 79x = 24,550,000 - 79(214,000),$$
$$49x = 7,644,000,$$
$$x = 156,000 \text{ square miles.}$$

Substituting this value for x in Eq. (1) yields

$$y = 58,000 \text{ square miles.}$$

(b) In 1970, the population of California was

$$128x = 128(156,000) = 19,968,000,$$

and the population of Georgia was

$$79y = 79(58,000) = 4,582,000.$$

Since two lines in a plane can (1) intersect in a point, (2) be parallel, or (3) be coincident, the solution set of a linear system can (1) consist of exactly one pair of real numbers, (2) be empty, or (3) contain infinitely many ordered pairs of real numbers, respectively.

1.5 LINEAR SYSTEMS AND THEIR APPLICATIONS

If the lines represented by the equations $Ax + By = E$ and $Cx + Dy = F$, are distinct and not parallel, then they will intersect in precisely one point and the solution set will contain exactly one pair of real numbers. If neither line is vertical, then the slopes of the lines are $-A/B$ and $-C/D$. Furthermore, since the lines are not parallel,

$$-\frac{A}{B} \neq -\frac{C}{D}, \quad AD \neq BC,$$

and

$$AD - BC \neq 0.$$

If one of the lines is vertical and if the two lines are not parallel, it is still true that $AD - BC \neq 0$. Consequently, the linear system

$$Ax + By = E,$$
$$Cx + Dy = F$$

has a unique solution provided $AD - BC \neq 0$.

For the linear system $Ax + By = E$ and $Cx + Dy = F$, if $AD - BC = 0$, then the lines have the same slope. If the lines have the same slope and are distinct, they do not intersect and there is no solution to the system. If $AD - BC = 0$ and the lines are not distinct, then one equation is a constant multiple of the other and there are infinitely many ordered pairs in the solution set, namely, the ordered pairs associated with the points on the line.

Stock portfolio

EXAMPLE 5 An investor bought two stocks, one selling at $30 a share and the other selling at $12 a share, for a total of $18,900. The dividend on the higher priced stock is $2.60 per share, and the dividend on the lower priced stock is $1.12 per share. If the investor expects to receive a total of $1694 in dividends from the two stocks in a given year, how many shares of each stock did he buy?

Solution: Let x be the number of shares bought of the higher priced stock and let y be the number of shares bought of the lower priced stock. Therefore, the cost of the higher priced stock was $30x$ dollars and the cost of the lower priced stock was $12y$ dollars. Consequently,

$$30x + 12y = 18{,}900. \quad \text{(Total investments)}$$

Similarly, we conclude from the information on dividends that

$$2.60x + 1.12y = 1694. \quad \text{(Total dividends)}$$

By dividing each side of the first equation by the common factor 6 and multiplying each side of the second equation by 100 to get rid of decimals, we obtain the equivalent system

$$\begin{aligned} 5x + 2y &= 3150, \\ 260x + 112y &= 169{,}400. \end{aligned} \quad \text{(System 1)}$$

Multiplying the first equation by 52 yields

$$\begin{aligned} 260x + 104y &= 163{,}800, \\ 260x + 112y &= 169{,}400. \end{aligned}$$

Subtracting the first equation from the second in the preceding system we obtain

$$\begin{aligned} 8y &= 5600, \\ y &= 700. \end{aligned}$$

Substituting $y = 700$ in the equation $5x + 2y = 3150$ from System 1, we find $x = 350$. Thus the investor bought 350 shares of the higher priced stock and 700 shares of the lower priced stock.

Gasohol mixture

EXAMPLE 6 A gasoline supplier has two large gasohol tanks, one containing 8% alcohol and the other containing 14% alcohol. If 1000 gallons of gasohol with a 10% alcohol content are needed, how many gallons should be taken from each tank in order to provide the proper mixture?

Solution: Let x be the number of gallons required of the 8% mixture and let y be the number of gallons required of the 14% mixture. Since there is to be a total of 1000 gallons,

$$x + y = 1000.$$

Now, $0.08x$ is the amount of alcohol in x gallons of the first mixture and $0.14y$ is the amount of alcohol in y gallons of the second mixture. Since there is to be

$$0.10(1000) = 100$$

1.5 LINEAR SYSTEMS AND THEIR APPLICATIONS

gallons of alcohol in the 1000 gallons,

$$0.08x + 0.14y = 100.$$

The problem therefore results in the following system:

$$x + y = 1000, \qquad (1)$$
$$8x + 14y = 10{,}000. \qquad (2)$$

The system can be solved (a) by multiplying the first equation by 8 and then eliminating x, or (b) by substitution. Let us use the second technique. From Eq. (1), we get

$$y = 1000 - x.$$

From Eq. (2) and the preceding expression for y, we obtain

$$8x + 14(1000 - x) = 10{,}000,$$
$$8x + 14{,}000 - 14x = 10{,}000,$$
$$-6x = -4000,$$

$$x = 666\frac{2}{3} \text{ gallons.}$$

Therefore, $y = 1000 - x = 333\frac{1}{3}$ gallons.

Size of a shipment

EXAMPLE 7 A retailer pays \$178.50 for a box containing lids and pans. He sells the lids for \$3.50 apiece and the pans for \$7.80 apiece and receives a total of \$547 for the sale of the shipment. If the lids cost \$1.05 apiece and the pans cost \$2.66 apiece, how many of each were in the shipment?

Solution: Let x be the number of lids in the shipment and let y be the number of pans in the shipment. From the given information, we obtain the following linear system:

$$3.50x + 7.80y = 547,$$
$$1.05x + 2.66y = 178.50.$$

The preceding linear system is equivalent to

$$35x + 78y = 5470,$$
$$105x + 266y = 17{,}850.$$

Multiplying the first equation in the preceding system by 3 yields

$$105x + 234y = 16{,}410,$$
$$105x + 266y = 17{,}850.$$

Subtracting the first equation in the preceding system from the second equation we get

$$32y = 1440,$$
$$y = 45.$$

Substituting $y = 45$ in the equation $35x + 78y = 5470$ yields 56. Therefore there were 56 lids and 45 pans in the shipment.

EXERCISES

Set A—Solving Linear Systems

For each linear system in Exercises 1 through 10, find the solution set.

1. $3x - 5y = 21$
   $4x + 7y = -13$

2. $2x + 4y = 14$
   $7x - 2y = 17$

3. $2x + 7y = 8$
   $4x - 9y = -7$

4. $4x - 3y = 2$
   $x + 3y = 3$

5. $5x - 7y = -12$
   $3x + 4y = 8$

6. $7x + 8y = 10$
   $2x - 3y = 12$

7. $\frac{3}{4}x + \frac{2}{3}y = -2$
   $\frac{1}{2}x - \frac{5}{6}y = 14$

8. $\frac{5}{8}x - \frac{3}{4}y = \frac{1}{4}$
   $\frac{3}{4}x + \frac{3}{2}y = \frac{3}{8}$

9. $1.5x - 3.2y = 6$
   $4.5x + 2.8y = 2.5$

10. $2.4x + 3.2y = 12.4$
    $1.5x - 5.4y = -1.5$

Set B—Applications Involving Linear Systems

In Exercises 11 through 25, use linear systems to solve the problems.

Size of shipment

11. A large department store ordered a shipment of two different types of shirts, one costing $6 apiece and another costing $9 apiece. The total cost of the shipment was $7680. A second order was placed later for the same number of shirts of each type. By this time the cost of the less expensive shirt had gone up to $6.25 and the cost of the more expensive shirt had gone up to $9.50. If the second shipment cost $8050, how many shirts of each type were in each shipment?

1.5 LINEAR SYSTEMS AND THEIR APPLICATIONS

Size of shipment

12. A store ordered a shipment of two different types of sweaters, one costing $12 apiece and another costing $20 apiece. The total cost of the shipment was $768. A second order was placed later for the same number of sweaters of each type. By this time the cost of the less expensive sweater had risen to $13 and the cost of the more expensive sweater had risen to $22. If the cost of the second shipment was $842, how many sweaters of each type were in each shipment?

Location of fire station

13. Consider the information given in Exercise 31, Section 1.3, and suppose a fire station is to be built equidistant from each of the three subdivisions. What should be its location in reference to City Hall?

Location of fire station

14. Consider the information given in Exercise 32, Section 1.3, and suppose a fire station is to be built equidistant from each of three subdivisions. What should be its location in reference to City Hall?

Stock portfolio

15. An investor bought two stocks, one selling at $20 a share and the other selling at $15 a share, for a net total of $16,500. The dividend on the higher priced stock is $1.60 per share and the dividend on the lower priced stock is $1.00 per share. If the investor expects to receive a total of $1300 in dividends from the two stocks in a given year, how many shares of each stock did he buy?

Stock portfolio

16. An investor bought two stocks, one selling at $18.75 a share and the other selling at $10.50 a share, for a net total of $9150. The dividend on the higher priced stock is $1.50 per share and the dividend on the lower priced stock is $1.00 per share. If the investor expects to receive a total of $800 in dividends from the two stocks in a given year, how many shares of each stock did he buy?

Gasohol mixture

17. A gasoline supplier has two large gasohol tanks, one containing 9% alcohol and the other containing 13% alcohol. If 2000 gallons of gasohol with a 10% alcohol content are needed, how many gallons should be taken from each tank in order to provide the proper mixture?

Gasohol mixture

18. A gasoline supplier has two large gasohol tanks, one containing 8% alcohol and the other containing 14% alcohol. If 3000 gallons of gasohol with a 10% alcohol content are needed, how many gallons should be taken from each tank in order to provide the proper mixture?

Size of shipment

19. A retailer pays $111.50 for a box containing Chewey and Crispy candy bars. He sells the Chewey bars for 75 cents apiece, the Crispy bars for 35 cents apiece, and takes in a total

of $171.25 for the sale of the shipment. If the Chewey bars cost 50 cents each and the Crispy bars cost 22 cents each, how many of each were in the shipment?

Size of shipment

20. A retailer pays $115 for a box containing nut and caramel candy bars. He sells the nut bars for 65 cents each, the caramel bars for 40 cents each, and takes in a total of $175.25 for the sale of the shipment. If the nut bars cost 45 cents apiece and the caramel bars cost 25 cents apiece, how many of each were in the shipment?

Acid solution mixture

21. The ABC Chemical Company wants to produce 500 gallons of an 18% acid solution by mixing a 25% acid solution with a 14% acid solution. How many gallons of each solution are needed?

Acid solution mixture

22. The ABC Chemical Company wants to produce 800 gallons of an 18% acid solution by mixing a 20% acid solution with a 14% acid solution. How many gallons of each solution are needed?

Stock portfolio

23. A person has $85,000 invested in stocks and bonds. If the return per year from stocks is 12% and that from bonds is 8% and if, in a given year, the income from these investments is $8680, what amount is invested in stocks?

Voting analysis

24. A political scientist obtained the following information for a mayoral race in which a Democrat named Smith won the election. The total number of Republicans and Democrats voting was 29,900. An exit poll found that 15% of the Republicans and 80% of the Democrats voted for Smith. If these two groups cast a total of 15,600 votes for Smith, how many Republicans voted? How many Democrats voted?

Dietary requirements

25. Suppose a person can obtain the minimum daily requirement of vitamin B_1, iron, and niacin from eating either one or both of two food products A and B. The following table indicates the percentage of the minimum daily requirement provided by one ounce of each food product.

Product	A, %	B, %
Vitamin B_1	25	13
Iron	10	15
Niacin	12	20

(a) How many ounces each of Product A and Product B would provide exactly 100% of the daily vitamin B_1 and iron requirements?

1.6 FUNCTIONS AND THEIR GRAPHS

Dietary requirement

(b) Would the amount of each food product calculated in part (a) provide the minimum daily requirement of niacin?

26. (a) Using the information in Exercise 25, determine how many ounces each of Product A and Product B would provide exactly 100% of the daily vitamin B_1 and niacin requirements?

(b) Would the amount of each food product calculated in part (a) provide the minimum daily requirement of iron?

1.6 FUNCTIONS AND THEIR GRAPHS

relation
graph
domain
range

As we know, a set of ordered pairs of real numbers can be paired with a set of points in the rectangular coordinate plane. Such a set is called a **relation** and the set of corresponding points in the plane is called the **graph** of the relation. If $f = \{(x, y)\}$ is such a relation, then its **domain** is the set of all first elements x in the ordered pairs and its **range** is the set of all second elements y. All of the following are examples of relations; their graphs are given in Figures 1.17 through 1.20.

$f_1 = \{(x, y) \mid y = 2x \text{ and } 1 \leq x \leq 3\}$
Domain of f_1: [1, 3]
 Range of f_1: [2, 6]
 Graph of f_1: Figure 1.17

$f_2 = \{(x, y) \mid y = x + 4, x \text{ a real number}\}$
Domain of f_2: set of real numbers
 Range of f_2: set of real numbers
 Graph of f_2: Figure 1.18

FIG. 1.17.

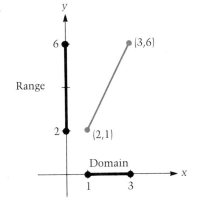

FIG. 1.18.

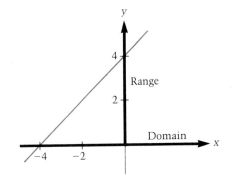

FIG. 1.19.

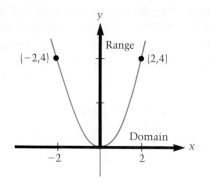

FIG. 1.20.

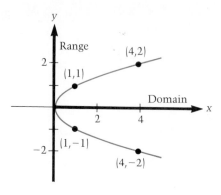

$f_3 = \{(x, y) \mid y = x^2, x \text{ a real number}\}$
  Domain of f_3: set of real numbers
  Range of f_3: set of nonnegative reals
  Graph of f_3: Figure 1.19

$f_4 = \{(x, y) \mid y^2 = x, x \geq 0\}$
  Domain of f_4: set of all nonnegative reals
  Range of f_4: set of all real numbers
  Graph of f_4: Figure 1.20

A special class of relations which are very useful in building mathematical models to solve practical problems is the class of functions. A **function** (of one variable) is a relation where no two ordered pairs have the same first element. Graphically, a relation is a function, provided no line parallel to the y-axis (range axis) intersects its graph in more than one point. For example, the relation defined by $y = x^2$, $x \geq 0$, is a function. (See Figure 1.21a.) However, the relation defined by $y^2 = x$, $x \geq 0$, is not a function since, for example, both ordered pairs $(4, 2)$ and $(4, -2)$ are elements in the relation. (See Figure 1.21b.)

function

FIG. 1.21. (a) $y = x^2$, $x \geq 0$. (b) $y^2 = x$, $x \geq 0$.

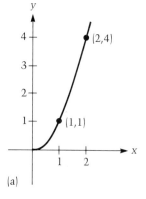

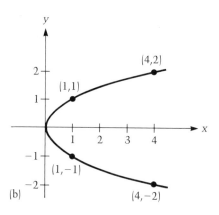

1.6 FUNCTIONS AND THEIR GRAPHS

The relation defined by the equation $Ax + By = C$, where x is any real number and $B \neq 0$, is a function. This function has the desirable property that the equation defining it can be solved for y in terms of x. Solving $Ax + By = C$ for y, we obtain

$$y = -\frac{A}{B}x + \frac{C}{B}.$$

linear function

Such an equation is said to give y *explicitly* in terms of x. Furthermore, this function, called a **linear function**, is often defined by the equation

$$y = mx + b.$$

As we know, the graph of this linear function is a nonvertical line in the coordinate plane having m as its slope and b as its y-intercept.

For any function f, the number y in the range paired with a given x in the domain is often denoted by $f(x)$ (read: "f of x," or "the value of f at x"). For example,

$$y = 3x + 5 \quad \text{and} \quad f(x) = 3x + 5$$

define the same function. Since $f(1)$ represents the number in the range paired with 1 in the domain, for the linear function $f(x) = 3x + 5$ we have $f(1) = 8$.

Linear function values

EXAMPLE 1 Consider the function defined by $f(x) = 5x - 6$, x any real number.

(a) What is $f(2)$?   (b) What is $f(-3)$?

(c) Is $f(2) \cdot f(-3) = f(-6)$?

Solution

(a) $f(2) = 5(2) - 6 = 4$.

(b) $f(-3) = 5(-3) - 6 = -21$.

(c) $f(-6) = 5(-6) - 6 = -36$. Since $f(2) \cdot f(-3) = -84$, the answer is "no."

independent variable
dependent variable

A letter, such as x, used to denote numbers in the domain of a function is called the **independent variable,** and a letter, such as y, used to denote numbers in the range is called the **dependent variable.** Generally, a function will be defined by an equation in the independent and dependent variables with the domain either stated or implied by the nature of the problem.

When functions are constructed to solve practical problems, it is often desirable to use letters other than x and y for the independent and dependent variables (domain and range variables). This practice helps to remember what each variable represents. Consider the following examples.

Area of a Circle: If A represents the area of a circle with radius r, then

$$A(r) = \pi r^2, \quad r > 0,$$

defines the corresponding area function; r is the independent variable and A is the dependent variable. (See Figure 1.22.)

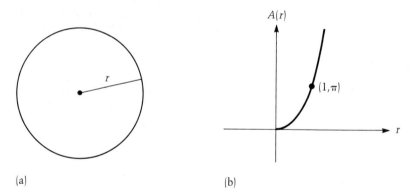

FIG. 1.22. (a) Circle with radius r. (b) $A(r) = \pi r^2$, $r \geq 0$.

(a)   (b)

Volume of a Sphere: If V represents the volume of a sphere with radius r, then

$$V(r) = \frac{4}{3}\pi r^3, \quad r > 0,$$

defines the corresponding volume function; r is the independent variable and V is the dependent variable. (See Figure 1.23.)

1.6 FUNCTIONS AND THEIR GRAPHS

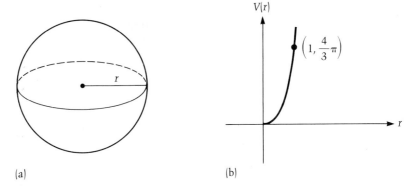

FIG. 1.23. (a) Sphere with radius r. (b) $V(r) = \dfrac{4}{3}\pi r^3$, $r \geq 0$.

Perimeter of a Square: If P represents the perimeter of a square with s as the length of one side, then

$$P(s) = 4s, \quad s > 0,$$

defines the corresponding perimeter function; s is the independent variable and P is the dependent variable. (See Figure 1.24.)

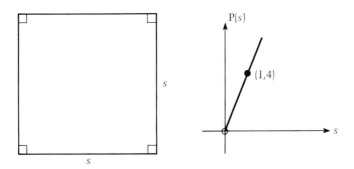

FIG. 1.24. $P(s) = 4s$.

The volume V of a right circular cylinder (tin can) having h as altitude and r as radius of the top and bottom is given by

$$V(r, h) = \pi r^2 h.$$

In this case, the dependent variable V is determined by the *two* independent variables r and h, and V is said to be a *function of two variables*. In Chapter 7, we shall investigate functions of more than one variable, but, unless stated otherwise, "function" will refer to a function of only one variable.

Consider the function defined by $f(x) = 2x - 4$, $3 \leq x \leq 8$. Since the graph of $f(x) = -2x - 4$ is a line, it is not hard to show that the graph of this function is the line segment given in Figure 1.25. Note that

$$f(3) \leq f(x) \quad \text{for each } x \text{ in } [3, 8]$$

and

$$f(8) \geq f(x) \quad \text{for each } x \text{ in } [3, 8].$$

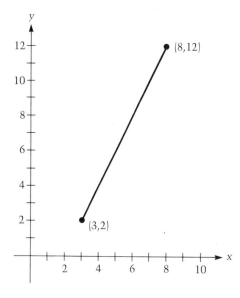

FIG. 1.25. $f(x) = 2x - 4$, $3 \leq x \leq 8$.

We call $f(3) = 2$ the *(absolute) minimum value* of the function and $f(8) = 12$ the *(absolute) maximum value* of the function. In general, for any function f, if there exists a number v in the domain such that $f(v) \geq f(x)$ for all x in the domain, then $f(v)$ is the **maximum value** of the function. Similarly, if there exists a number u in the domain such that $f(u) \leq f(x)$ for all x in the domain, then $f(u)$ is the **minimum value** of the function.

maximum value

minimum value

Maximum and minimum values

EXAMPLE 2 Consider the function defined by $f(x) = x^2$, $-2 \leq x \leq 2$. (See Figure 1.26.)

(a) What is its maximum value?

(b) What is its minimum value?

1.6 FUNCTIONS AND THEIR GRAPHS

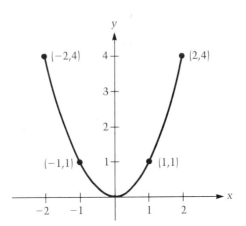

FIG. 1.26. $f(x) = x^2$, $-2 \leq x \leq 2$.

Solution

(a) It is clear from Figure 1.26 that 4 is the maximum value of the function.

(b) Since $x^2 \geq 0$ for any real number x and since $f(0) = 0$, the minimum value of the function is 0. (This can also be seen in Figure 1.26.)

PRACTICE PROBLEM

(a) Consider the function defined by $f(x) = x^2 + 4$, $-1 \leq x \leq 3$. What is its maximum value?

(b) What is its minimum value?

(c) What is its range?

Answer: (a) 13   (b) 4   (c) [4, 13] ∎

Although a function need not have an absolute maximum value (or an absolute minimum value), if it has one, it cannot have more than one. However, as in Example 2, a function can have a maximum value at more than one point in its domain. In fact, for the constant function defined by $f(x) = 6$, which has a horizontal line as its graph, the maximum and minimum value 6 is attained as each number in its domain.

Maximum and minimum values

EXAMPLE 3

(a) Graph the function defined by $f(x) = 2x + 4$, $-3 < x \leq 2$.

(b) Give the maximum value of the function if it exists.

(c) Give the minimum value of the function if it exists.

Solution

(a) See Figure 1.27.

(b) The maximum value of the function is $f(2) = 8$.

(c) The function has no minimum value. The range consists of all real numbers y such that $-2 < y \leq 8$; this set contains no least number.

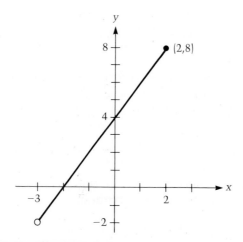

FIG. 1.27. $f(x) = 2x + 4$, $-3 < x \leq 2$.

Average cost function

EXAMPLE 4 The cost in dollars of producing q units of a certain product is given by

$$C(q) = 2q^2 + 4q + 50 \quad \text{for} \quad q \geq 1.$$

(a) What is the cost of producing one unit?

(b) What is the cost of producing the first 10 units?

(c) What is the average unit cost of producing the first 10 units?

(d) What is the average cost function?

(e) Graph the average cost function.

1.6 FUNCTIONS AND THEIR GRAPHS

(f) How many units will minimize the average cost per unit? (What is the minimum value of the average cost function?)

Solution

(a) $C(1) = 2(1)^2 + 4(1) + 50 = \$56$.

(b) $C(10) = 2(10)^2 + 4(10) + 50 = \$290$.

(c) Since $C(10) = \$290$ is the cost of the first 10 units,

$$\frac{C(10)}{10} = \$29$$

is the average unit cost.

average cost function

(d) In general, for a cost function $C(q)$, the **average cost function** is denoted by $\overline{C}(q)$ and defined by

$$\overline{C}(q) = \frac{C(q)}{q}.$$

Since $C(q) = 2q^2 + 4q + 50$,

$$\overline{C}(q) = \frac{2q^2 + 4q + 50}{q} = 2q + 4 + \frac{50}{q}.$$

(e) For "large" q, note that

$$2q + 4 + \frac{50}{q} \doteq 2q + 4.$$

Thus we see that for "large" q the function values for the average cost function are approximated by the function values of the linear function defined by

$$L(q) = 2q + 4.$$

This straight line with slope 2 and y-intercept 4 is called an *asymptote* for the graph of $\overline{C}(q)$; its graph approximates the graph of $\overline{C}(q)$ for large q. (See Figure 1.28.) Following is a table of values for the average cost function.

Average Cost Function

q	1	2	3	4	5	6	7	8	9
$\overline{C}(q)$	56	33	$26\frac{2}{3}$	$24\frac{1}{2}$	24	$24\frac{1}{3}$	$25\frac{1}{7}$	$26\frac{1}{4}$	$27\frac{5}{9}$

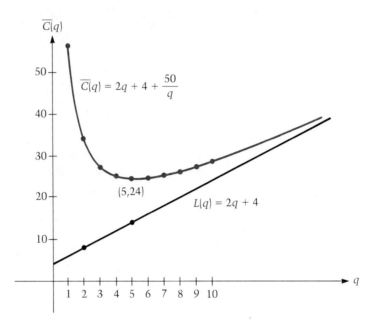

FIG. 1.28.

(f) It appears from the table of values and the graph of the average cost function that $\overline{C}(5) = \$24$ is the minimum average cost. If this is the case, producing 5 units would minimize average cost. (In Chapter 3 we shall find a way to prove that $\overline{C}(5) = 24$ is the minimum without resorting to a table of values or a graph of the function. This is just one type of application of differential calculus.)

Before turning to the exercises, let us point out that we will continue to learn more about functions, graphs of functions, and applications of functions throughout the remainder of the text.

EXERCISES

Set A–Function Values

In Exercises 1 through 8, answer the following questions.

1.6 FUNCTIONS AND THEIR GRAPHS

(a) What is $f(2)$? (b) What is $f(3)$? (c) Is $f(2) + f(3) = f(5)$? (d) Is $f(2) \cdot f(3) = f(6)$?

1. $f(x) = 8x + 3$, x a real number
2. $f(x) = 5x - 7$, x a real number
3. $f(x) = 3x^2$, x a real number
4. $f(x) = -x^2$, x a real number
5. $f(x) = x^2 + 4$, x a real number
6. $f(x) = x^2 - x$, x a real number
7. $f(x) = x^3$, x a real number
8. $f(x) = x^3 - 1$, x a real number
9. If f is the function defined by $f(x) = 3x - 7$, what numbers in the domain are paired with numbers in the range less than 15?
10. If f is the function defined by $f(x) = -7x + 3$, what numbers in the domain are paired with numbers in the range less than 12?
11. If g is the function defined by $g(x) = 5x - 3$, what numbers in the domain are paired with numbers in the range greater than -2?
12. If g is the function defined by $g(x) = -3x + 11$, what numbers in the domain are paired with numbers in the range greater than 3?
13. If $f(x) = 3x - 17$ and $g(x) = -3x + 4$, for what x is $f(x) \le g(x)$?
14. If $f(x) = -8x + 2$ and $g(x) = 12x - 8$, for what x is $f(x) < g(x)$?
15. If $f(x) = -7x + 13$ and $g(x) = 9x - 3$, for what x is $f(x) > g(x)$?
16. If $f(x) = 11x - 4$ and $g(x) = 13x + 8$, for what x is $f(x) \ge g(x)$?

Set B—Maximum and Minimum Values of Functions and Their Graphs

In Exercises 17 through 30, do each of the following. (a) Graph the function. (b) Give the maximum value of the function if one exists. (c) Give the minimum value of the function if one exists.

17. $f(x) = 2x - 5$, $-1 \le x \le 8$
18. $f(x) = 3x + 4$, $-3 \le x \le 1$
19. $f(x) = -3x + 2$, $-2 < x < 4$
20. $f(x) = -5x + 8$, $-4 < x < 5$

21. $f(x) = -4x + 7, -3 \leq x < 5$
22. $f(x) = -3x + 2, -2 < x \leq 6$
23. $f(x) = x^2 + 4, -1 \leq x \leq 6$
24. $f(x) = x^2 + 1, -3 \leq x \leq 8$
25. $f(x) = -x^2 + 9, -3 < x \leq 3$
26. $f(x) = -x^2 - 3, -2 \leq x < 4$
27. $f(x) = x + \dfrac{1}{x}, 1 \leq x \leq 8$
28. $f(x) = 4 + \dfrac{2}{x}, 1 \leq x \leq 10$
29. $f(x) = 4x + 2 + \dfrac{36}{x}$, x a positive real number
30. $f(x) = 2x + 3 + \dfrac{72}{x}$, x a positive real number

1.7 APPLICATIONS OF LINEAR FUNCTIONS

Suppose a new automobile costing $7200 depreciates at a constant amount of $800 per year. Its *initial value* $V(0)$ is $7200, and at the end of two years its value in dollars is

$$\begin{aligned} V(2) &= 7200 - 2(800) \\ &= 7200 - 1600 \\ &= 5600. \end{aligned}$$

At the end of t years, its value in dollars is

$$V(t) = 7200 - 800t.$$

Since $V(t)$ must be greater than or equal to zero,

$$\begin{aligned} 7200 - 800t &\geq 0, \\ -800t &\geq -7200, \\ t &\leq 9. \end{aligned}$$

Therefore, the domain of the function V is $0 \leq t \leq 9$. (See Figure 1.29.) If we assume that this formula holds for any number in the interval [0, 9], we find, for example, that

1.7 APPLICATIONS OF LINEAR FUNCTIONS

$$V\left(\frac{7}{2}\right) = 7200 - 800\left(\frac{7}{2}\right)$$
$$= 7200 - 2800$$
$$= \$4400$$

linear depreciation

is the value of the car after it is $3\frac{1}{2}$ years old. (*Linear depreciation* is a standard type of depreciation often used by businesses for machinery and other equipment.)

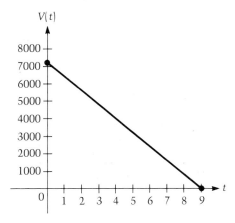

FIG. 1.29. *Note:* Different scales are used on axes. $V(t) = 7200 - 800t$, $0 \le t \le 9$.

Now let us consider the cost function C for buying x light bulbs selling for \$3.00 apiece. It is given by

$$C(x) = 3x$$

where x is a positive integer. The cost $C(x)$ is said to be *directly proportional* to x, the number of bulbs bought. If x is doubled, for example, then $C(x)$ is doubled. In general, if

$$y = kx,$$

directly proportional

then y is said to be **directly proportional** to x, and k is called the *constant of proportionality*. The graph of such a direct proportionality function is a straight line through the origin with slope k.

The average cost $\overline{C}(x)$ of x items, such as the \$3.00 light bulbs, is given by

$$\overline{C}(x) = \frac{C(x)}{x},$$

where $C(x)$ is the cost of x items. If $C(x) = 3x$, then $\overline{C}(x) = 3$, which, of course, is the expected result.

Suppose that the company from which $3.00 light bulbs are to be purchased will ship any number of the bulbs for an $8.00 packaging and mailing fee. The cost function for the person who plans to purchase the bulbs and have them delivered by mail is

$$C(x) = 3x + 8 \text{ dollars.}$$

We should observe the following things:

1. Although the cost function is a linear function, the cost is not directly proportional to the number of bulbs bought. For example, $C(1) = 11$ and $C(2) = 14$, and we see that although the number of bulbs is doubled, the cost is not doubled.

2. The average cost function is defined by

$$\overline{C}(x) = \frac{C(x)}{x} = 3 + \frac{8}{x},$$

which is neither a constant function nor a linear function.

3. We can see that the average cost decreases, since as x increases $8/x$ decreases. However, since $8/x$ is positive for positive x,

$$\overline{C}(x) = 3 + \frac{8}{x} > 3,$$

and the average cost is always greater than $3. In fact, $\overline{C}(x)$ "approaches" 3 as x gets large, and $y = 3$ is a horizontal asymptote for the graph of the average cost function.

4. Since the domain is the set of positive integers, we conclude that $C(1) = 11$ is the minimum value of $C(x)$ and that the cost function has no maximum value. (See Figure 1.30.) However, for the average cost function, $\overline{C}(1) = 11$ is its maximum value but it has no minimum value.

5. Although the domain of $C(x)$ and that of $\overline{C}(x)$ are the set of positive integers, it is common practice to graph such functions as though each domain were the set of real numbers x, where $x \geq 1$.

1.7 APPLICATIONS OF LINEAR FUNCTIONS

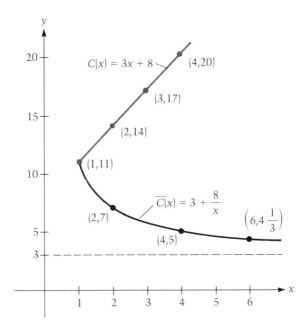

FIG. 1.30. $C(x) = 3x + 8$ and $\overline{C}(x) = 3 + \dfrac{8}{x}$ where $x \geq 1$.

Suppose a college organization decides to sell popcorn at a local fair in order to raise money for charity. If space at the fair rents for $60 and if the cost to produce each sack of popcorn is 20 cents, then the cost C of producing x sacks of popcorn is

$$C(x) = 0.20x + 60 \text{ dollars,}$$

where x is an integer greater than or equal to zero. Since

$$C(100) = 0.20(100) + 60 = 80,$$

the cost of producing 100 sacks of popcorn is $80. Although the graph of this function contains only those points on the line having 60 as y-intercept and $0.20 = \frac{1}{5}$ as slope for which x is an integer, this fact is ignored when we graph the function. (See Figure 1.31.) Of course, when solving any practical problem associated with such a graph, it should not be forgotten that the domain contains only integers.

Now suppose each sack of popcorn is sold for 50 cents. The total revenue R from selling x sacks is

$$R(x) = \$0.50x.$$

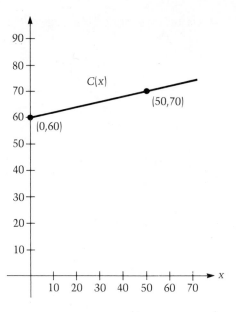

FIG. 1.31.
$C(x) = 0.20x + 60$, $x \geq 0$.

Therefore, for example, $R(100) = 0.5(100) = \$50$ would be received for selling 100 sacks of popcorn. The profit P on selling x sacks of popcorn is the difference between the revenue and cost for x sacks; it is given by

$$P(x) = R(x) - C(x).$$

If $R(x) > C(x)$, then $P(x) > 0$ and a profit is made. If $R(x) < C(x)$, then $P(x) < 0$ and a loss is incurred. For example, since

$$P(100) = R(100) - C(100) = 50 - 80 = -30,$$

the organization loses $30 if only 100 sacks of popcorn are sold.

break-even point

The point at which $P(x) = 0$ is called the **break-even point;** that is, the break-even point is the value of x for which

$$R(x) = C(x).$$

1.7 APPLICATIONS OF LINEAR FUNCTIONS

If the revenue and cost functions are graphed in the same coordinate system, the break-even point is where their graphs intersect. (See Figure 1.32.) For the organization selling popcorn at the fair, the break-even point is found as follows. Since $R(x) = 0.50x$ and $C(x) = 0.20x + 60$, $R(x) = C(x)$ if

$$0.50x = 0.20x + 60,$$
$$0.30x = 60,$$
$$x = 200.$$

Therefore, to make a profit, the organization would have to sell more than 200 sacks of popcorn. If the organization wants to make a profit of $300, it would require that $P(x) = 300$. Solving $R(x) - C(x) = 300$, we obtain

$$0.50x - (0.20x + 60) = 300,$$
$$0.30x = 360,$$
$$x = 1200.$$

Therefore, a profit of $300 is realized by selling 1200 sacks of popcorn.

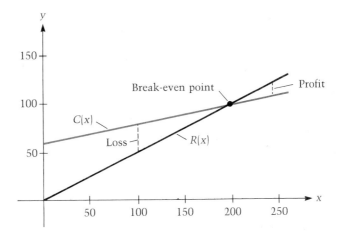

FIG. 1.32.
$P(x) = R(x) - C(x)$.

PRACTICE PROBLEM Suppose the organization selling popcorn at the fair could rent space for $50. If popcorn costing 15 cents per sack to produce could be sold for 35 cents, what would be the break-even point?

Answer: 250. ■

Sales function

EXAMPLE 1

(a) A manufacturer of television sets concludes from previous introductions of new television models that 15,000 sets can be sold with no advertising. Also based on past sales records, it is assumed that for each $1000 in advertising an additional 50 sets will be sold. How many sets will be sold with an advertising budget of x thousand dollars?

Solution: If N is the number of sets sold, then

$$N(x) = 15{,}000 + 50x.$$

Since, for example, $N(0) = 15{,}000$ and $N(3) = 15{,}150$, this means that 15,000 sets will be sold with no advertising budget and 15,150 sets will be sold with a $3000 advertising budget.

Revenue function

(b) Suppose each television set sells for $650. What is the revenue function R in terms of x?

Solution: Since $N(x)$ is the number sold with an advertising budget of x thousand dollars and since each set sells for $650,

$$\begin{aligned} R(x) &= 650 N(x) \\ &= 650(50x + 15{,}000) \\ &= 32{,}500x + 9{,}750{,}000. \end{aligned}$$

Note that the total revenue with no advertising budget is

$$R(0) = \$9{,}750{,}000.$$

Cost function

(c) Suppose that the initial (fixed) costs to produce the television sets were $4,500,000 and that each set costs $500 to manufacture. What is the total cost function C in terms of x?

Solution: The total cost would include
    the fixed costs: $4,500,000,
    the costs to manufacture $N(x)$ sets: $\$500 N(x)$,
    the costs for x thousand dollars of advertising: $\$1000x$.
Consequently,

$$\begin{aligned} C(x) &= 1000x + 500(15{,}000 + 50x) + 4{,}500{,}000 \\ &= 1000x + 7{,}500{,}000 + 25{,}000x + 4{,}500{,}000 \\ &= 26{,}000x + 12{,}000{,}000. \end{aligned}$$

1.7 APPLICATIONS OF LINEAR FUNCTIONS

Note that if $x = 0$, then no advertising is done and the cost to produce the first 15,000 sets would be

$$C(0) = \$12,000,000.$$

Profit function

(d) Find the profit function P in terms of x.

Solution: Since the profit function P is given by $P(x) = R(x) - C(x)$,

$$P(x) = 32,500x + 9,750,000 - 26,000x - 12,000,000$$
$$= 6500x - 2,250,000.$$

Since $P(0) = -\$2,250,000$, a loss of $\$2,250,000$ would result if the company did not advertise.

Break-even point

(e) What is the break-even point?

Solution: The break-even point is at the x such that $P(x) = 0$; it is found as follows.

$$6500x - 2,250,000 = 0,$$
$$6500x = 2,250,000,$$
$$x \doteq 346.15385.$$

Since x is in thousands of dollars, an advertising budget of approximately \$346,154 is required to break even; that is, the break-even point is reached when $N(346.2) = 32,310$ sets are sold.

Profit from advertising

(f) What would be the profit with an advertising budget of \$1,000,000?

Solution: Since $1,000,000 = 1000$ thousands, the profit from a million dollar advertising budget is

$$P(1000) = 6500(1000) - 2,250,000$$
$$= \$4,250,000.$$

---

demand function

Assume $D(p)$ is the number of units of a certain product that consumers will demand at a certain price p. If we assume D is determined solely by p, then a **demand function** with p as independent variable can be constructed as a model for market demand. Since consumers are reluctant to pay high prices, it is reasonable to expect that the demand decreases as price p increases. On the other side of this market are the producers who supply the product. Since their goal is to make a profit, it is reasonable to expect that as price increases, profit increases and the number of units S the producers are willing to supply increases. If supply S is considered to be determined solely by price p, then a **supply function** $S(p)$ with p as independent variable can be constructed as a model for market supply. For a given price p, if supply $S(p)$ is less than demand $D(p)$, then a shortage exists. If supply $S(p)$ is greater than demand $D(p)$, then a surplus results. The value $\bar{p}$ such that $D(\bar{p}) = S(\bar{p})$ is called the **equilibrium price**, $\bar{q} = D(\bar{p}) = S(\bar{p})$ is called the **equilibrium quantity**, and $(\bar{p}, \bar{q})$ is called the **market equilibrium point.** (See Figure 1.33.)

supply function

equilibrium price
equilibrum quantity

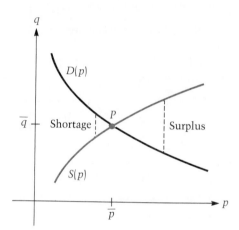

FIG. 1.33. P = equilibrium point; $\bar{p}$ = equilibrium price.

Demand function

EXAMPLE 2 Suppose the demand function for a certain product is given by

$$D(p) = 1000(82 - 8p) \text{ units,}$$

where $p \geq \$4.50$.

(a) How many units will be sold if the price is $4.50?

(b) How many units will be sold if the price is $6.00?

(c) At what price would the consumer be unwilling to buy the product?

1.7 APPLICATIONS OF LINEAR FUNCTIONS

Solution

(a) The number of units that will be sold at $4.50 is

$$D(4.50) = 1000[82 - 8(4.5)]$$
$$= 1000(82 - 36)$$
$$= 46{,}000 \text{ units.}$$

(b) The number of units that will be sold at $6.00 is

$$D(6) = 1000(82 - 48) = 34{,}000 \text{ units.}$$

(c) The price at which the consumer will be unwilling to buy the product is when $D(p) = 0$. Since $D(p) = 0$ when

$$1000(82 - 8p) = 0,$$
$$82 - 8p = 0,$$
$$-8p = -82,$$
$$p = 10.25,$$

demand will be zero if the price is $10.25.

Supply function

EXAMPLE 3 Suppose the supply function for the product discussed in Example 2 is given by

$$S(p) = 2000p \text{ units,}$$

where $p \geq \$4.50$.

(a) How many units will be supplied if the price is $4.50?
(b) How many units will be supplied if the price is $6.00?
(c) What is the equilibrium price for the product?
(d) What is the equilibrium quantity?

Solution

(a) $S(4.50) = 2000(4.50) = 9000$ units will be supplied if $p = \$4.50$.

(b) $S(6) = 2000(6) = 12{,}000$ units will be supplied if $p = \$6.00$.

(c) Since the equilibrium price is reached when $S(p) = D(p)$,

$$2000p = 1000(82 - 8p),$$
$$2p = 82 - 8p,$$
$$10p = 82,$$
$$p = 8.2.$$

The equilibrium price is $8.20.

(d) The equilibrium quantity is $S(8.2) = 2000(8.2) = 16{,}400$ units.

EXERCISES

Set A—Applications of Linear Functions

Linear depreciation

1. (a) Assume an automobile costs $8000 and assume it depreciates linearly at $1200 per year. Express the value V of the car in terms of t years.
    (b) What is $V(2)$?
    (c) For what value of t is $V(t) = 0$?
    (d) What is the value of the car after 30 months?

Linear depreciation

2. (a) Assume an automobile costs $6000 and assume it depreciates linearly at $800 per year. Express the value V of the car in terms of t years.
    (b) What is $V(2)$?
    (c) For what value of t is $V(t) = 0$?
    (d) What is the value of the car after 33 months?

Demand function

3. (a) Suppose the demand function for a certain product is given by

    $$D(p) = 1000(76 - 8p) \text{ units}$$

    if $p \geq \$3.50$. How many units will be sold if the price is $3.50?
    (b) How many units will be sold if the price is $6.00?
    (c) At what price would the consumer be unwilling to buy the product?
    (d) What is $D(p + 1) - D(p)$? What does this number represent?

Demand function

4. (a) Suppose the demand function for a certain product is given by

    $$D(p) = 200(84 - 6p) \text{ units}$$

    if $p \geq \$2.50$. How many units will be sold if the price is $2.50?

1.7 APPLICATIONS OF LINEAR FUNCTIONS

(b) How many units will be sold if the price is $5.00?
(c) At what price would the consumer not be willing to buy the product?
(d) What is $D(p + 1) - D(p)$? What does this number represent?

Supply function

5. (a) Suppose the supply function for the product discussed in Exercise 3 is given by

$$S(p) = 8000p \text{ units,}$$

if $p \geq \$3.50$. How many units will be supplied if the price is $3.50?
(b) How many units will be supplied if the price is $6.00?
(c) What is the equilibrium price for the product?
(d) What is the equilibrium quantity?

Supply function

6. (a) Suppose the supply function for the product discussed in Exercise 4 is given by

$$S(p) = 1200p \text{ units,}$$

if $p \geq \$2.50$. How many units will be supplied if the price is $2.50?
(b) How many units will be supplied if the price is $5.00?
(c) What is the equilibrium price for the product?
(d) What is the equilibrium quantity?

Repair charges

7. (a) An automobile repair agency states that its repair charges C are directly proportional to the time t in hours spent working on a car. If it charges $43.50 for $1\frac{1}{2}$ hours of work, what is the constant of proportionality?
(b) What does the constant of proportionality represent in this problem?
(c) What is the function associated with the repair charges?
(d) What would it cost for a repair job that took $4\frac{3}{4}$ hours?

Production function

8. (a) Assume that the number of units N of a product that can be produced by a given machine is directly proportional to the number of hours t the machine is allowed to run. If the machine produces 4880 units in 8 hours, what is the constant of proportionality?
(b) What does the constant of proportionality represent in this problem?
(c) What is the production function in terms of the number of hours t that the machine runs?
(d) How many units would the machine produce in 100 hours?

(e) How many hours would it take for the machine to produce 128,100 units?

Charges for checking account

9. (a) A bank charges $3.50 per month for a checking account and 15 cents for each check. What would be the charges for writing 20 checks in one month?
   (b) If x checks are written in one month, what would be the charges C in terms of x?
   (c) Use part (b) to find the charges for writing 35 checks.
   (d) What is the average cost function per month for writing checks?
   (e) What is the average cost for writing 35 checks in one month?
   (f) Graph the cost and average cost functions in the same coordinate system.

Charges for checking account

10. (a) A bank charges $5.00 per month for a checking account and 10 cents for each check. What would be the charges for writing 20 checks in one month?
    (b) If x checks are written in one month, what would be the charges C in terms of x?
    (c) Use part (b) to find the charges for writing 35 checks.
    (d) What is the average cost function per month for writing checks?
    (e) What is the average cost of writing 35 checks in one month?
    (f) Graph the cost and average cost functions in the same coordinate system.

Break-even point

11. Given the checking account charges discussed in Exercises 9 and 10, how many checks would have to be written to result in equal charges for the two accounts?

Political campaign

12. (a) A political scientist concludes that in a particular election a Republican candidate in a given district will receive not only 8000 votes from registered Republicans but also 40 percent of any additional votes cast. Express the number of votes V the Republican receives in terms of the number x of additional votes cast.
    (b) What is the maximum number of additional votes that could be cast so that the Republican would have a majority of the votes?

Average adult weight

13. (a) It is estimated that the average weight W, in pounds, of adults in the United States who are h inches in height is given by $W(h) = (11/3)h - 110$, where $60 \leq h \leq 84$. What would be the average weight of an adult who is 5 feet 5 inches tall?

1.7 APPLICATIONS OF LINEAR FUNCTIONS

(b) What would be the expected height of an adult who weighs 198 pounds?

Use the following information in Exercises 14 through 19. A manufacturer of television sets decides from previous experiences in introducing new television models that 12,000 sets of a given model can be sold with no advertising. Also, based on past sales records, it is anticipated that for each $1000 in advertising an additional 40 sets will be sold.

Sales function

14. (a) How many sets N will be sold with an advertising budget of x thousand dollars?
    (b) Use part (a) to determine how many sets will be sold with an advertising budget of $90,000.

Revenue function

15. (a) Suppose each television set sells for $600. What is the total revenue function R in terms of x?
    (b) Use part (a) to determine the revenue received with an advertising budget of $90,000.

Cost function

16. (a) Suppose that the fixed costs to produce the television sets are $3,500,000 and that each set costs $450 to manufacture. What is the total cost function in terms of x?
    (b) Use part (a) to determine the total cost with an advertising budget of $90,000.

Profit function

17. (a) Find the profit function in terms of x.
    (b) Use part (a) to determine the profit (or loss) with an advertising budget of $90,000.

Break-even point

18. (a) What advertising budget is required to reach the break-even point?
    (b) How many sets need to be sold to break even?

Profit from advertising

19. What would the profit (loss) be with an advertising budget of $1,000,000?

Use the following information in Exercises 20 through 25. A manufacturer of televisions decides from previous experiences in introducing new television models that 8000 sets can be sold with no advertising. Also, based on past sales records, it is anticipated that for each $1000 in advertising an additional 30 sets will be sold.

Sales function

20. (a) How many sets will be sold with an advertising budget of x thousand dollars?
    (b) Use part (a) to determine how many sets will be sold with an advertising budget of $100,000.

Revenue function

21. (a) Suppose each television set sells for $500. What is the total revenue function R in terms of x?

Cost function

      (b) Use part (a) to determine the revenue received with an advertising budget of $100,000?

22. (a) Suppose the fixed costs to produce the television sets were $2,500,000 and suppose each set costs $400 to manufacture. What is the total cost function in terms of x?
    (b) Use part (a) to determine the total cost with an advertising budget of $100,000.

Profit function

23. (a) Find the profit function in terms of x.
    (b) Use part (a) to determine the profit with an advertising budget of $100,000.

Break-even point

24. (a) What advertising budget is required to break even?
    (b) How many sets need to be sold to break even?

Profit from advertising

25. What would be the profit (loss) with an advertising budget of $1,000,000?

1.8 SUMMARY AND REVIEW EXERCISES

1. BASIC INEQUALITY PROPERTIES

(a) If $a < b$ and $b < c$, then $a < c$.

(b) If $a < b$, then $a + c < b + c$ and $a - c < b - c$.

(c) If $a < b$ and $c > 0$, then $ac < bc$ and $\dfrac{a}{c} < \dfrac{b}{c}$.

(d) If $a < b$ and $c < 0$, then $ac > bc$ and $\dfrac{a}{c} > \dfrac{b}{c}$.

(e) If $a < b$ and $c < d$, then $a + c < b + d$.

2. MIDPOINT AND DISTANCE

(a) The **midpoint** of the line segment with (x_1, y_1) and (x_2, y_2) as endpoints is given by

$$\left(\frac{x_1 + x_2}{2}, \frac{y_1 + y_2}{2}\right).$$

(b) The **distance** between the two points (x_1, y_1) and (x_2, y_2) in the coordinate plane is given by

$$d = \sqrt{(x_2 - x_1)^2 + (y_2 - y_1)^2}.$$

3. SLOPES OF LINES

(a) If (x_1, y_1) and (x_2, y_2) are two points in the coordinate plane such that $x_1 \neq x_2$, then

1.8 SUMMARY AND REVIEW EXERCISES

$$m = \frac{y_2 - y_1}{x_2 - x_1}$$

is the **slope** of the line containing the two points.

(b) Two nonvertical lines with slopes m_1 and m_2 are **parallel** if and only if $m_1 = m_2$.

(c) Two nonvertical lines with slopes m_1 and m_2 are **perpendicular** if and only if $m_1 = -1/m_2$.

4. EQUATIONS OF LINES

(a) An equation of a line containing the point (x_1, y_1) having slope m is

$$y - y_1 = m(x - x_1). \quad \text{(Point-slope form)}$$

(b) An equation of the line with m as slope and b as y-intercept is

$$y = mx + b. \quad \text{(Slope-intercept form)}$$

(c) Any *linear equation in two variables* can be written in the form $Ax + By = C$. If $B \neq 0$, its *slope* is $-A/B$.

5. FUNCTIONS

(a) A **function** is a set of ordered pairs of real numbers such that no two ordered pairs have the same first element. The **domain** is the set of all first elements and the **range** is the set of all second elements.

(b) A letter, such as x, used to denote numbers in the domain of a function is called the **independent variable,** and a letter, such as y, used to denote numbers in the range is called the **dependent variable.**

(c) For any function f, if there exists a number v in the domain such that $f(v) \geq f(x)$ for each x in the domain, then $f(v)$ is the **maximum value** of f. If there exists a number u in the domain such that $f(u) \leq f(x)$ for each x in the domain, then $f(u)$ is the **minimum value** of f.

6. LINEAR FUNCTIONS

(a) A **linear function** is defined by $y(x) = mx + b$, where x is any real number. Its graph is the line having m as slope and b as y-intercept.

(b) For the linear function defined by $y(x) = kx$, where k is a constant, $y(x)$ is said to be **directly proportional** to x and k is called the **constant of proportionality**.

7. SOME GENERAL FUNCTIONS

(a) If $C(x)$ is the total cost function for x units of a given product, then

$$\overline{C}(x) = \frac{C(x)}{x}$$

is the **average cost function**.

(b) If $R(x)$ is the revenue function for x units of a given product and $C(x)$ is the total cost function, then $P(x) = R(x) - C(x)$ is the **profit function**. The **break-even point** is where $P(x) = 0$.

(c) If $D(p)$ is a demand function and $S(p)$ is a supply function for a certain commodity, then the value $\overline{p}$ such that $D(\overline{p}) = S(\overline{p})$ is called the **equilibrium price**, $\overline{q} = D(\overline{p}) = S(\overline{p})$ is called the **equilibrium quantity**, and $(\overline{p}, \overline{q})$ is the **market equilibrium point**.

REVIEW EXERCISES

1. Find the solution set:

$$\frac{4x - 3}{5} - \frac{2x + 1}{3} \geq \frac{5x + 6}{4}.$$

2. Find the solution set:

$$1.35x + 4.0 < 5.45x - 8.3.$$

3. Find an equation of the line containing $(-2, 3)$ and $(4, 7)$.

4. Find an equation of the line containing $(-5, 2)$ and $(1, -3)$.

5. Find an equation of the perpendicular bisector of the line segment with $(-4, 7)$ and $(8, 5)$ as endpoints.

6. Find an equation of the perpendicular bisector of the line segment with $(-3, -4)$ and $(5, 8)$ as endpoints.

7. Solve the linear system:

$$2x - 3y = 9,$$
$$4x + 7y = -8.$$

1.8 SUMMARY AND REVIEW EXERCISES

8. Solve the linear system:

$$5x + 3y = 6,$$
$$9x - 4y = -55.$$

9. Find the maximum and minimum values of the function defined by $f(x) = -6x + 2$, $1 \le x \le 5$.

10. Find the maximum and minimum values of the function defined by

$$f(x) = x^2 + 2x + 5, \quad -2 \le x \le 6.$$

Hint: $x^2 + 2x + 5 = (x^2 + 2x + 1) + 4$.

11. A large department store receives a shipment of 225 radios containing two makes. One costs $35 apiece and the other costs $62 apiece. If the total bill for the shipment is $11,088, how many radios priced at $35 are in the shipment?

12. A total of $28.25 is collected from a parking meter containing 419 coins which are nickels and dimes. How many dimes were in the meter?

13. A person bought a total of 76 gallons of gasoline for two different cars in one month and received a bill for $91.44. If one car uses "regular" costing $1.17 a gallon and the other car uses "unleaded" costing $1.23 a gallon, how many gallons of "regular" were bought?

14. Two trucks must make deliveries over a route totaling 600 miles. Truck A gets 12 miles per gallon and Truck B gets 9 miles per gallon. If the total fuel consumption for the deliveries cannot exceed 56 gallons, what is the minimum number of miles that Truck A must travel?

15. A company has three outlet stores which are located in relation to company headquarters as follows: Outlet A is 300 miles east and 400 miles south. Outlet B is 500 miles west and 200 miles north. Outlet C is 100 miles east and 400 miles north. At what location in relation to company headquarters should a warehouse be located to be equidistant from each of the outlet stores?

16. A retailer pays $67.50 for a shipment containing nut bars and chocolate bars. He sells the nut bars for 65 cents apiece, the chocolate bars for 45 cents apiece, and takes in a total of $99.50 for the sale of the shipment. If the nut bars cost 45

cents apiece and the chocolate bars cost 30 cents apiece, how many of each were in the shipment?

17. A salesman sells two types of motors selling for $328 and $450 apiece. In addition to a salary, he gets a 5% commission on each $328 motor he sells and 7% commission on each $450 motor. In one month his combined sales for the two motors are $20,190 and his total commission is $1216.50. How many of each type of motor did he sell?

18. The cost function for x units of a particular product is $C(x) = 3x^2 + 2x + 15$ dollars and the revenue function is $R(x) = 48x$ dollars. Find the break-even point for this product.

19. An automobile repair shop states that its repair charges C are directly proportional to the time t in hours worked on a car. If it charges $62.50 for a job of 1 hour and 40 minutes, what is the repair function C in terms of t?

20. For the cost function $C(x) = 3x^2 + 2x + 27$, what is the average cost function $\overline{C}(x)$? Determine what integer x minimizes $\overline{C}(x)$. What is the minimum average cost?

BIOGRAPHICAL SKETCH

The Bettmann Archive

René Descartes (1596–1650)

René Descartes was born on March 31, 1596, near Tours, France. He is ranked, along with Fermat and Pascal, as one of the foremost French mathematicians of the seventeenth century, the century which produced such other great figures in the arts and sciences as Shakespeare, Milton, Galileo, Newton, and Kepler. In addition to Descartes' contribution to mathematics, he wrote a text on physiology, studied psychology, and greatly influenced the development of modern philosophy.

In mathematics Descartes is most widely known for establishing the field of analytic geometry. His treatise *La géométrie*, published in 1637, contains his important results in this field and gained for him the title of "father of analytic geometry." John Stuart Mill said that "analytic geometry constitutes the greatest single step ever made in the progress of the exact sciences"; although this might be an overstatement, he was correct when he remarked that the development of analytic geometry "immortalized the name of Descartes." The creation of analytic geometry brought about a unification of algebra and geometry into a reservoir of important ideas from which the concepts of the calculus would flow naturally.

Although Descartes was neither very healthy nor physically strong, he indulged in an active life for many years. Besides spending some rakish years in Paris, he chose on several occasions to become involved in the wars of the period. In 1628 he moved to Holland where for the next 20 years he led a quieter life and produced much of his most important work. In 1649 Queen Christina of Sweden tried for a year to persuade Descartes to join her retinue of intellectuals, but he managed to ignore the invitation until she flattered him by providing a battleship for him to make the journey. Just 11 weeks after his arrival in Stockholm and exposure to bitter winter weather, he contracted influenza and died in 1650 at the age of 54.

CHAPTER 2

DIFFERENTIAL CALCULUS AND ITS APPLICATIONS

2

2.1 INTRODUCTION
2.2 THE LIMIT OF A FUNCTION AND CONTINUITY
2.3 THE NEWTON QUOTIENT
2.4 THE DERIVATIVE OF A FUNCTION
2.5 PRODUCT, QUOTIENT, AND GENERAL POWER RULES FOR DIFFERENTIATION
2.6 THE DIFFERENTIAL AND MARGINAL ANALYSIS
2.7 SUMMARY AND REVIEW EXERCISES

2.1 INTRODUCTION

Some ideas related to the calculus date back as far as the time of Archimedes (287–213 B.C.), but the subject was not developed as a branch of mathematics until the seventeenth century. Much of this was done independently by the English mathematician, Sir Isaac Newton (1642–1727), and the German mathematician, Wilhelm Leibniz (1646–1716). It is remarkable that the calculus evolves quite naturally from two rather elementary problems:

(1) *finding the slopes of tangents to curves in the coordinate plane and*

(2) *finding the areas of certain types of regions in the coordinate plane.*

Differential calculus was developed to solve the first problem and *integral calculus* to solve the second.

In this chapter we find how "tools" of *differential calculus* can be used to solve such problems as:

1. Find the slope of the tangent to the graph of $f(x) = x^{2/3} - 2$ at the point where $x = 8$. (See Figure 2.1a and Exercise 18, Section 2.4.)

2. Find an equation of the line tangent to the graph of $f(x) = x^2 + 3x + 1$ and parallel to the line $2x - y = 2$. (See Figure 2.1b and Exercise 23, Section 2.4.)

3. Suppose that after x seconds the distance in feet traveled by a bullet shot vertically upward is given by $s(x) = 400x - 16x^2$.

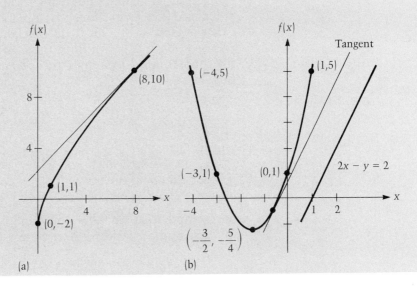

FIG. 2.1.
(a) $f(x) = 3x^{2/3} - 2$.
(b) $f(x) = x^2 + 3x + 1$.

What is its initial velocity? What is the velocity of the bullet at the end of 2 seconds? When does it reach its maximum height? How far has the bullet traveled when the maximum height is reached? (See Exercise 32, Section 2.5.)

4. If the cost function in dollars for x units of a given product is $C(x) = 4000 + 15x + 8x^{1/2}$, at what level of production would one additional unit cost $15.50? (See Exercise 18, Section 2.6.)

After developing the basic techniques of differential calculus in this chapter, we shall be prepared to solve a much wider variety of problems in Chapter 3. However, it is important to consider first what is called the *limit of a function*. The limit concept is not only the basis for both differential and integral calculus but also for a very large branch of mathematics called *analysis*. ■

2.2 THE LIMIT OF A FUNCTION AND CONTINUITY

limit

For a function f, let t be a real number such that each open interval which contains t also contains at least one number x different from t in the domain of f. Note we do not insist that t be in the domain of f, only that there exist numbers in the domain of f "close" to t. (Of course, the statement does not preclude t from being in the domain.) A number L is called the **limit** of f at t if and only if $f(x)$ is close to L for each x different from t in the domain of f that is close to t. This definition, though not mathematically precise enough to prove limit theorems, should become clear in the context of the examples. If the limit of f at t is L, we express this by

$$\lim_{x \to t} f(x) = L. \quad \text{(The limit of } f \text{ as } x \text{ approaches } t \text{ is } L.\text{)}$$

or

$$f(x) \to L \text{ as } x \to t. \quad (f(x) \text{ approaches } L \text{ as } x \text{ approaches } t.)$$

For example, the limit of $f(x) = 3x + 2$ at $t = 5$ (or as x approaches 5) is denoted by $\lim_{x \to 5}(3x + 2)$. To find this limit we note, for example, that $f(4.99) = 16.97$ and $f(5.001) = 17.003$ and that for each number x near 5, $f(x)$ is near 17. Therefore

$$\lim_{x \to 5}(3x + 2) = 17.$$

2.2 THE LIMIT OF A FUNCTION AND CONTINUITY

Note for this function that $f(5) = 17$.

To find the limit of

$$f(x) = \frac{x^2 - 25}{x - 5} \quad \text{at} \quad t = 5$$

is somewhat more complicated than in the previous example since 5 is not in the domain of the function. (See Figure 2.2.) Since

$$\frac{x^2 - 25}{x - 5} = \frac{(x + 5)(x - 5)}{x - 5} = x + 5$$

for each $x \neq 5$, $f(x) = x + 5$ for $x \neq 5$ and we find that for each number close to (but different from 5) $f(x)$ is close to 10. Therefore

$$\lim_{x \to 5} \frac{x^2 - 25}{x - 5} = 10.$$

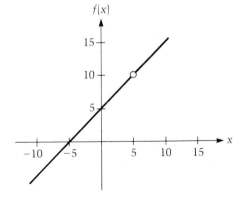

FIG. 2.2. $f(x) = \dfrac{x^2 - 25}{x - 5}$, $x \neq 5$.

PRACTICE PROBLEM Find

$$\lim_{x \to 7} \frac{x^2 - 49}{x - 7}$$

Answer: 14. ∎

Limit of a function

EXAMPLE 1 Let $f(x) = 8x + 1$. Then, for example,

$$\lim_{x \to 3} f(x) = 25.$$

In fact, for any real number t,

$$\lim_{x \to t} f(x) = 8t + 1.$$

polynomial function

For any **polynomial function** $P(x) = a_0 x^n + a_1 x^{n-1} + \cdots + a_n$, it can be proved from a precise definition of the limit of a function that

$$\lim_{x \to t} P(x) = P(t).$$

In fact, for any real number r,

$$\lim_{x \to t} [P(x)]^r = [P(t)]^r,$$

provided t is in the domain of $[P(x)]^r$. Furthermore, for any two polynomials $P(x)$ and $Q(x)$, where $Q(t) \neq 0$,

$$\lim_{x \to t} \frac{P(x)}{Q(x)} = \frac{P(t)}{Q(t)}.$$

Limit of a function

EXAMPLE 2 Find the limit of $f(x) = \dfrac{x^2 + 1}{5x - 3}$ at $t = 4$.

Solution: $\lim\limits_{x \to 4} \dfrac{x^2 + 1}{5x - 3} = \dfrac{17}{17} = 1.$

PRACTICE PROBLEM Find $\lim\limits_{x \to 1} \dfrac{x^2 - 3}{3x + 7}$.

Answer: $-\frac{1}{5}$. ■

2.2 THE LIMIT OF A FUNCTION AND CONTINUITY

Limit of a function

EXAMPLE 3 Find the limit of $f(x) = (4x + 1)^{3/2}$ at $t = 2$.

Solution: $\lim_{x \to 2} (4x + 1)^{3/2} = 9^{3/2} = 27$.

Limit of a function

EXAMPLE 4 Find $\lim_{x \to -2} \dfrac{(x^3 + 8)}{(x + 2)}$.

Solution: Since -2 is not in the domain of the function, we could determine the function values for such numbers as -1.99 and -2.01 in order to find (guess) the limit, if it exists. However, the following algebraic approach demonstrates a convincing and reliable analysis. Note that $x^3 + y^3 = (x + y)(x^2 - xy + y^2)$. Hence

$$\lim_{x \to -2} \frac{x^3 + 8}{x + 2} = \lim_{x \to -2} \frac{(x + 2)(x^2 - 2x + 4)}{x + 2}$$
$$= \lim_{x \to -2} (x^2 - 2x + 4)$$
$$= 12.$$

PRACTICE PROBLEM Find $\lim_{x \to 2} \dfrac{x^4 - 16}{x - 2}$.

Answer: 32. ∎

As indicated earlier, the limit of a function f at a particular t may not exist. For example,

$$\lim_{x \to 0} \frac{1}{x^2} \quad \text{does not exist.}$$

This can be seen by observing that $f(x) = 1/x^2$ becomes "large" as x approaches 0 and thus cannot be close to any given real number L.

For a given function, it may be very difficult to find whether or not the limit exists at a given t. Even if it is known that the limit of a function f exists at some t, it still is often difficult (or impossible) to determine the limit exactly. As an example, consider the function f defined by

$$f(x) = (1 + x)^{1/x}, \quad x > -1 \quad \text{and} \quad x \neq 0.$$

This function is not defined at zero and there is no simple algebraic manipulation that will determine whether or not it has a limit at zero. A careful sketch of the graph of f will strongly suggest that the limit does exist at zero. Furthermore, some numerical computations will provide an approximation of the limit. Let us do this.

First let us make several observations about the graph of $f(x) = (1 + x)^{1/x}$ where $x > -1$ and $x \neq 0$. We see, for example, that

$$f(-1/2) = 4, \quad f(-99/100) = (100)^{100/99} > 100, \quad \text{and} \quad f(1) = 2.$$

For x "close" to -1, the function values are very large. If x is very large, the function values are "close" to (but greater than) 1. Thus the graph approaches the lines $x = -1$ and $y = 1$. These lines are asymptotes for the graph of f. (They are lines which "eventually" approximate the graph of f.) The function values of f, which are given in the following table, clearly indicate what happens to the graph of f for x near 0. (See Figure 2.3.)

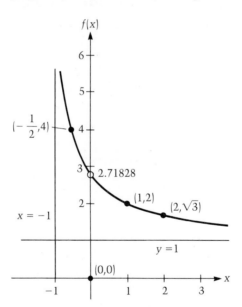

FIG. 2.3. $f(x) = (1 + x)^{1/x}$, $x > -1$, $x \neq 0$.

Table of Function Values for
$f(x) = (1 + x)^{1/x}$

$f(0.1) = 2.594$	$f(-0.1) = 2.868$
$f(0.01) = 2.705$	$f(-0.01) = 2.732$
$f(0.001) = 2.716$	$f(-0.001) = 2.720$
$f(0.0001) = 2.718$	$f(-0.0001) = 2.719$

2.2 THE LIMIT OF A FUNCTION AND CONTINUITY

It is apparent that the graph of f approaches the same point on the y-axis from each side of the axis. As indicated in the table of function values, the y-coordinate of the point is between 2.718 and 2.719. In fact, it can be shown that

$$\lim_{x \to 0} (1 + x)^{1/x}$$

exists. This limit (number) is 2.71828, correct to five decimal places, and is so important in calculus that a special symbol is used for its designation. We let

$$e = \lim_{x \to 0} (1 + x)^{1/x}.$$

Using e to designate this limit is much the same as using π to designate 3.14159..., another important number in mathematics. We shall find that e, like π, plays a significant role in many applied problems.

Most functions whose limits we seek fall into one of two categories. The first type, and the easier one with which to deal, is made up of what are called *continuous functions*. A function f is **continuous** at t if and only if

continuous

(1) t is in the domain of f,
(2) $\lim_{x \to t} f(x)$ exists, and
(3) $\lim_{x \to t} f(x) = f(t)$.

A function is said to be *continuous in an interval* if and only if it is continuous at each point in the interval.

The function f defined by $f(x) = x^2$ is continuous at 2 since 2 is in the domain of f and since

$$\lim_{x \to 2} x^2 = 4 = f(2).$$

In fact, $f(x) = x^2$ is continuous on the set of real numbers.

For any polynomial $P(x) = a_0 x^n + a_1 x^{n-1} + \cdots + a_n$, since

$$\lim_{x \to t} P(x) = P(t)$$

for each real number t, every polynomial function is continuous at each real number. It is also true that the sum, difference, and product of two polynomial functions P and Q are continuous functions on the set of real numbers. Furthermore, the quotient

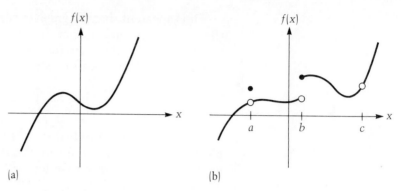

FIG. 2.4. (a) Continuous function: *domain:* all real numbers. (b) Discontinuous function at $x = a$, $x = b$, and $x = c$. *Note:* c is not in the domain.

P/Q of two polynomials is continuous in any interval where $Q(x) \neq 0$. Geometrically, the graph of a continuous function has no "jumps," "holes," or "breaks." (See Figure 2.4.)

Continuity

EXAMPLE 5 Since the function

$$f(x) = \frac{3x^2 - 5x + 1}{3x - 4}$$

is the quotient of two polynomials, it is continuous at each real number x where (its denominator) $3x - 4 \neq 0$, that is, for each $x \neq 4/3$. (It is *discontinuous* at $x = 4/3$.)

The second type of function for which we shall need to find the limit is discussed in Section 2.4. We will see that for a given function f it is often necessary to obtain the limit at t of the function F defined by

$$F(x) = \frac{f(x) - f(t)}{x - t}.$$

Although t must be in the domain of f for F to be defined, t is not in the domain of F since the denominator $x - t$ is 0 at $x = t$. To find the limit of F at t, it is necessary to rely heavily either on algebraic techniques or on some basic theorems; generally, we shall be prudent and take the latter course.

2.2 THE LIMIT OF A FUNCTION AND CONTINUITY

EXERCISES

Set A–Limit of a Function

In Exercises 1 through 20 find the indicated limits.

1. $\lim_{x \to 1} (3x - 5)$
2. $\lim_{x \to 3} (5x + 7)$
3. $\lim_{x \to 2} (x^2 - 3x - 1)$
4. $\lim_{x \to 6} (x^2 - 5x - 6)$
5. $\lim_{x \to 2} (x^3 + 3x - 7)$
6. $\lim_{x \to 2} (5x^3 + x^2 - 10)$
7. $\lim_{x \to -1} (2x^4 - 5x + 17)$
8. $\lim_{x \to 0} (6x^5 + 9x^3 - x + 4)$
9. $\lim_{x \to \sqrt{3}} \dfrac{3x^2 + 1}{x^2 + 4}$
10. $\lim_{x \to 1} \dfrac{x^2 - 2x + 1}{3x^4 + 9}$
11. $\lim_{x \to -2} \dfrac{3x - 4}{8x + 3}$
12. $\lim_{x \to 3} \dfrac{5x + 13}{3x - 12}$
13. $\lim_{x \to 3} \dfrac{x^2 - 9}{x - 3}$
14. $\lim_{x \to -8} \dfrac{x^2 - 64}{x + 8}$
15. $\lim_{x \to 3/2} \dfrac{4x^2 - 9}{2x - 3}$
16. $\lim_{x \to 1/2} \dfrac{6x^2 + 7x - 5}{2x - 1}$
17. $\lim_{x \to 3} \dfrac{x^2 - x - 6}{x - 3}$
18. $\lim_{x \to 4} \dfrac{2x^2 - 11x + 12}{x - 4}$
19. $\lim_{x \to 2} \dfrac{x^3 - 8}{x - 2}$
20. $\lim_{x \to -4} \dfrac{x^3 + 64}{x + 4}$

Set B–Continuity of a Function

In each of Exercises 21 through 30 state for what real numbers the function is continuous.

21. $f(x) = 3x^2 + 2x - 5$
22. $f(x) = 4x^3 - x + 11$
23. $f(x) = x^8 + 3x^2 + x$
24. $f(x) = x^9 - 6x^5 + 4$
25. $f(x) = \dfrac{3x + 6}{5x - 7}$
26. $f(x) = \dfrac{8x - 3}{6x + 7}$
27. $f(x) = \dfrac{2x^2 - 8}{x^2 + 4}$
28. $f(x) = \dfrac{3x^2 + x + 3}{x^2 - 2x - 8}$
29. $f(x) = \dfrac{x^2 - 1}{x^2 - 6x - 20}$
30. $f(x) = \dfrac{x - 1}{x^2 + 1}$

2.3 THE NEWTON QUOTIENT

In this section we define the *Newton quotient*, or *difference quotient*, of a function. But first let us consider an important practical problem which leads up to this fundamental concept in calculus.

Suppose the total cost C in dollars to produce x units of a given product is expressed by

$$C(x) = 2x^2 + 3x + 840$$

where $x \geq 0$. After the manufacturer has made 4 units, what is the total cost of the next 6 units? Since

$$C(4) = 2(16) + 3(4) + 840 = 884,$$

the cost of producing the first 4 units is $884. (See Figure 2.5.) Since the cost of producing the first 10 units is

$$C(10) = 2(100) + 3(10) + 840 = \$1070,$$

the cost of producing the 6 additional units is

$$C(10) - C(4) = 1070 - 884 = \$186.$$

The average cost per unit of the 6 additional units is

$$\frac{C(10) - C(4)}{6} = \frac{186}{6} = \$31.$$

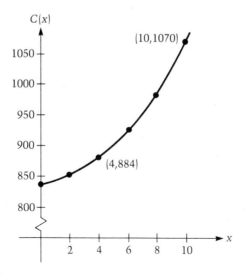

FIG. 2.5.
$C(x) = 2x^2 + 3x + 840$.

2.3 THE NEWTON QUOTIENT

This is called the *average marginal cost* for the 6 additional units.

In general, if $C(x)$ is the total cost of producing x units of a given product, then $C(x + \Delta x) - C(x)$ is the cost of producing Δx (read, "delta x") additional units. If we let ΔC denote this corresponding *change in cost* of production, then the **average marginal cost** at x units for Δx additional units is defined by

average marginal **cost**

$$\textbf{average marginal cost} = \frac{\Delta C}{\Delta x}$$

$$= \frac{\text{change in total cost}}{\text{change in production}}$$

$$= \frac{C(x + \Delta x) - C(x)}{\Delta x}.$$

In the preceding example where the cost function is $C(x) = 2x^2 + 3x + 840$, $x = 4$, $\Delta x = 6$, $C(4) = 884$, $C(x + \Delta x) = C(4 + 6) = C(10) = 1070$, and

$$\frac{\Delta C}{\Delta x} = \$31.$$

Note: The symbol Δ is frequently used in mathematics to represent "the change in." It is not a variable itself. The symbol Δx is a single variable representing "the change in x."

Recall that if $C(x)$ is the cost of producing x units of a given product then the *average cost* $\overline{C}(x)$ of the first x units is defined by

$$\overline{C}(x) = \frac{C(x)}{x}.$$

Therefore, if $C(x) = 2x^2 + 3x + 840$, then the average cost of the first 4 units is

$$\overline{C}(4) = \frac{C(4)}{4} = \frac{884}{4} = \$221.$$

However, the average *marginal* cost of the first 4 units is

$$\frac{C(4) - C(0)}{4} = \frac{884 - 840}{4} = \$11.$$

The difference between the average cost and the average marginal cost for the first 4 units results from the value $C(0) = 840$, the initial costs. Although average cost is affected by initial costs, generally marginal cost is not.

Average cost and average marginal cost

EXAMPLE 1

(a) If $C(x) = 2x^2 + 3x + 140$ is the total cost of producing x units of a particular product, what is the average marginal cost for one additional unit after 20 units have been produced? *Note:* This is the cost of the 21st unit.

(b) What is the average cost of the first 20 units?

(c) What is the average marginal cost of the first 20 units?

Solution

(a) $C(21) - C(20) = 1085 - 1000 = \$85$.

(b) $\overline{C}(20) = \dfrac{C(20)}{20} = \dfrac{1000}{20} = \$50$.

(c) $\dfrac{C(20) - C(0)}{20} = \dfrac{1000 - 140}{20} = \$43$.

For any function f, if we let Δx denote a change in the domain value at a given x, then the corresponding change in the range value, denoted by Δy, is given by

$$\Delta y = f(x + \Delta x) - f(x).$$

In Figure 2.6 we see that

$$\frac{\Delta y}{\Delta x} = \frac{f(x + \Delta x) - f(x)}{\Delta x}$$

slope

is the **slope** of the line containing the points P and Q on the graph of f. Observe the similarity in the preceding formula and that for marginal cost. If $y = f(x)$, we call

$$\frac{\Delta y}{\Delta x} = \frac{f(x + \Delta x) - f(x)}{\Delta x}$$

2.3 THE NEWTON QUOTIENT

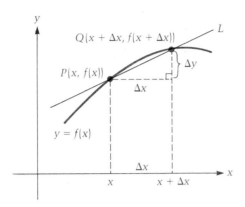

FIG. 2.6. Slope of L:
$$\frac{\Delta y}{\Delta x} = \frac{f(x + \Delta x) - f(x)}{\Delta x}.$$

Newton quotient

the **Newton quotient** (or **difference quotient**) of f at x with increment Δx. Although Δx can be positive or negative, it cannot equal zero.

Newton quotient

EXAMPLE 2 Find the Newton quotient at x with increment Δx for the function $f(x) = 3x + 6$.

Solution

$$\frac{\Delta y}{\Delta x} = \frac{f(x + \Delta x) - f(x)}{\Delta x}$$

$$= \frac{[3(x + \Delta x) + 6] - [3x + 6]}{\Delta x}$$

$$= \frac{3x + 3(\Delta x) + 6 - 3x - 6}{\Delta x}$$

$$= \frac{3(\Delta x)}{\Delta x}$$

$$= 3.$$

The Newton quotient for the linear function $f(x) = 3x + 6$ is the constant 3; the graph of f is a line with slope 3. For nonlinear functions, the Newton quotient will depend upon x and Δx.

PRACTICE PROBLEM Find the Newton quotient at x with increment Δx for the function $f(x) = 7x - 5$.

Answer: $\dfrac{\Delta y}{\Delta x} = 7.$ ∎

Newton quotient and slope

EXAMPLE 3

(a) Find the Newton quotient of $f(x) = x^2 - 4$. (See Figure 2.7.)

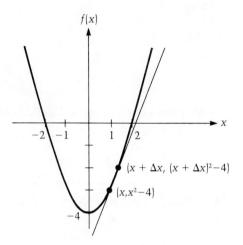

FIG. 2.7. $f(x) = x^2 - 4$.

(b) Use part (a) to find the slope of the line containing the points $(3, f(3))$ and $(3.02, f(3.02))$.

Solution

(a)
$$\frac{\Delta y}{\Delta x} = \frac{f(x + \Delta x) - f(x)}{\Delta x}$$
$$= \frac{[(x + \Delta x)^2 - 4] - [x^2 - 4]}{\Delta x}$$
$$= \frac{x^2 + 2x(\Delta x) + (\Delta x)^2 - 4 - x^2 + 4}{\Delta x}$$
$$= \frac{2x(\Delta x) + (\Delta x)^2}{\Delta x}$$
$$= \frac{\Delta x(2x + \Delta x)}{\Delta x}$$
$$= 2x + \Delta x.$$

2.3 THE NEWTON QUOTIENT

(b) To find the slope of the line through the two points using the Newton quotient, we can let $x = 3$ and $\Delta x = 0.02$. Therefore,

$$\text{slope} = 2(3) + 0.02 = 6.02.$$

Note: We could have let $x = 3.02$ and $\Delta x = -0.02$ and obtained the result.

PRACTICE PROBLEM

(a) Find the Newton quotient of $f(x) = 3x^2 + 5$.

(b) Find the slope of the line containing $(2, f(2))$ and $(1.98, f(1.98))$, using the Newton quotient.

Answer: (a) $\Delta y / \Delta x = 6x + 3(\Delta x)$. (b) 11.94. ∎

Newton quotient for $f(x) = x^{1/2}$

EXAMPLE 4 Find the Newton quotient for the function $f(x) = x^{1/2}$ where $x > 0$. (See Figure 2.8.)

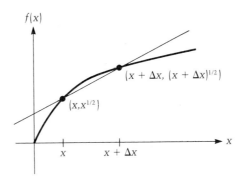

FIG. 2.8. $f(x) = x^{1/2}$.

Solution

$$\frac{\Delta y}{\Delta x} = \frac{(x + \Delta x)^{1/2} - x^{1/2}}{\Delta x}.$$

Although this is one form of the answer, it will be useful in the next section to have the answer written in a different form, one obtained by multiplying the numerator and denominator of the preceding fraction by $(x + \Delta x)^{1/2} + x^{1/2}$. Using the fact that

$(a^{1/2} - b^{1/2})(a^{1/2} + b^{1/2}) = a - b$ yields

$$\frac{\Delta y}{\Delta x} = \frac{x + \Delta x - x}{(\Delta x)[(x + \Delta x)^{1/2} + x^{1/2}]}$$

$$= \frac{1}{(x + \Delta x)^{1/2} + x^{1/2}}.$$

Newton quotient

EXAMPLE 5 For $f(x) = 5/(x + 5)$ find the Newton quotient at $x = 3$ with increment Δx.

Solution: At $x = 3$,

$$\frac{\Delta y}{\Delta x} = \frac{f(3 + \Delta x) - f(3)}{\Delta x}$$

$$= \frac{\dfrac{5}{(3 + \Delta x) + 5} - \dfrac{5}{8}}{\Delta x}$$

$$= \left(\frac{5}{8 + \Delta x} - \frac{5}{8}\right) \cdot \frac{1}{\Delta x}$$

$$= \frac{5(8) - 5(8 + \Delta x)}{8(\Delta x)(8 + \Delta x)}$$

$$= \frac{-5(\Delta x)}{8(\Delta x)(8 + \Delta x)}$$

$$= \frac{-5}{8(8 + \Delta x)}.$$

PRACTICE PROBLEM Find the Newton quotient at $x = 2$ with increment Δx for $f(x) = 4/(2x - 1)$.

Answer: $-8/[3(3 + 2\Delta x)]$. ∎

The Newton quotient $\Delta y/\Delta x$ has many applications since it represents the *average change* in the dependent variable y corresponding to a change Δx in the independent variable x. Consider the following example.

Average heartbeat

EXAMPLE 6 Suppose a person's heart beats $f(t) = 130t - 2t^2$ times during the first t minutes, $0 \le t \le 16$, after completing a very strenuous exercise.

2.3 THE NEWTON QUOTIENT

(a) After 2 minutes have passed, what is the person's average heartbeat during the next 4 minutes?

(b) After 10 minutes have passed, what is the person's average heartbeat during the next 4 minutes?

Solution

(a) The person's heart beats $f(2) = 260 - 8 = 252$ times during the first 2 minutes and $f(6) = 780 - 72 = 708$ times during the first 6 minutes. Since the person's heart beats

$$f(6) - f(2) = 708 - 252 = 456$$

times during this 4-minute interval, the average heartbeat for this period is $\frac{456}{4} = 114$ times per minute. This, of course, is the Newton quotient at $t = 2$ with increment $\Delta t = 4$.

(b) The answer to the question in part (b) is the Newton quotient of $y = 130t - t^2$ at $t = 10$ with increment $\Delta t = 4$. It is

$$\frac{\Delta y}{\Delta t} = \frac{f(10 + 4) - f(10)}{4}$$

$$= \frac{1428 - 1100}{4}$$

$$= 82 \text{ times per minute.}$$

Let us conclude this section by considering one more very important application of the Newton quotient. If $s(x)$ is the distance in feet a body travels in x seconds, then the average velocity in feet per second during the time interval from x to $x + \Delta x$ seconds is given by the Newton quotient of the distance function. That is,

average velocity

$$\textbf{average velocity} = \frac{\text{change in distance}}{\text{change in time}} = \frac{s(x + \Delta x) - s(x)}{\Delta x}.$$

It can be shown that the distance s a free-falling body falls from rest in x seconds is given by

$$s(x) = 16x^2 \text{ feet.}$$

Therefore the average velocity in the time interval from x to $x + \Delta x$ seconds is given by

$$\frac{\Delta s}{\Delta x} = \frac{16(x + \Delta x)^2 - 16x^2}{\Delta x}$$

$$= \frac{16[x^2 + 2x\Delta x + (\Delta x)^2] - 16x^2}{\Delta x}$$

$$= \frac{32x(\Delta x) + 16(\Delta x)^2}{\Delta x}$$

$$= 32x + 16(\Delta x).$$

Note that if $x = 2$ and $\Delta x = 0.1$, then this represents the time interval from 2 to 2.1 seconds and the average velocity during this time interval is

$$\frac{\Delta s}{\Delta x} = 32(2) + 16(0.1) = 64 + 1.6 = 65.6 \text{ ft/sec}.$$

Furthermore, we can see that if $x = 2$ and Δx is "very small," then

$$32x + 16(\Delta x) = 64 + 16(\Delta x) \doteq 64 \text{ ft/sec}.$$

This prompts us to *define* the velocity at the end of 2 seconds to be 64 ft/sec.

Newton quotient and velocity

EXAMPLE 7

(a) If a bullet is fired directly upward with an initial velocity of 128 ft/sec, then its height h in feet above the ground at the end of x seconds is given by $h(x) = 128x - 16x^2$. Find the Newton quotient of the function h in terms of x and Δx.

(b) Evaluate the Newton quotient when $x = 2$ and $\Delta x = 0.01$. Give a physical interpretation for your answer.

(c) Evaluate the Newton quotient when $x = 3.9$ and $\Delta x = 0.1$.

(d) Evaluate the Newton quotient when $x = 4$ and $\Delta x = 0.1$.

(e) Interpret the answers to parts (c) and (d) with regard to their signs. (See Figure 2.9.)

2.3 THE NEWTON QUOTIENT

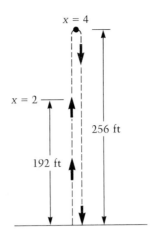

FIG. 2.9.

Solution

(a) $\dfrac{\Delta h}{\Delta x} = \dfrac{h(x + \Delta x) - h(x)}{\Delta x}$

$= \dfrac{128(x + \Delta x) - 16(x + \Delta x)^2 - 128x + 16x^2}{\Delta x}$

$= \dfrac{128(\Delta x) - 32x(\Delta x) - 16(\Delta x)^2}{\Delta x}$

$= 128 - 32x - 16(\Delta x).$

(b) If $x = 2$ and $\Delta x = 0.01$, then

$$\dfrac{\Delta h}{\Delta x} = 128 - 64 - 0.16 = 63.84 \text{ ft/sec}.$$

The bullet averages 63.84 ft/sec in the time interval from 2 to 2.01 seconds.

(c) If $x = 3.9$ and $\Delta x = 0.1$, then

$$\dfrac{\Delta h}{\Delta x} = 128 - 124.8 - 1.6 = 1.6 \text{ ft/sec}.$$

(d) If $x = 4$ and $\Delta x = 0.1$, then

$$\dfrac{\Delta h}{\Delta x} = 128 - 128 - 1.6 = -1.6 \text{ ft/sec}.$$

(e) Since Δx is positive in both cases, for $\Delta h/\Delta x$ to be positive, as in part (c), Δh must be positive. Therefore, $h(4) - h(3.9) > 0$ and the height of the bullet above the ground is greater at 4 seconds than at 3.9 seconds; that is, it is going up. For $\Delta h/\Delta x$ to be negative, as in part (d), Δh must be negative. Therefore, $h(4.1) - h(4) < 0$ and the height of the bullet above the ground is less at 4.1 seconds than at 4 seconds; that is, it is coming down. (A reasonable guess is that the bullet reaches its maximum height at the end of 4 seconds. This can be verified by noting that $h(8) = 0$ and knowing that it will take the bullet as long to go up as to come down.)

EXERCISES

Set A—Newton Quotient

In Exercises 1 through 14 find the Newton quotient at x with increment Δx for each function and simplify.

1. $f(x) = 7x + 5$
2. $f(x) = 4x - 6$
3. $g(x) = 8x - \frac{2}{3}$
4. $g(x) = -6x + 11$
5. $f(x) = x^2 + 3$
6. $f(x) = x^2 - 5$
7. $s(x) = 7 - 2x^2$
8. $C(x) = -3x^2 + 2$
9. $f(x) = 4x^2 - 3x$
10. $g(x) = 3x^2 + 8x$
11. $f(x) = x^2 + 5x - 7$
12. $f(x) = 9 - 3x - 5x^2$
13. $f(x) = x^3$
14. $g(x) = 2x^3 - 8$

In Exercises 15 through 20 find the Newton quotient at $x = 2$ with increment Δx for each function and simplify.

15. $g(x) = 2x^3 - 4x$
16. $h(x) = \sqrt{x + 2}$
17. $f(x) = \dfrac{1}{x + 1}$
18. $f(x) = \dfrac{3}{x + 5}$
19. $f(x) = \dfrac{x}{3x + 7}$
20. $f(x) = \dfrac{x}{2x - 3}$

Set B—Average Marginal Cost

21. Suppose the total cost function for producing x units of a given product is $C(x) = 38x + 600$ dollars.
    (a) What is $C(0)$? What does it represent?
    (b) What does $C(15) - C(14)$ represent?
    (c) Find the average marginal cost for producing 10 additional units when 25 have been produced.

(d) Find the average marginal cost in terms of x and Δx.

22. Suppose the total cost function for producing x units of a given product is $C(x) = 42x + 550$ dollars.
    (a) What is $C(0)$? What does it represent?
    (b) What does $C(23) - C(22)$ represent?
    (c) Find the average marginal cost for producing 6 additional units when 22 have been produced.
    (d) Find the average marginal cost in terms of x and Δx.

23. (a) The cost of producing x units of a certain product is given by $C(x) = 2x^2 + 4x + 8$ dollars. Find the average marginal cost for producing 5 additional units after 10 have been produced.
    (b) Find the average marginal cost in terms of x and Δx.
    (c) Use part (b) to check your answer in part (a).
    (d) Find the average marginal cost of the first 20 units.
    (e) Find the average cost of the first 20 units.

24. (a) The cost of producing x units of a certain product is given by $C(x) = 0.02x^2 + 2.5x + 24$ dollars. Find the average marginal cost for producing 5 additional units after 10 have been produced.
    (b) Find the average marginal cost in terms of x and Δx.
    (c) Use part (b) to check your answer in part (a).
    (d) Find the average marginal cost of the first 20 units.
    (e) Find the average cost of the first 20 units.

25. (a) The cost of producing x thousand units of a certain product is given by $C(x) = (2x^2 + 4x + 500)100$ dollars where $x \geq 1$. Find the average marginal cost for producing 500 additional units after 4000 have been produced.
    (b) Find the average marginal cost in terms of x and Δx.
    (c) Use part (b) to check your answer in part (a).

26. (a) The cost in thousands of dollars for producing x thousand units of a particular product is given by $C(x) = 0.08x^2 + 1.2x + 12$, where $x \geq 2$. Find the average marginal cost for producing 500 additional units after 4000 have been produced.
    (b) Find the average marginal cost in terms of x and Δx.
    (c) Use part (b) to check your answer in part (a).

Set C—Newton Quotient and Slope

27. (a) Find the Newton quotient for the function $f(x) = x^2$.
    (b) Use part (a) to find the slope of the line through the two

points on the graph of f where the x-coordinates are 3 and 3.2.

(c) Use part (a) to find the slope of the line through the two points on the graph of f where the x-coordinates are 3 and 3.1.

(d) What number does the Newton quotient get close to at 3 as Δx gets close to 0?

28. (a) Find the Newton quotient for the function $f(x) = x^2 + 3x$.
    (b) Use part (a) to find the slope of the line through the two points on the graph of f where the x-coordinates are 2 and 2.01.
    (c) What number does the Newton quotient get close to at 2 as Δx gets close to 0?

29. (a) Find the Newton quotient for the function $f(x) = 1/x$.
    (b) Use part (a) to find the slope of the line through the two points on the graph of f where the x-coordinates are 4 and 3.99.
    (c) What number does the Newton quotient get close to at $x = 4$ as Δx gets close to 0?

30. (a) Find the Newton quotient for the function $f(x) = x/(3x + 2)$.
    (b) Use part (a) to find the slope of the line through the two points on the graph of f where the x-coordinates are 1 and 1.01.
    (c) What number does the Newton quotient get close to at $x = 1$ as Δx gets close to 0?

Set D—Miscellaneous Applications

Average velocity

31. (a) If a bullet is shot directly upward with an initial velocity of 96 ft/sec, then its height in feet above the ground at the end of x sec is given by $h(x) = 96x - 16x^2$. Find the Newton quotient of h in terms of x and Δx. What does it represent?
    (b) Evaluate the Newton quotient when $x = 2$ and $\Delta x = 0.1$.
    (c) Evaluate the Newton quotient when $x = 2.9$ and $\Delta x = 0.1$.
    (d) Evaluate the Newton quotient when $x = 3$ and $\Delta x = 0.1$.
    (e) Interpret the answers to parts (c) and (d) with regard to their signs.

Average marginal revenue

32. (a) **Average marginal revenue** is defined to be the change in total revenue per unit of change in production at a given level of output. Hence for a revenue function $R(x)$,

2.3 THE NEWTON QUOTIENT

$$\textbf{average marginal revenue} = \frac{\Delta R}{\Delta x} = \frac{R(x + \Delta x) - R(x)}{\Delta x},$$

where Δx is the change in production at a given level x of output. If the revenue in thousands of dollars from x thousand units of a particular product is given by $R(x) = -0.08x^2 + 2.8x - 4$, where $x \geq 2$, what is the average marginal revenue in terms of x and Δx.

(b) Find the average marginal revenue from the sale of an additional 1000 units after 10,000 units have been sold.

Average marginal profit

33. (a) **Average marginal profit** is defined to be the change in total profit per unit change in production at a given level of output. Hence, for a profit function $P(x)$,

$$\textbf{average marginal profit} = \frac{\Delta P}{\Delta x} = \frac{P(x + \Delta x) - P(x)}{\Delta x},$$

where Δx is the change in production at a given level x of output. If the profit in thousands of dollars on x thousand units is given by $P(x) = -0.16x^2 + 4.0x - 16$, where $x \geq 2$, what is the average marginal profit in terms of x and Δx.

(b) Find the average marginal profit when $x = 11$ and $\Delta x = 1$.
(c) Find the average marginal profit when $x = 12$ and $\Delta x = 1$.
(d) Find the average marginal profit when $x = 13$ and $\Delta x = 1$.
(e) Interpret the answers to parts (b), (c), and (d).

Average acceleration

34. (a) **Average acceleration** during a given time interval from x to $x + \Delta x$ of a body moving in a straight line is defined as the change in velocity divided by the time elapsed. For a velocity function $v(x)$,

$$\textbf{average acceleration} = \frac{\Delta v}{\Delta x} = \frac{v(x + \Delta x) - v(x)}{\Delta x}.$$

If the velocity function of a free-falling body is given by $v(x) = 32x$ ft/sec, what is the average acceleration for $x > 0$ and $\Delta x > 0$.

(b) Interpret the physical significance of your answer to part (a).

2.4 THE DERIVATIVE OF A FUNCTION

Now that we have considered the limit of a function and the Newton quotient of a function, we are prepared to investigate the *derivative* of a function, one of the most powerful mathematical tools for solving practical problems. Consider the Newton quotient of $f(x) = x^2 - 4$ at $x = 1$. It is

$$\frac{\Delta y}{\Delta x} = \frac{f(1 + \Delta x) - f(1)}{\Delta x}$$

$$= \frac{[(1 + \Delta x)^2 - 4] - (1 - 4)}{\Delta x}$$

$$= \frac{1 + 2(\Delta x) + (\Delta x)^2 - 4 - (-3)}{\Delta x}$$

$$= \frac{2(\Delta x) + (\Delta x)^2}{\Delta x}$$

$$= 2 + \Delta x.$$

As we know, $2 + \Delta x$ is the slope of the line L containing the two points $P(1, f(1))$ and $Q(1 + \Delta x, f(1 + \Delta x))$ on the graph of f. (See Figure 2.10.) As $\Delta x \to 0$ (Δx approaches zero), the line L "turns into the tangent" to the graph of f at $(1, f(1)) = (1, -3)$. Since

$$\frac{\Delta y}{\Delta x} = 2 + \Delta x$$

approaches 2 as $\Delta x \to 0$, we define the *slope of the tangent to f at $x = 1$* to be 2. In symbols,

$$\lim_{\Delta x \to 0} \frac{\Delta y}{\Delta x} = \lim_{\Delta x \to 0} (2 + \Delta x) = 2.$$

For any function f, if the Newton quotient

$$\frac{\Delta y}{\Delta x} = \frac{f(1 + \Delta x) - f(1)}{\Delta x}$$

2.4 THE DERIVATIVE OF A FUNCTION

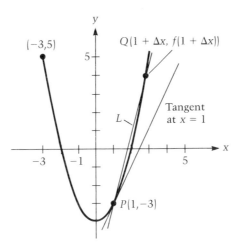

FIG. 2.10. $f(x) = x^2 - 4$.

can be made arbitrarily close to exactly one real number as $\Delta x \to 0$, it is called the *limit* of $\Delta y/\Delta x$ as $\Delta x \to 0$. This limit is also called the *derivative of f at* 1 and is denoted by $f'(1)$. Hence, for the function defined by

$$f(x) = x^2 - 4,$$

the derivative of f at 1 is $f'(1) = 2$.

In Example 3, Section 2.3, we found that if

$$f(x) = x^2 - 4,$$

then for any x and any $\Delta x \neq 0$,

$$\frac{\Delta y}{\Delta x} = 2x + \Delta x.$$

To express the fact that the limit of this Newton quotient as $\Delta x \to 0$ is $2x$, we write

$$\lim_{\Delta x \to 0} \frac{\Delta y}{\Delta x} = \lim_{\Delta x \to 0} (2x + \Delta x) = 2x.$$

The limit is the *derivative of* $f(x) = x^2 - 4$ *at* x and is denoted by $f'(x)$; thus $f'(x) = 2x$. This derivative evaluated at $x = a$ is $2a$; it is the slope of the line tangent to the graph of $f(x) = x^2 - 4$ at the point $(a, f(a))$. Note that $f'(0) = 0$ and observe in Figure 2.10 that 0 is the slope of the line tangent to $f(x) = x^2 - 4$ at the low point

(0, −4) on its graph. In general, the derivative is useful in finding high points and low points on the graphs of functions. (This will make it possible to solve problems such as those listed in the introduction to this chapter.)

derivative of f at x

> **DEFINITION** For any function $y = f(x)$, if the limit of the Newton quotient $\Delta y / \Delta x$ exists as Δx approaches zero, it is called the *(first)* **derivative** *of f at x*. Symbolically,
>
> $$f'(x) = \lim_{\Delta x \to 0} \frac{\Delta y}{\Delta x} = \lim_{\Delta x \to 0} \frac{f(x + \Delta x) - f(x)}{\Delta x}.$$

If the derivative of f exists at x, then f is said to be **differentiable** at x. Furthermore, we define $f'(a)$ to be the **slope** of the line tangent to the graph of f at $(a, f(a))$. (See Figure 2.11.)

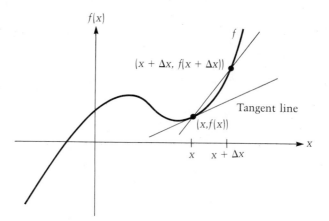

FIG. 2.11. Slope of tangent at x is $f'(x) = \lim_{\Delta x \to 0} \frac{f(x + \Delta x) - f(x)}{\Delta x}$.

Derivative from the definition

EXAMPLE 1 Given that $f(x) = x^2$, find the derivative of f at x using the definition of the derivative.

Solution

$$f'(x) = \lim_{\Delta x \to 0} \frac{f(x + \Delta x) - f(x)}{\Delta x}$$

2.4 THE DERIVATIVE OF A FUNCTION

$$= \lim_{\Delta x \to 0} \frac{(x + \Delta x)^2 - x^2}{\Delta x}$$

$$= \lim_{\Delta x \to 0} \frac{x^2 + 2x(\Delta x) + (\Delta x)^2 - x^2}{\Delta x}$$

$$= \lim_{\Delta x \to 0} \frac{2x(\Delta x) + (\Delta x)^2}{\Delta x}$$

$$= \lim_{\Delta x \to 0} (2x + \Delta x)$$

$$= 2x.$$

Note: The derivative of $F(x) = x^2 - 4$ is the same as the derivative of $f(x) = x^2$. For these two functions, which differ by a constant, this is seen geometrically if we observe that the graph of F is identical to the graph of f except that it is 4 units lower in the coordinate plane and that the tangent line at $(a, F(a))$ will be parallel to the tangent line at $(a, f(a))$. (Recall that parallel lines have the same slope and see Figure 2.12.)

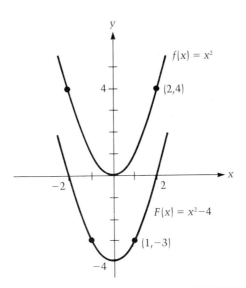

FIG. 2.12. $f(x) = x^2$ and $F(x) = x^2 - 4$.

PRACTICE PROBLEM Use the definition of the derivative of f at x to find the derivative of $f(x) = x^2 + 3x$.

Answer: $f'(x) = 2x + 3$. ∎

Derivative from the definition

EXAMPLE 2 Use the definition to find the derivative of $f(x) = x^3$ at x.

Solution

$$f'(x) = \lim_{\Delta x \to 0} \frac{(x + \Delta x)^3 - x^3}{\Delta x}$$

$$= \lim_{\Delta x \to 0} \frac{x^3 + 3x^2(\Delta x) + 3x(\Delta x)^2 + (\Delta x)^3 - x^3}{\Delta x}$$

$$= \lim_{\Delta x \to 0} \frac{3x^2(\Delta x) + 3x(\Delta x)^2 + (\Delta x)^3}{\Delta x}$$

$$= \lim_{\Delta x \to 0} [3x^2 + 3x(\Delta x) + (\Delta x)^2]$$

$$= 3x^2.$$

Derivative from the definition

EXAMPLE 3 Use the definition to find the derivative of $f(x) = 1/x$ at x. (See Figure 2.13.)

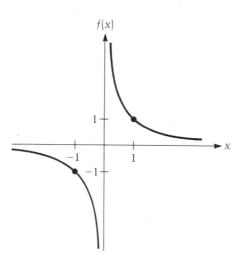

FIG. 2.13. $f(x) = \dfrac{1}{x}$.

Solution

$$f'(x) = \lim_{\Delta x \to 0} \frac{1/(x + \Delta x) - 1/x}{\Delta x}.$$

Multiplying the numerator and the denominator of the complex fraction by $x(x + \Delta x)$ yields

2.4 THE DERIVATIVE OF A FUNCTION

$$f'(x) = \lim_{\Delta x \to 0} \frac{x - (x + \Delta x)}{x(\Delta x)(x + \Delta x)}$$

$$= \lim_{\Delta x \to 0} \frac{-\Delta x}{x(\Delta x)(x + \Delta x)}$$

$$= \lim_{\Delta x \to 0} \frac{-1}{x(x + \Delta x)}$$

$$= -\frac{1}{x^2}$$

PRACTICE PROBLEM Use the definition of the derivative of f at x to find the derivative of $f(x) = 1/(x + 2)$.

Answer: $f'(x) = -1/(x + 2)^2$. ∎

Derivative of constant function

EXAMPLE 4 Use the definition to find the derivative of $f(x) = c$, where c is a constant.

Solution

$$f'(x) = \lim_{\Delta x \to 0} \frac{c - c}{\Delta x} \quad \text{from the definition.}$$

Since the numerator is always zero, the fraction is zero for any $\Delta x \neq 0$. Hence

$$f'(x) = 0$$

for each x in the domain of f. (This example shows that **the derivative of a constant function is zero at each x in its domain.**)

Derivative from the definition

EXAMPLE 5 Use the definition to find the derivative of $f(x) = x^{1/2}$, where $x \geq 0$.

Solution: In Example 4, Section 2.3, we found for $f(x) = x^{1/2}$ that

$$\frac{\Delta y}{\Delta x} = \frac{1}{(x + \Delta x)^{1/2} + x^{1/2}}.$$

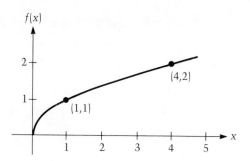

FIG. 2.14. $f(x) = x^{1/2}$.

Therefore,

$$f'(x) = \lim_{\Delta x \to 0} \frac{1}{(x + \Delta x)^{1/2} + x^{1/2}}$$

$$= \frac{1}{2x^{1/2}}$$

$$= \frac{1}{2} x^{-1/2}.$$

Note that although $f(x) = x^{1/2}$ is defined at $x = 0$ (that is, 0 is in the domain of f), the derivative of f does not exist at $x = 0$. Geometrically, the tangent line to the graph of $f(x) = x^{1/2}$ becomes vertical at $x = 0$ and slope is undefined for such lines. (See Figure 2.14.)

There are three major areas of investigation for us to pursue. They are:

(1) to find the derivatives of specific functions such as $f(x) = x^{1/3}$, $g(x) = \log x$, etc.;

(2) to derive general differentiation theorems so that we can find the derivative, for example, of $F(x) = (x^{1/3})(\log x)$ by using known derivatives of $f(x) = x^{1/3}$ and $g(x) = \log x$; and

(3) to consider many of the important applications of the derivative.

We begin our investigations by listing three important theorems; they are stated without proof. (The proofs follow directly from the definition of the derivative and some limit theorems which we have chosen not to discuss.)

2.4 THE DERIVATIVE OF A FUNCTION

power rule

THEOREM 1 If $f(x) = x^r$ where r is any real number, then

$$f'(x) = rx^{r-1}.$$

Note: The student should check to see that this theorem has already been verified for $r = 2, 3, -1, 0,$ and $1/2$ in Examples 1, 2, 3, 4, and 5.

PRACTICE PROBLEM Find the derivative of $f(x) = x^{-3/2}$.

Answer: $f'(x) = -\frac{3}{2}x^{-5/2}$. ∎

Equation of tangent line

EXAMPLE 6 Find an equation of the line tangent to the graph of $f(x) = x^{2/3}$ at the point $(8, 4)$ on the graph of f.

Solution: Since $f'(x) = \frac{2}{3}x^{-1/3}$, the slope of the line tangent to the graph of f at $x = 8$ is

$$f'(8) = \frac{2}{3(8)^{1/3}} = \frac{2}{6} = \frac{1}{3}.$$

Using the point $(8, 4)$ and the slope $\frac{1}{3}$, we find that an equation of the tangent line at $(8, 4)$ is $x - 3y = -4$. [Although $f(0)$ exists, note that $f'(0)$ does not exist.]

THEOREM 2 If g is a differentiable function at x and if $f(x) = cg(x)$, where c is a constant, then

$$f'(x) = cg'(x).$$

(The derivative of a constant times a function is the constant times the derivative of the function.)

Illustration. Since the derivative of $g(x) = x^3$ is $g'(x) = 3x^2$, it follows from Theorem 2 that the derivative of $f(x) = 5x^3$ is $f'(x) = 5(3x^2) = 15x^2$.

Slope of tangent

EXAMPLE 7 What is the slope of the line tangent to the graph of $f(x) = 6x^{3/2}$ at $x = 4$?

Solution: Since

$$f'(x) = 6\left(\frac{3}{2}\right)x^{1/2} = 9x^{1/2},$$

the slope of the line tangent to the graph of f at $x = 4$ is $f'(4) = 18$.

PRACTICE PROBLEM What is the slope of the line tangent to the graph of $f(x) = 6x^{5/3}$ at $x = 8$? What is the equation of the tangent line at that point.

Answer: $m = 40$. $40x - y = 128$. ∎

THEOREM 3 If f and g are differentiable functions at x and if $F(x) = f(x) + g(x)$, then

$$F'(x) = f'(x) + g'(x).$$

(The derivative of the sum of two functions is the sum of their derivatives.)

If $F(x) = f(x) - g(x)$, then $F(x) = f(x) + (-1)g(x)$. By Theorems 2 and 3,

$$F'(x) = f'(x) + (-1)g'(x) = f'(x) - g'(x).$$

Therefore, **the derivative of the difference of two functions is the difference of their derivatives.** For example,

2.4 THE DERIVATIVE OF A FUNCTION

$$\text{if} \quad F(x) = 2x^3 - 5x^2$$
$$\text{then} \quad F'(x) = 6x^2 - 10x.$$

We shall find later that the derivative of the (first) derivative of a function f also has many useful applications. It is called the **second derivative** of f and is denoted by $f''(x)$. For example,

second derivative

$$\text{if} \quad f(x) = 6x^3 - 4x^2 + 2x + 5,$$
$$\text{then} \quad f'(x) = 18x^2 - 8x + 2$$
$$\text{and} \quad f''(x) = 36x - 8.$$

First and second derivatives

EXAMPLE 8 Find the first and second derivatives of $f(x) = 3x^5 + 8x^{-3/2}$.

Solution

$$f'(x) = 15x^4 - 12x^{-5/2}$$

and

$$f''(x) = 60x^3 + 30x^{-7/2}.$$

PRACTICE PROBLEM Find the first and second derivatives of $f(x) = -4x^{-3/2} + 2x^7$.

Answer: $f'(x) = 6x^{-5/2} + 14x^6$ and $f''(x) = -15x^{-7/2} + 84x^5$. ∎

Inequalities involving derivatives

EXAMPLE 9

(a) Given the function $2x^3 - 4x^2$, for what x is $f'(x) = 0$?

(b) For what x is $f'(x) < 0$?

(c) For what x is $f''(x) = 0$?

(d) For what x is $f''(x) < 0$? (See Figure 2.15.)

Solution

(a) Since $f'(x) = 6x^2 - 8x = 2x(3x - 4)$, $f'(x) = 0$ at $x = 0$ and $x = \frac{4}{3}$.

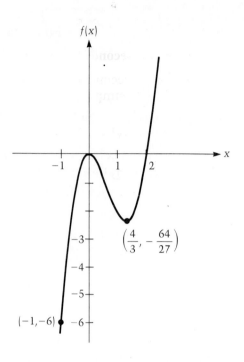

FIG. 2.15.
$f(x) = 2x^3 - 4x^2$.

(b) $f'(x) < 0$ if and only if $2x(3x - 4) < 0$, or $x(3x - 4) < 0$. For the product $x(3x - 4)$ to be negative

either	or
$x < 0$ and $3x - 4 > 0$ $x < 0$ and $x > \frac{4}{3}$.	$x > 0$ and $3x - 4 < 0$ $x > 0$ and $x < \frac{4}{3}$.
Since there are no real numbers less than 0 *and* greater than $\frac{4}{3}$, this case yields no solution.	This means all real numbers between 0 and $\frac{4}{3}$ make the product $2x(3x - 4)$ less than 0.

Therefore, $f'(x) < 0$ provided $0 < x < \frac{4}{3}$.

(c) Since $f''(x) = 12x - 8$, $f''(x) = 0$, where $12x - 8 = 0$, $12x = 8$, or $x = \frac{2}{3}$.

(d) $f''(x) < 0$, where $12x - 8 < 0$, $12x < 8$, or $x < \frac{2}{3}$.

EXERCISES

Set A—First and Second Derivatives of Functions

Find the first and second derivatives of each function in Exercises 1 through 12 and simplify.

1. $f(x) = 6x + 7$
2. $f(x) = 8x - 3$
3. $g(x) = 3x^4 - 5x^2 + 7x + 6$
4. $F(x) = 8x^4 - 2x^3 - 3x + 5$
5. $G(x) = 8x^{10} - 4x^3 + 5x$
6. $g(x) = 11x^4 + 7x + 2$
7. $f(x) = 3x^{4/3} + 2x^{1/2} + 1$
8. $g(x) = 4x^{3/4} - 4x^{3/2} - 18$
9. $g(x) = 6/x^{1/3} - 7/x^{3/5} + x$
10. $f(x) = 6/x^{2/3} - 8/x^{1/4} - 2x$
11. $f(x) = x^3 - 3x + 1/x$
12. $f(x) = 2x^2 - 1/x + 2/x^2$
13. $f(x) = 2x^3 + 3x + 5 - 1/x$; find $f'(1)$ and $f''(1)$.
14. $f(x) = 1/x^2 - 3/x$; find $f'(2)$ and $f''(2)$.
15. $f(x) = 6x^{2/3} - 3x$; find $f'(8)$ and $f''(8)$.
16. $g(x) = 4x^{1/2} - 5x^2$; find $g'(4)$ and $g''(4)$.

Set B—Slopes and Equations of Tangent Lines

17. Find the slope of the tangent to the graph of $f(x) = x^{3/2} - x$ at $x = 4$.
18. Find the slope of the tangent to the graph of $f(x) = 3x^{2/3} - 2$ at $x = 8$.
19. Find an equation of the tangent to the graph of $f(x) = 3x^2 - 5x + 2$ at $x = 1$.
20. Find an equation of the tangent to the graph of $f(x) = x^2 - 3x + 1$ at $x = 2$.
21. Find an equation of the tangent to the graph of $f(x) = 1/x + x^2$ at $x = 2$.
22. Find an equation of the tangent to the graph of $f(x) = 3x^3 - 1/x^2$ at $x = 1$.
23. Find an equation of the line tangent to the graph of $f(x) = x^2 + 3x + 1$ and parallel to the line $2x - y = 2$.
24. Find an equation of the line tangent to the graph of $f(x) = 2x^2 - 2x + 3$ and perpendicular to the line $x + 6y = 5$.

Set C—Inequalities Involving the First and Second Derivatives

25. Given the function $f(x) = x^3$, for what x is
    (a) $f'(x) = 0$? (b) $f'(x) > 0$? (c) $f'(x) < 0$?
    (d) $f''(x) > 0$? (e) $f''(x) < 0$?

26. Given the function $f(x) = x^4$, for what x is
    (a) $f'(x) = 0$?   (b) $f'(x) > 0$?   (c) $f'(x) < 0$?
    (d) $f''(x) > 0$?   (e) $f''(x) < 0$?

27. Given the function $f(x) = 4x^2 - 12x + 3$, for what x is
    (a) $f'(x) = 0$?   (b) $f'(x) > 0$?   (c) $f'(x) < 0$?
    (d) $f''(x) > 0$?   (e) $f''(x) < 0$?

28. Given the function $f(x) = -5x^2 + 15x - 3$, for what x is
    (a) $f'(x) = 0$?   (b) $f'(x) > 0$?   (c) $f'(x) < 0$?
    (d) $f''(x) > 0$?   (e) $f''(x) < 0$?

29. Given the function $g(x) = x^3 - x^2$, for what x is
    (a) $g'(x) = 0$?   (b) $g'(x) > 0$?   (c) $g'(x) < 0$?
    (d) $g''(x) > 0$?   (e) $g''(x) < 0$?

30. Given the function $g(x) = 2x^3 + x^2 - 1$, for what x is
    (a) $g'(x) = 0$?   (b) $g'(x) > 0$?   (c) $g'(x) < 0$?
    (d) $g''(x) > 0$?   (e) $g''(x) < 0$?

2.5 PRODUCT, QUOTIENT, AND GENERAL POWER RULES FOR DIFFERENTIATION

Now let us look at three more general differentiation theorems. With these, the three from Section 2.4, and one more given in Section 4.2, we shall have all the general differentiation theorems we will need.

THEOREM 4 *(General power rule)* If g is a differentiable function and if $F(x) = [g(x)]^r$, where r is any given real number, then

general power rule

$$F'(x) = r[g(x)]^{r-1} g'(x).$$

Note: If $g(x) = x$, then $g'(x) = 1$ and Theorem 1 is a special case of Theorem 4. (In fact, this power formula is a special case of the "chain rule," the final differentiation theorem discussed in Section 4.2.)

Power of a function

EXAMPLE 1 Find the derivative of $F(x) = (x^2 + 1)^{3/2}$.

2.5 PRODUCT, QUOTIENT, AND GENERAL POWER RULES FOR DIFFERENTIATION

Solution

$$F'(x) = \left(\frac{3}{2}\right)(x^2 + 1)^{1/2}(2x) = 3x(x^2 + 1)^{1/2}.$$

PRACTICE PROBLEM Find the derivative of $F(x) = (2x^3 + 6x)^{4/3}$.

Answer

$$F'(x) = \frac{4}{3}(2x^3 + 6x)^{1/3}(6x^2 + 6) = 8(2x^3 + 6x)^{1/3}(x^2 + 1). \blacksquare$$

Power of a function

EXAMPLE 2 Find the derivative of $f(x) = (x^3 + 2)^2$ by Theorem 4 and check your answer by finding the derivative of the function expressed as a polynomial.

Solution

$$f'(x) = 2(x^3 + 2)(3x^2) = 6x^5 + 12x^2. \quad \text{(Theorem 4)}$$

Since

$$f(x) = x^6 + 4x^3 + 4,$$

$$f'(x) = 6x^5 + 12x^2. \quad \text{(Theorems 3 and 2)}$$

PRACTICE PROBLEM Find the derivative of $f(x) = (2x^3 - 3x)^2$ by Theorem 4 and check your answer by finding the derivative of the function expressed as a polynomial.

THEOREM 5 If f and g are differentiable functions and if $F(x) = f(x)g(x)$ then

product rule

$$F'(x) = f(x)g'(x) + g(x)f'(x).$$

(The derivative of the product of two functions is the first (function) times the derivative of the second plus the second times the derivative of the first. This is called the *product rule*.)

Although we do not prove Theorem 5, it should seem more reasonable if we verified it for a simple example. If
$$f(x) = x^3, \quad g(x) = x^4, \quad \text{and} \quad F(x) = f(x)g(x),$$
then
$$F(x) = x^7 \quad \text{and} \quad F'(x) = 7x^6.$$

Applying Theorem 5, we get
$$F'(x) = (x^3)(4x^3) + (x^4)(3x^2) = 4x^6 + 3x^6 = 7x^6.$$

(It is worth emphasizing that the derivative of a product of two functions is *not* the product of their derivatives.)

Derivative of a product

EXAMPLE 3 Find the derivative of $f(x) = x^2(3x + 1)^2$ by using Theorems 4 and 5 and check your answer by finding the derivative of the function expressed as a polynomial.

Solution

$$\begin{aligned}
f'(x) &= x^2(2)(3x + 1)(3) + 2x(3x + 1)^2 \\
&= 6x^2(3x + 1) + 2x(9x^2 + 6x + 1) \\
&= 18x^3 + 6x^2 + 18x^3 + 12x^2 + 2x \\
&= 36x^3 + 18x^2 + 2x.
\end{aligned}$$

Since
$$f(x) = x^2(9x^2 + 6x + 1) = 9x^4 + 6x^3 + x^2,$$

it follows that
$$f'(x) = 36x^3 + 18x^2 + 2x.$$

Derivative of a product

EXAMPLE 4 Find the derivative of $f(x) = x^4(x^3 + 1)^{5/3}$.

Solution

$$\begin{aligned}
f'(x) &= x^4\left[\frac{5}{3}(x^3 + 1)^{2/3}(3x^2)\right] + (x^3 + 1)^{5/3}(4x^3) \\
&= 5x^6(x^3 + 1)^{2/3} + 4x^3(x^3 + 1)^{5/3}.
\end{aligned}$$

By factoring $x^3(x^3 + 1)^{2/3}$ from the sum we could write this derivative in the following simplified form:

2.5 PRODUCT, QUOTIENT, AND GENERAL POWER RULES FOR DIFFERENTIATION

$$f'(x) = x^3(x^3 + 1)^{2/3}(5x^3 + 4x^3 + 4)$$
$$= x^3(x^3 + 1)^{2/3}(9x^3 + 4).$$

PRACTICE PROBLEM Find the derivative of $f(x) = x^3(x^2 + 4)^{3/2}$.

Answer: $f'(x) = 6x^2(x^2 + 2)(x^2 + 4)^{1/2}$. ∎

Now that we have theorems for the derivative of the power of a function and for the sum, difference, and product of two functions, let us state the theorem giving the rule for differentiating the quotient of two functions.

quotient rule

THEOREM 6 If f and g are differentiable functions and if $F(x) = f(x)/g(x)$, then

$$F'(x) = \frac{g(x)f'(x) - f(x)g'(x)}{[g(x)]^2},$$

provided $g(x) \neq 0$. (**The derivative of the quotient of two functions is the denominator times the derivative of the numerator minus the numerator times the derivative of the denominator all divided by the square of the denominator.** This is called the *quotient rule*.)

Derivative of a quotient

EXAMPLE 5 Use Theorem 6 to find the derivative of the following function:

$$F(x) = \frac{3x + 2}{x^2 + 1}.$$

Solution

$$F'(x) = \frac{(x^2 + 1)(3) - (3x + 2)(2x)}{(x^2 + 1)^2}$$
$$= \frac{-3x^2 - 4x + 3}{(x^2 + 1)^2}.$$

PRACTICE PROBLEM Find the derivative of $F(x) = (x^3 + 1)/(2x + 5)$.

Answer: $F'(x) = (4x^3 + 15x^2 - 2)/(2x + 5)^2$. ∎

Derivative of a quotient

EXAMPLE 6 Use the quotient rule to find the derivative of

$$f(x) = \frac{x}{(3x - 1)^{1/2}}.$$

Solution

$$f'(x) = \frac{(3x - 1)^{1/2} - x(1/2)(3x - 1)^{-1/2}(3)}{3x - 1}.$$

One of the easiest ways to simplify this answer is to multiply the numerator and denominator of the fraction by $2(3x - 1)^{1/2}$. (In order to make similar simplifications, the student should carefully analyze the algebra involved.) We find that

$$f'(x) = \frac{2(3x - 1) - 3x}{2(3x - 1)^{3/2}} = \frac{3x - 2}{2(3x - 1)^{3/2}}.$$

Let us turn our attention from differentiation theorems to an important idea associated with both the derivative and the graph of a function. If f is a differentiable function at x, then

$$f'(x) = \lim_{\Delta x \to 0} \frac{f(x + \Delta x) - f(x)}{\Delta x}$$

exists. If we let $u = x + \Delta x$, then as Δx approaches 0, we see that u approaches x. This leads to an *equivalent definition* of the derivative of f at x. (See Figure 2.16.) It is

alternative definition of the derivative

$$f'(x) = \lim_{u \to x} \frac{f(u) - f(x)}{u - x}.$$

2.5 PRODUCT, QUOTIENT, AND GENERAL POWER RULES FOR DIFFERENTIATION

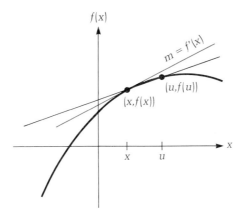

FIG. 2.16.
$f'(x) = \lim_{u \to x} \dfrac{f(u) - f(x)}{u - x}.$

From this definition, we see that as u approaches x, then $u - x$ approaches 0. Furthermore, the numerator $f(u) - f(x)$ must also approach 0; for, otherwise, the quotient would "get large" and not have a limit. Hence, for a differentiable function, $f(u)$ approaches $f(x)$ as u approaches x. In terms of limits, this means that

$$\lim_{u \to x} f(u) = f(x).$$

Consequently, we see that *any differentiable function is continuous.*

Although every differentiable function is continuous, note that we did not say that every continuous function is differentiable. In fact, this is not true. For example

$$f(x) = x^{1/3}$$

is continuous at $x = 0$ but its derivative,

$$f'(x) = \frac{1}{3x^{2/3}},$$

does not exist at $x = 0$. (The line tangent to the graph of $f(x) = x^{1/3}$ becomes vertical as x approaches 0. See Figure 2.17.)

We conclude this section by considering another application of the derivative. (More will be considered later.) In Section 2.3 we found that if $s(x)$ is the distance in feet a body travels in x seconds, then the average velocity in feet per second during the time interval from x to $x + \Delta x$ seconds is given by the Newton

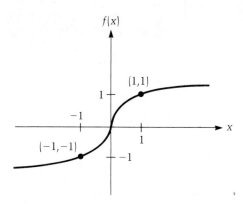

FIG. 2.17. $f(x) = x^{1/3}$.

quotient of the distance function. That is,

$$\text{average velocity} = \frac{\Delta s}{\Delta x} = \frac{s(x + \Delta x) - s(x)}{\Delta x}.$$

If the limit of the Newton quotient exists as Δx approaches zero, it is defined to be the *(instantaneous)* **velocity** in feet per second of the object at the end of x seconds. Therefore, the *velocity function* $v(x)$ is defined by

velocity

velocity function

$$v(x) = s'(x) = \lim_{\Delta x \to 0} \frac{\Delta s}{\Delta x}.$$

velocity

EXAMPLE 7 If $s(x) = 16x^2$ is the distance a free-falling body falls from rest in x seconds, what is its velocity at the end of 2 seconds.

Solution: Since $s'(x) = 32x$, $s'(2) = 64$ ft/sec is its velocity at the end of 2 seconds.

EXERCISES

Set A–Derivative of the Power of a Function and the Product and Quotient of Two Functions

In Exercises 1 through 16, find the first derivative of each function and simplify.

1. $f(x) = (x^3 + 6x)^{5/2}$
2. $f(x) = (x^2 + 8x)^{3/2}$
3. $f(x) = (2x + 1)^{3/2}$
4. $f(x) = (2x^6 + 4)^{1/4}$

5. $F(x) = (2x^3 - 3x)^{2/7}$
6. $G(x) = (5x^3 - 4x)^{2/5}$
7. $g(x) = (3 - x)^{4/5}$
8. $g(x) = (4 - x^2)^{1/2}$
9. $f(x) = x(2x + 3)^{1/2}$
10. $f(x) = 3x(x^3 + 1)^{4/3}$
11. $f(x) = x^2(x + 1)^{3/2}$
12. $f(x) = x^2(4x + 1)^{1/2}$
13. $f(x) = \dfrac{2x + 1}{3x - 4}$
14. $f(x) = \dfrac{5x - 3}{4x + 7}$
15. $f(x) = \dfrac{x}{x^2 + 1}$
16. $f(x) = \dfrac{x}{(x^3 + 1)^2}$

17. (a) Find the derivative of $f(x) = (x + 1)/x$ by using the quotient theorem for differentiation.
   (b) Use the fact that $f(x) = 1 + 1/x = 1 + x^{-1}$ to find the derivative. Compare your answer to the one obtained in part (a).

18. (a) Find the derivative of $f(x) = (3x^2 - 2x + 5)/x$ by using the quotient theorem for differentiation.
   (b) Use the fact that $f(x) = 3x - 2 + 5x^{-1}$ to find the derivative. Compare your answer to the one obtained in part (a).

19. (a) Find the derivative of $f(x) = x(x^2 + 3)^2$ by using the product and power rules for differentiation.
   (b) Use multiplication to express f as a polynomial and then find the derivative. Compare your answer to the one obtained in part (a).

20. (a) Find the derivative of $f(x) = (3x^2 + 7)(4x + 9)$ by using the product theorem for differentiation.
   (b) Use multiplication to express f as a polynomial and then find the derivative. Compare your answer to the one obtained in part (a).

Set B—Slopes of Tangent Lines

In each of Exercises 21 through 26, find the slope of the tangent to the graph of f at the given x.

21. $f(x) = \dfrac{4x - 3}{x^2 + 1}$, $x = 2$
22. $f(x) = \dfrac{7x - 5}{x^3 + 4}$, $x = -2$
23. $f(x) = x(x + 3)^{1/2}$, $x = 1$
24. $f(x) = x(2x + 1)^{3/2}$, $x = 4$
25. $f(x) = x^2(4x + 1)^{5/2}$, $x = 2$
26. $f(x) = x^3(5x^2 + 3)^{2/3}$, $x = 1$

Set C—Equations of Tangent Lines

In each of Exercises 27 through 30, find the equation of the line tangent to the graph of f at the given x.

27. $f(x) = \dfrac{3x - 4}{x^2 + 1}$, $x = 1$
28. $f(x) = \dfrac{5x + 4}{x^3 - 1}$, $x = 2$

29. $f(x) = x(2x + 7)^{1/2}$, $x = 1$ 30. $f(x) = x(3x - 2)^{4/3}$, $x = 1$

Set D–Velocity and Acceleration

31. Suppose an object is propelled upward so that its height in feet above the ground at the end of x seconds is given by $h(x) = 80x - 16x^2$.
    (a) Find $h'(0)$ and explain its physical significance.
    (b) What is the object's velocity at the end of 2 seconds?
    (c) For what x is $h'(x) = 0$? Explain the physical significance.
    (d) How long is the object in the air?

Velocity

32. Suppose that the distance (in feet) traveled in x seconds by a bullet shot vertically upward is given by $s(x) = 400x - 16x^2$.
    (a) What is the bullet's initial velocity?
    (b) What is the velocity of the bullet at the end of 2 seconds?
    (c) When does it reach its maximum height?
    (d) How far has the bullet traveled when the maximum height is attained?

Acceleration

33. In Exercise 34, Section 2.3, average acceleration was defined to be $\Delta v/\Delta x$ where $v(x)$ is the velocity function for an object in rectilinear motion. **Acceleration** *at time x* is defined by

$$\text{acceleration} = \lim_{\Delta x \to 0} \frac{\Delta v}{\Delta x} = v'(x).$$

Since $v(x) = s'(x)$, it follows that $v'(x) = s''(x)$. Therefore, the second derivative of the distance function is the acceleration function. Use the fact that $s(x) = 16x^2$ is the distance a free-falling body falls from rest in x seconds to find its acceleration at any time x.

2.6 THE DIFFERENTIAL AND MARGINAL ANALYSIS

Recall that if f is a differentiable function, then

$$f'(x) = \lim_{\Delta x \to 0} \frac{\Delta y}{\Delta x} = \lim_{\Delta x \to 0} \frac{f(x + \Delta x) - f(x)}{\Delta x}.$$

As indicated earlier, if Δx is close to zero, then

$$\frac{f(x + \Delta x) - f(x)}{\Delta x} \doteq f'(x).$$

2.6 THE DIFFERENTIAL AND MARGINAL ANALYSIS

Consequently,

$$f(x + \Delta x) - f(x) \doteq f'(x)\Delta x.$$

Since $\Delta y = f(x + \Delta x) - f(x)$, it follows that for Δx close to zero,

$$\Delta y \doteq f'(x)\Delta x.$$

We call $f'(x)\Delta x$ the **differential** *of f at x with increment* Δx; it is denoted by dy. Hence

differential of a function

$$dy = f'(x)\Delta x.$$

For symmetry in notation, we let $dx = \Delta x$ and use these two symbols interchangeably. Therefore, the differential dy is also defined by

$$dy = f'(x)dx.$$

The differential dy approximates the change Δy in y resulting from a small change Δx in x. Since $f(x + \Delta x) - f(x) \doteq dy$,

$$f(x + \Delta x) \doteq f(x) + dy$$

and $f(x) + dy$ is often used to approximate $f(x + \Delta x)$ when $f(x)$ and $f'(x)$ are known, or are easily obtained.

Differential approximation

EXAMPLE 1 Let $f(x) = x^{1/2}$ and use the differential of f at 25 to approximate $f(27) = \sqrt{27}$.

Solution: Since $f'(x) = \frac{1}{2}x^{-1/2}$ at $x = 25$ and $\Delta x = 2$, we get

$$f(27) \doteq f(25) + [f'(25)](2) = 5 + \frac{1}{5} = 5.2.$$

Consequently, 5.2 is an approximation of $\sqrt{27}$. (In fact, $\sqrt{27} = 5.196$, correct to three decimal places.)

PRACTICE PROBLEM Let $f(x) = x^{1/2}$ and use the differential of f at 9 to approximate $f(11) = \sqrt{11}$.

Answer: $f(11) \doteq 3\frac{1}{3}$. ∎

In Figure 2.18, we see that the difference between Δy and dy is small provided Δx is small. If we let $E = \Delta y - dy$, then

$$E = [f(x + \Delta x) - f(x)] - dy = f(x + \Delta x) - [f(x) + dy],$$

and E is the **error** resulting from using $f(x) + dy$ to approximate $f(x + \Delta x)$. It can be proved for a twice differentiable function that there exists a number t between x and $x + \Delta x$ such that

error formula

$$E = \frac{1}{2} f''(t)(\Delta x)^2.$$

Furthermore, the sign of the error tells us whether the approximation $f(x) + dy$ is too large or too small; if the error is positive, the approximation is too small, and if the error is negative, the approximation is too large.

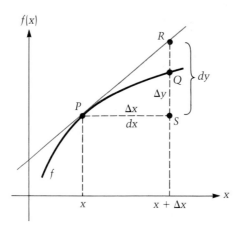

FIG. 2.18. $dy = f'(x) dx$,
$|SR| = dy$, $|SQ| = \Delta y$,
$E = \Delta y - dy$.

2.6 THE DIFFERENTIAL AND MARGINAL ANALYSIS

Error approximation

EXAMPLE 2 Use the error formula to find an upper bound on the error in the differential approximation of $\sqrt{27}$ obtained in Example 1.

Solution: Since $f(x) = x^{1/2}$,

$$f'(x) = \frac{1}{2}x^{-1/2} \quad \text{and} \quad f''(x) = -\frac{1}{4}x^{-3/2}.$$

In the approximation, $x = 25$ and $\Delta x = 2$; therefore,

$$E = \frac{1}{2}\left(\frac{-1}{4t^{3/2}}\right)(2)^2 \quad \text{where} \quad 25 < t < 27.$$

To find an upper bound on the error, we ignore the negative sign and conclude that if $25 < t < 27$, then

$$E < \frac{1}{2(25)^{3/2}} = \frac{1}{2(125)} = \frac{1}{250} = 0.004.$$

Consequently, the approximation was too large but the error was less than 0.004.

PRACTICE PROBLEM Use the error formula to find an upper bound on the error in the differential approximation of $\sqrt{11}$ determined in the Practice Problem following Example 1.

Answer: $\frac{1}{54} \doteq .0185$. ■

In Section 2.3 we found that if $C(x)$ is the total cost of producing x units of a given product, then the average marginal cost is given by

$$\frac{\Delta C}{\Delta x} = \frac{C(x + \Delta x) - C(x)}{\Delta x},$$

where Δx is the number of additional units produced after x units have been produced. Again, for Δx close to zero,

$$C'(x) \doteq \frac{\Delta C}{\Delta x},$$

and

$$C'(x)\Delta x \doteq \Delta C = C(x + \Delta x) - C(x).$$

In particular, if $\Delta x = 1$, then

$$C'(x) \doteq C(x + 1) - C(x).$$

marginal cost

The **marginal cost** of production at x is defined to be $C'(x)$; it approximates the cost of producing one additional unit after x units have been produced. In other words, the marginal cost is the differential of $C(x)$ at x where $\Delta x = 1$. Similarly, if $R(x)$ is the revenue function for x units of a given product, then $R'(x)$ is defined to be the **marginal revenue** at the level of output x; it approximates the revenue from the $(x + 1)$st unit. If $P(x)$ is the profit function for x units of a given product, then $P'(x)$ is defined to be the **marginal profit** at the level of output x; it approximates the profit from the $(x + 1)$st unit. The use of the derivative to find the marginal cost, marginal revenue, or marginal profit at x is called *marginal analysis*.

marginal revenue

marginal profit

marginal analysis

Marginal cost

EXAMPLE 3

(a) For the cost function $C(x) = 2x^2 + 3x + 840$ given in Example 1, Section 2.3, we found that the cost of producing one additional unit after 20 units had been produced was $85. Find the marginal cost at 20 units to approximate the cost of the 21st unit.

(b) Use the error formula for the differential to check the accuracy of your answer.

Solution

(a) Since $C'(x) = 4x + 3$, $C'(20) = 83$. The marginal cost at 20 units is $83; that is, the approximate cost of the 21st unit is $83.

(b) Since $C''(x) = 4$ for *any* x, we find by using the differential error formula that

$$E = \frac{1}{2} C''(t)(\Delta x)^2 = \frac{1}{2} (4)(1)^2 = \$2$$

Note that since $C''(t) = 4$ is a constant, the actual error is determined. Therefore, the 21st unit costs $83 + $2 = $85, the answer obtained in Example 1, Section 2.3.

2.6 THE DIFFERENTIAL AND MARGINAL ANALYSIS

Marginal cost

EXAMPLE 4

(a) According to Exercise 25, Section 2.3, the cost of producing x thousand units of a certain product is

$$C(x) = (2x^2 + 4x + 500)100 \text{ dollars}$$

where $x \geq 1$. How much would the actual cost be to produce one additional unit after 4000 units have been produced?

(b) Use the differential to approximate the answer to the question in part (a).

Solution

(a) Since x *represents thousands of units*, $x = 4$ and $\Delta x = 0.001$. Hence, the actual cost of producing the 4001st unit is

$$C(4.001) - C(4) = 54{,}802.0002 - 54{,}800 = 2.0002,$$

that is, $2 to the nearest cent.

(b) Since $C'(x) = (4x + 4)(100)$, the cost of producing the 4001st unit is approximately $dC = C'(x)(\Delta x)$, where $x = 4$ and $\Delta x = 0.001$; that is,

$$C'(4)(0.001) = 20(100)(0.001) = \$2.$$

Note: The increased accuracy of this answer as compared to the one in Example 3 might have been anticipated since Δx is nearer 0.

The accuracy of this approximation can be verified without part (a) by using the error formula for the differential approximation. Since $C'(x) = (4x + 4)(100)$, it follows that $C''(x) = 400$ for *any* x. Therefore

$$E = \frac{1}{2} C''(t)(\Delta x)^2$$

$$= \frac{1}{2}(400)(0.001)^2$$

$$= 0.002.$$

Again, the error formula gives the actual error. (In general, for any quadratic function, since the second derivative is a constant, the formula for E gives the actual error in the differential approximation.)

Suppose a manufacturer's total cost function in dollars for producing x radios is

$$C(x) = 5000 + 20x + 8x^{1/2}.$$

Let us analyze and review various conclusions that can be obtained from this cost function.

(1) The initial (fixed) costs are $C(0) = \$5000$.

(2) The average cost of producing the first 100 radios is

$$\overline{C}(100) = \frac{C(100)}{100} = \frac{5000 + 2000 + 80}{100} = \$70.80.$$

(3) The cost of producing the 101st radio is approximated by $C'(100)$. Since

$$C'(x) = 20 + 4x^{-1/2},$$

the approximate cost of the 101st radio is

$$C'(100) = 20 + \frac{4}{10} = \$20.40.$$

(4) Since $C''(x) = -2x^{-3/2}$ is negative, the approximation in part (3) is too large. Furthermore, since

$$E = \frac{1}{2}C''(t)(\Delta x)^2 = -t^{-3/2}(1)^2 \quad \text{for} \quad 100 < t < 101,$$

the error E in absolute value is less than

$$\frac{1}{(100)^{3/2}} = \frac{1}{(10)^3} = 0.001.$$

Consequently, marginal analysis approximates the cost of the 101st radio to the nearest cent.

(5) Finally, as x gets large,

$$C'(x) = 20 + \frac{1}{x^{1/2}}$$

decreases and approaches 20; therefore the cost of each radio decreases and approaches $20 per radio.

2.6 THE DIFFERENTIAL AND MARGINAL ANALYSIS

Since there are many different notations for the derivative which are frequently used, it is important not only to be familiar with them but also to be able to use them effectively. For a function $y = f(x)$, each of the following is a notation for the first derivative:

$$f'(x), \frac{dy}{dx}, y', D_x y, \frac{d}{dx} f(x), Df(x), Df.$$

The "prime" notation $f'(x)$ and y' and the *Leibniz notation* dy/dx are the most widely used and are the ones we shall continue to use. Note that

$$\frac{dy}{dx} = f'(x)$$

is consistent with the differential notation $dy = f'(x)dx$. Furthermore, instead of using the notation $f'(3)$ to denote the derivative of $y = f(x)$ at $x = 3$, we can use

$$\left. \frac{dy}{dx} \right|_{x=3}.$$

For example, if $y = 3x^2 + 5x$ then

$$\frac{dy}{dx} = 6x + 5 \quad \text{and} \quad \left. \frac{dy}{dx} \right|_{x=2} = 17.$$

The notations used most often for the second derivative are

$$f''(x) \quad \text{and} \quad \frac{d^2 y}{dx^2}.$$

Other commonly used notations are

$$y'', \; y''(x), \; D_x^2 y, \; \frac{d^2}{dx^2} f(x), \; D^2 f(x), \; D^2 f.$$

Since *higher-order derivatives* of a function have useful applications, it is important to have a notation for the nth derivative; notations used are

$$f^{(n)}(x), \; \frac{d^n y}{dx^n}, \; y^{(n)}(x), \; D_x^n y, \; \frac{d^n}{dx^n} f(x), \; D^n f(x), \; D^n f.$$

Leibniz notation

EXAMPLE 5 If $y = 3x^4 - 5x^2 + 11x - 6$,

then

$$\frac{dy}{dx} = 12x^3 - 10x + 11$$

and

$$\frac{d^2 y}{dx^2} = 36x^2 - 10.$$

(Note the offset of "2" in the Leibniz notation; it should be understood that "2" is *not* an exponent.) Furthermore,

$$\frac{d^3 y}{dx^3} = 72x \quad \text{and} \quad \frac{d^4 y}{dx^4} = 72.$$

Each higher derivative of the polynomial function is zero.

PRACTICE PROBLEM Find the first, second, and third derivatives of $y = 7x^3 - 8x^{5/2} - 1/x$.

2.6 THE DIFFERENTIAL AND MARGINAL ANALYSIS

Answer

$$\frac{dy}{dx} = 21x^2 - 20x^{3/2} + 1/x^2,$$

$$\frac{d^2y}{dx^2} = 42x - 30x^{1/2} - 2/x^3,$$

and

$$\frac{d^3y}{dx^3} = 42 - 15x^{-1/2} + 6/x^4. \blacksquare$$

EXAMPLE 6 If $C = 3 - 8p + p^2 + p^3$, then

$$\frac{dC}{dp} = -8 + 2p + 3p^2, \qquad \frac{d^2C}{dp^2} = 2 + 6p, \qquad \text{and} \qquad \frac{d^3C}{dp^3} = 6.$$

Each higher-order derivative of the polynomial function is zero.

EXERCISES

Set A–Derivatives and the Leibniz Notation

1. Let $y = 4x^2 - 6x + 1/x$. Find

$$\frac{dy}{dx} \quad \text{and} \quad \frac{d^2y}{dx^2}.$$

2. Let $y = 18x^{5/3} + 17x$. Find

$$\frac{dy}{dx} \quad \text{and} \quad \frac{d^2y}{dx^2}.$$

3. Let $y = 4x^{-1/2} + 6x^{2/3}$. Find

$$\frac{dy}{dx} \quad \text{and} \quad \frac{d^2y}{dx^2}.$$

4. Let $C = 4x^3 - 5x + 2$. Find

$$\frac{dC}{dx} \quad \text{and} \quad \frac{d^2C}{dx^2}.$$

5. Let $C = 4 - 3p + p^2 - 2p^3$. Find
$$\frac{dC}{dp} \quad \text{and} \quad \frac{d^2C}{dp^2}.$$

6. Let $u = v^4 - 7v^2 + 13v - 2$. Find
$$\frac{du}{dv} \quad \text{and} \quad \frac{d^2u}{dv^2}.$$

7. Let $q = 5p^3 - 2p + 3/p$. Find
$$\frac{dq}{dp} \quad \text{and} \quad \frac{d^2q}{dp^2}.$$

8. Let $q = 6p + 5/p - 7/p^2$. Find
$$\frac{dq}{dp} \quad \text{and} \quad \frac{d^2q}{dp^2}.$$

9. Let $V = \frac{4}{3}\pi r^3$. Find
$$\frac{dV}{dr} \quad \text{and} \quad \frac{d^2V}{dr^2}.$$

10. Let $h = 88t - 32t^2$. Find
$$\frac{dh}{dt} \quad \text{and} \quad \frac{d^2h}{dt^2}.$$

Set B–Marginal Analysis

Marginal cost

11. (a) The cost of producing x units per week of a given product is $C(x) = 0.02x^2 + 6x + 300$ dollars. Use marginal analysis to approximate the cost of the 51st unit.
    (b) Find $C(51) - C(50)$. What does this represent? Compare your answer with that to part (a).
    (c) Use the error formula to verify the difference in the answers to parts (a) and (b).
    (d) Find the marginal cost at $x = 100$.
    (e) Suppose when the level of production is 100 units per week that the manufacturer can sell the units for $10 each. If the selling price is not raised, would it be wise for the manufacturer to increase the level of production? Justify your answer. [See part (d).]

2.6 THE DIFFERENTIAL AND MARGINAL ANALYSIS

Marginal cost

12. (a) The cost of producing x units per week of a given product is $C(x) = 0.04x^2 + 8x + 500$ dollars. Use marginal analysis to approximate the cost of the 101st unit.
    (b) Find $C(101) - C(100)$. What does this represent? Compare your answer with that to part (a).
    (c) Use the error formula to verify the difference in the answers to parts (a) and (b).
    (d) Find the marginal cost at $x = 150$.
    (e) Suppose that when the level of production is 150 units per week, the manufacturer can sell the units for $20 apiece. If the selling price is not raised, would it be wise for the manufacturer to increase the level of production? Justify your answer. [See part (d).]

Differential approximation of cost

13. (a) Suppose the cost in thousands of dollars for producing x thousand units of a particular product is given by

$$C(x) = 0.8x^2 + 1.2x + 12, \quad x \geq 2.$$

Use the differential to approximate the cost of producing one additional unit after 5000 have been produced.
    (b) Check your answer by finding the actual cost of the 5001st unit.
    (c) Use the error formula to verify the difference in your answers to parts (a) and (b).

Differential approximation of cost

14. (a) Suppose the cost in thousands of dollars of producing x thousand units of a particular product is given by

$$C(x) = 0.1x^2 + 3.7x + 63, \quad 2 \leq x \leq 40.$$

Use the differential to approximate the cost of producing one additional unit after 9000 have been produced.
    (b) Check your answer by finding the actual cost of the 9001st unit.
    (c) Use the error formula to verify the difference in your answers to parts (a) and (b).

Marginal revenue

15. (a) For the revenue function $R(x) = -0.08x^2 + 2.8x - 4$, $x \geq 2$, find the marginal revenue at $x = 11$.
    (b) Find $R(12) - R(11)$ and compare your result with the answer to part (a).

Marginal profit

16. (a) For the profit function $P(x) = 0.16x^2 + 4.0x - 16$, $x \geq 2$, find the marginal profit at $x = 11$.
    (b) Find $P(12) - P(11)$ and compare your result with the answer to part (a).

Marginal analysis

17. Prove that marginal profit is zero when marginal cost equals marginal revenue.

Marginal cost

18. If the cost function, in dollars, for x units of a given product is $C(x) = 4000 + 15x + 8x^{1/2}$, for what level of production would the marginal cost be $15.50?

Marginal cost

19. If the cost function, in dollars, for x units of a given product is $C(x) = 0.02x^2 + 6x + 300$, for what level of production would the marginal cost be $22?

Differential approximation

20. (a) Suppose the cost function, in dollars, for producing x radios is $C(x) = 5000 + 20x + 8x^{1/2}$. Use the differential to approximate the cost of producing 4 additional radios after 100 have been produced.
    (b) Use the error formula to analyze the possible error in the answer to part (a).

Set C—Differential Approximations of Function Values

21. (a) For the function $f(x) = x^{1/5}$, let $x = 32$, $\Delta x = 2$ and use the differential to approximate $\sqrt[5]{34}$.
    (b) Use the error formula to find an upper bound on the error in the approximation.

22. Use the differential of $f(x) = x^{2/5}$ at 32 to approximate $f(33)$.

23. Use the differential of $f(x) = 1/x$ at 4 to approximate $f(4.01)$.

24. Use the differential of $f(x) = 1/x^2$ at 2 to approximate $f(2.03)$.

2.7 SUMMARY AND REVIEW EXERCISES

1. THE NEWTON QUOTIENT OF A FUNCTION

For a function f given by $y = f(x)$, the **Newton quotient** is

$$\frac{\Delta y}{\Delta x} = \frac{\text{change in } y}{\text{change in } x} = \frac{f(x + \Delta x) - f(x)}{\Delta x}.$$

(a) For a function f, the Newton quotient is the **slope** of the line containing the points $(x, f(x))$ and $(x + \Delta x, f(x + \Delta x))$ on the graph of f.

(b) If f is the cost function for x units of a given product, then the Newton quotient is the **average marginal cost** at x units for Δx additional units. If $\Delta x = 1$, it is the cost of the $(x + 1)$st unit.

(c) If f is a distance function, the Newton quotient is the **average velocity** during the time interval x to $x + \Delta x$.

2.7 SUMMARY AND REVIEW EXERCISES

(d) If f is a revenue function, the Newton quotient is the **average marginal revenue** at x units for Δx additional units.

(e) If f is a profit function, the Newton quotient is the **average marginal profit** at x units for Δx additional units.

(f) If f is a velocity function, the Newton quotient is the **average acceleration** during the time interval x to $x + \Delta x$.

2. THE DERIVATIVE OF A FUNCTION

For a function f given by $y = f(x)$, the **derivative** of f at x is

$$\frac{dy}{dx} = f'(x) = \lim_{\Delta x \to 0} \frac{f(x + \Delta x) - f(x)}{\Delta x}.$$

(a) For a function f, $f'(x)$ is the **slope** of the line tangent to the graph of f at $(x, f(x))$.

(b) If f is a cost function, $f'(x)$ is the **marginal cost** at x; it is the approximate cost of one additional unit after x units have been produced.

(c) If f is a distance function, $f'(x)$ is the **velocity** at time x.

(d) If f is a revenue function, $f'(x)$ is the **marginal revenue** at x; it is the approximate revenue from one additional unit after x units have been produced.

(e) If f is a profit function, $f'(x)$ is the **marginal profit** at x; it is the approximate profit from one additional unit after x units have been produced.

(f) If f is a velocity function, $f'(x)$ is the **acceleration** at time x.

3. DIFFERENTIATION THEOREMS

(a) If $f(x) = c$, c a constant, then $f'(x) = 0$.

(b) If $f(x) = x^r$, r a real number, then $f'(x) = rx^{r-1}$.

(c) If $F(x) = [g(x)]^r$, r a real number, then $F'(x) = r[g(x)]^{r-1}g'(x)$.

(d) Let f and g be differentiable functions.
If $F(x) = cf(x)$, c a constant, then $F'(x) = cf'(x)$.
If $F(x) = f(x) + g(x)$, then $F'(x) = f'(x) + g'(x)$.
If $F(x) = f(x)g(x)$, then $F'(x) = f(x)g'(x) + f'(x)g(x)$.
If $F(x) = f(x)/g(x)$, then

$$F'(x) = \frac{g(x)f'(x) - f(x)g'(x)}{[g(x)]^2}.$$

4. THE DIFFERENTIAL OF A FUNCTION

(a) For a function f given by $y = f(x)$, the **differential** of f at x with increment dx (or Δx) is defined by

$$dy = f'(x)dx.$$

It is the approximate change in y at x with an increment dx; that is,

$$dy \doteq f(x + dx) - f(x).$$

(b) An upper bound on the **error** in using dy to approximate Δy, the actual change in y, can be found from the fact that

$$\Delta y - dy = \frac{1}{2}f''(t)(dx)^2$$

for some t between x and $x + dx$.

5. REVIEW EXERCISES

In Exercises 1 and 2, find the given limits.

1. $\lim\limits_{x \to 3} \dfrac{3x^2 - 7x + 1}{5x + 11}$

2. $\lim\limits_{x \to -1} \dfrac{x^2 - 6x - 7}{2x^2 - 3x - 5}$

In Exercises 3 and 4, state for what x the functions are continuous.

3. $f(x) = \dfrac{x^2 - x - 6}{x^2 + 1}$

4. $f(x) = \dfrac{x^2 + 1}{x^2 - x - 6}$

5. Find the Newton quotient of

$$f(x) = \frac{x}{3x + 4}$$

at x with increment Δx.

6. Let $C(x) = 6x^3 - 5x^2 + 3x + 2$. Find $C'(x)$ and $C''(x)$.

7. Let $y = -2x^{5/2} + 3/x^2$. Find

$$\frac{dy}{dx} \quad \text{and} \quad \frac{d^2y}{dx^2}.$$

8. Let $f(x) = (3x^2 + 5)^{2/3}$. Find $f'(x)$.

2.7 SUMMARY AND REVIEW EXERCISES

9. Let $y = (3x + 1)/(x^2 + 4)$. Find
$$\frac{dy}{dx}.$$

10. Let $u = v/(3v + 1)$. Find
$$\frac{du}{dv} \quad \text{and} \quad \frac{d^2u}{dv^2}.$$

11. The cost of producing x units of a certain product is given by $C(x) = 0.04x^2 + 3.5x + 36$ dollars.
    (a) Find the average marginal cost for producing 4 additional units after 10 have been produced.
    (b) Find the average marginal cost in terms of x and Δx.
    (c) Use part (b) to check your answer in part (a).

12. Use the Newton quotient of $f(x) = x/(3x + 4)$ at x with increment Δx to find the slope of the line through the two points on the graph of f where $x = 5$ and $x = 5.01$. (See Exercise 5.)

13. Find an equation of the line tangent to the graph of $f(x) = 2x^{3/2} - 4/x$ at $x = 4$.

14. (a) If the cost of producing x units of a certain product is given by $C(x) = 0.04x^2 + 3.5x + 36$ dollars, find the cost of the 11th unit.
    (b) Use marginal analysis to approximate the cost of the 11th unit.

15. (a) For the profit function $P(x) = 0.12x^2 + 5.2x - 10$, where $x \geq 4$, find the marginal profit at $x = 20$.
    (b) Find $P(21) - P(20)$ and compare with the answer to part (a).

16. If the cost of producing x units of a certain product is given by $C(x) = 0.04x^2 + 3.5x + 36$ dollars, for what level of production would the marginal cost be $6.30?

17. Suppose the cost function in dollars for producing x units of a given product is $C(x) = 2500 + 20x + 4x^{1/2}$.
    (a) Use the differential to approximate the cost of producing 4 additional units after 100 have been produced.
    (b) Use the error formula to analyze the possible error in the answer to part (a).
    (c) Use a calculator to approximate $C(104) - C(100)$.

18. (a) Use the differential of $f(x) = 1/(x^2 + 3)$ at $x = 3$ to approximate $f(3.01)$.
    (b) Use a calculator to approximate $f(3.01)$ and compare your answer with that in part (a).

19. A bullet fired directly upward reaches a height of $h(x) = 112x - 16x^2$ feet above the ground at the end of x seconds.
    (a) Find its initial velocity.
    (b) Find its velocity at the end of 2 seconds.
    (c) What is its maximum height above the ground?

20. Find an equation for each of the lines tangent to the graph of $f(x) = x^3 - 6x^2 + 2$ and parallel to the line $15x - y = 11$.

BIOGRAPHICAL SKETCH

Isaac Newton (1642–1727)

The Bettmann Archive

Isaac Newton was born at Woolsthorpe, England, on Christmas day in 1642, the day Galileo died. Isaac Barrow, his teacher at Cambridge, was the first to recognize Newton's genius. Barrow had a great influence on Newton's academic growth, and in 1669 he resigned the Lucasian chair of mathematics in favor of Newton, quite a rare and magnanimous act.

Newton's discovery of the universal law of gravitation and his invention of differential calculus were made during the two years 1664–1665. It was during these two years, before he was 25 years old and when the University was closed due to the Great Plague, that he first exhibited his great creative ability. L. T. More said, "there are no other examples of achievement in the history of science to compare with that of Newton during those two golden years." Unfortunately, Newton often waited an inordinate amount of time before publishing his discoveries. For example, he waited 20 years to publish his universal law of gravitation; it appeared in his monumental work *Principia*, published in 1686. He delayed even longer the publication of his results on the calculus that he had obtained at the age of 23.

In his development of the calculus, Newton relied heavily on his remarkable intuition. He was more interested in the development of new ideas and in methods for solving practical problems related to them than he was in a rigorous verification of the new theories and techniques. Newton did establish mechanics on an axiomatic basis, made contributions to chemistry, optics, and also studied theology and philosophy. He once said that if he had seen further than others, it was because he had "stood on the shoulders of giants." Although he was a sickly child, he was fortunate in having been spared the poverty that plagued many of the other great mathematicians in the seventeenth, eighteenth, and nineteenth centuries. His creative ability is best described by Leibniz's remark, "Taking mathematics from the beginning of the world to the time of Newton, what he has done is much the better half."

CHAPTER

3

FURTHER APPLICATIONS OF THE DERIVATIVE

3

3.1 INTRODUCTION
3.2 INCREASING AND DECREASING FUNCTIONS
3.3 RELATIVE MAXIMA AND RELATIVE MINIMA
3.4 MORE APPLIED MAXIMUM AND MINIMUM PROBLEMS
3.5 RELATED RATES (OPTIONAL)
3.6 SUMMARY AND REVIEW EXERCISES

3.1 INTRODUCTION

In this chapter, we discuss additional applications of the derivative and learn how to solve such problems as the following.

1. A psychologist finds that on the average a student in the first t hours of study of a foreign language can learn the meaning of

$$w(t) = \frac{100t^2}{t^2 + 27} + 10$$

   new words provided $1 \leq t \leq 8$. Use the derivative to show that $w(t)$ is an increasing function on $[1, 8]$. During what time interval is the *rate of increase* in learning new words increasing? (See Example 7, Section 3.2.)

2. Suppose a large soup company plans to manufacture a can (right circular cylinder) that will contain K cubic units (a fixed volume) of soup. What should the ratio of the height to the radius be in order to minimize the can's surface area and thus its cost? (See Example 9, Section 3.4.)

3. A book publisher plans to produce a textbook with each page containing 48 in^2 of print. If the top and bottom margins are each to be one inch wide and if the side margins are each to be $\frac{3}{4}$ inch wide, what page dimensions would require the minimum amount of paper? (See Exercise 15, Section 3.4.)

4. A company has an order for 12,500 units of a special product. It can rent machines to produce the units for $100 each and pay an individual $8 per hour to monitor all the machines used. If there are no additional costs and if each machine can produce 40 units per hour, how many machines should be rented to minimize the cost of production? (See Exercise 10, Section 3.4.) ■

3.2 INCREASING AND DECREASING FUNCTIONS

Suppose a manufacturer determines that if x represents the number of days a worker has been on a new job, then the number of units $w(x)$ produced by the worker on the xth day is approximated by

$$w(x) = \frac{800x^2}{x^2 + 2352} + 200.$$

For example, since

$$w(10) = \frac{800(100)}{100 + 2352} + 200 \doteq 233,$$

a worker will produce 233 units on the 10th day. Similarly, since

$$w(30) = \frac{800(900)}{900 + 2352} + 200 \doteq 421,$$

a worker will produce 421 units on the 30th day.

Using the differential of w at $x = 10$ with $\Delta x = 1$, we know that

$$dw = w'(10) \doteq w(11) - w(10).$$

Therefore $w'(10)$ approximates the change in the number of units produced from the 10th day to the 11th day. Since the derivative of $w(x)$ is given by

$$w'(x) = \frac{(w^2 + 2352)(1600x) - 800x^2(2x)}{(x^2 + 2352)^2}$$

$$= \frac{(2352)(1600)x}{(x^2 + 2352)^2},$$

we find that

$$w'(10) = \frac{37{,}632{,}000}{6{,}012{,}304} \doteq 6.$$

Since the answer is positive, it follows that there is an approximate *increase* in production of 6 units on the 11th day over that of the 10th day. Since 233 units are produced on the 10th day, 239 units should be produced on the 11th day. In fact, we find that

3.2 INCREASING AND DECREASING FUNCTIONS

$$w(11) = \frac{96{,}800}{2473} + 200 \doteq 239.14.$$

Three questions arise quite naturally when considering this production function. (1) Will a worker continue to produce more units each day? (2) Is there an upper bound on how many units a worker can produce in a day? (3) On what day will a worker's production increase by the greatest number of units? Although all three questions could be answered by arithmetic or algebraic means, let us answer only the first two by such techniques. Note that if we divide the numerator and denominator of the fraction in

$$w(x) = \frac{800x^2}{x^2 + 2352} + 200$$

by x^2, we get

$$w(x) = \frac{800}{1 + 2352/x^2} + 200.$$

As x increases, $2352/x^2$ approaches 0 and $1 + 2352/x^2$ *decreases* and gets close to (but stays greater than) 1. Therefore $w(x)$ *increases* but is always less than

$$800 + 200 = 1000 \text{ units.}$$

A sketch of the graph of $w(x)$ is given in Figure 3.1. We see that the line $y = 1000$ approximates the curve for large x; that is, it is an asymptote for the curve. Although the function continues to increase, a worker's production could not increase beyond 999 units per day according to the model. (We leave as an exercise

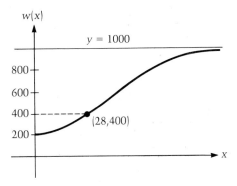

FIG. 3.1.
$w(x) = \dfrac{800x^2}{x^2 + 2352} + 200.$

for the student to determine how many days it would take a worker to meet this maximum production level.)

Often it is not so simple to determine algebraically whether or not a function is increasing. For example, it would not be easy to determine either algebraically or arithmetically for what x the derivative $w'(x)$ is an increasing function. Let us therefore turn our attention to a general discussion of increasing and decreasing functions.

increasing

A function f is said to be **increasing** *in an interval* if for each pair of numbers u and v in the interval where $u < v$, we have $f(u) < f(v)$. It is geometrically obvious that if throughout some interval the tangent lines to the graph of f have positive slopes, then the graph "moves up" as x "moves to the right" and the function is increasing. (See Figure 3.2.) Similarly, a function is said to be **decreasing** *in an interval* if for each pair of numbers u and v in the interval where $u < v$ it follows that $f(u) > f(v)$. If throughout some interval the tangent lines to the graph of f have negative slopes, then the graph "moves down" as x "moves to the right" and the function is decreasing. This leads to the following theorem which we state without proof.

decreasing

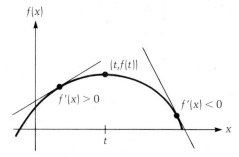

FIG. 3.2. $f'(x) > 0$ for $x < t$ and $f'(x) < 0$ for $x > t$.

increasing and decreasing functions

THEOREM 1 Assume a function f is differentiable in an open interval with a and b as endpoints and assume f is continuous at a and b.

(a) If $f'(x) > 0$ for $a < x < b$, then f is **increasing** on $[a, b]$.

(b) If $f'(x) < 0$ for $a < x < b$, then f is **decreasing** on $[a, b]$.

Notes:

1. As we shall find in Example 2, $f'(x) > 0$, for example, is a sufficient condition (enough) for a function to be increasing, but it is not necessary.

3.2 INCREASING AND DECREASING FUNCTIONS

> 2. Assume f is differentiable for $x > c$ and continuous at c. If $f'(x) > 0$ [$f'(x) < 0$] for $x > c$, then f is increasing (decreasing) where $x \geq c$.
>
> Similar remarks hold for $x < c$.

For the production curve $w(x)$, it is obvious that

$$w'(x) = \frac{(2352)(1600)x}{(x^2 + 2352)^2} > 0 \quad \text{for} \quad x \geq 1;$$

hence, $w(x)$ is an increasing function for $x \geq 1$ as a consequence of Theorem 1.

Note we did not say that if a differentiable function f is increasing in an interval, then $f'(x) > 0$ for each x in the interval. In fact, this need not be true. Consider f defined by $f(x) = x^3$. For any real numbers u and v, if $u < v$, then $u^3 < v^3$; therefore, f is increasing on the set of real numbers. However, $f'(x) = 3x^2$ and $f'(0) = 0$. Of course, if the function is increasing, we know that the derivative is *not negative*. Let us now consider several examples where we find intervals in the domain of a function for which the function is **monotonic** *(increasing or decreasing)*.

monotonic

Intervals where function is monotonic

EXAMPLE 1 Let f be defined by $f(x) = x^{2/3}$, where $-3 \leq x \leq 3$. The function is continuous on $[-3, 3]$ and

$$f'(x) = \frac{2}{3}x^{-1/3} = \frac{2}{3x^{1/3}} \quad \text{for} \quad x \neq 0.$$

Since $f'(x) > 0$, where $0 < x < 3$, the function increases on $[0, 3]$.
Since $f'(x) < 0$, where $-3 < x < 0$, the function decreases on $[-3, 0]$.
Since $f(0) = 0$ and since f is positive on either side of zero, f is not monotonic in any open interval containing zero. (See Figure 3.3a.) *Note:* Although 0 is in the domain of f, $f'(0)$ does not exist.

Intervals where function is monotonic

EXAMPLE 2 Let f be defined by $f(x) = x^{1/3}$, where $-3 \leq x \leq 3$. The function is continuous on $[-3, 3]$ and

$$f'(x) = \frac{1}{3}x^{-2/3} = \frac{1}{3x^{2/3}} \quad \text{for} \quad x \neq 0.$$

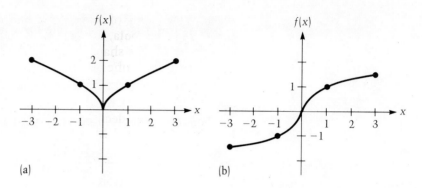

FIG. 3.3. (a) $f(x) = x^{2/3}$, $-3 \le x \le 3$. (b) $f(x) = x^{1/3}$, $-3 \le x \le 3$.

Since $f'(x) > 0$, where $0 < x < 3$, the function increases on $[0, 3]$. Since $f'(x) > 0$, where $-3 < x < 0$, the function also increases on $[-3, 0]$. Although 0 is not in the domain of f', it is in the domain of f. Since $f(x) < f(0)$ for $x < 0$ and $f(x) > f(0)$ for $x > 0$, f is increasing in any interval containing 0. Consequently, f increases on its domain $[-3, 3]$. (See Figure 3.3b.)

Quadratic function

EXAMPLE 3 Prove that if $f(x) = ax^2 + bx + c$ and $a > 0$, then f is an increasing function where $x \ge -b/2a$.

Solution: $f'(x) = 2ax + b$. Furthermore, $f'(x) > 0$ if and only if

$$2ax + b > 0,$$
$$2ax > -b,$$
$$x > -\frac{b}{2a}.$$

Since the derivative is positive for $x > -b/2a$, f is increasing on the set of real numbers for which $x \ge -b/2a$. (See Figure 3.4.)

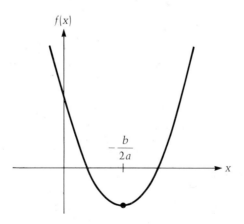

FIG. 3.4. $f(x) = ax^2 + bx + c$, $a > 0$.

3.2 INCREASING AND DECREASING FUNCTIONS

parabola

Note: The graph of a *quadratic function* $f(x) = ax^2 + bx + c$, $a \neq 0$, is a **parabola,** an important curve in many applications. (For example, the shape of light reflectors and satellite receivers are usually parabolic.)

PRACTICE PROBLEM Prove that if $f(x) = ax^2 + bx + c$ and $a > 0$, then f is a decreasing function where $x \leq -b/2a$. ∎

Intervals where function is monotonic

EXAMPLE 4

(a) For the function $f(x) = x^3 + 4x^2 - 3x + 1$, $-8 \leq x \leq 10$, find where the function is increasing.

(b) Find where it is decreasing.

Solution

(a) Since $f'(x) = 3x^2 + 8x - 3$, $f'(x) > 0$, provided

$$3x^2 + 8x - 3 > 0,$$
$$(3x - 1)(x + 3) > 0.$$

The preceding product is positive if both $3x - 1 > 0$ and $x + 3 > 0$; that is, if $x > \frac{1}{3}$ and $x > -3$. Therefore, if $x > \frac{1}{3}$, then $f'(x) > 0$ and f is increasing on the interval $[\frac{1}{3}, 10]$. Also, $(3x - 1)(x + 3)$ is positive if both $3x - 1 < 0$ and $x + 3 < 0$; that is, if $x < \frac{1}{3}$ and $x < -3$. Therefore, if $x < -3$, then $f'(x) > 0$ and f is also increasing on the interval $[-8, -3]$. (See Figure 3.5.)

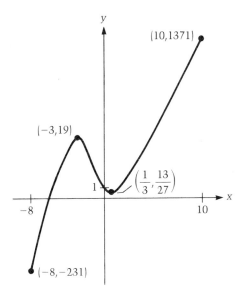

FIG. 3.5.
$f(x) = x^3 + 4x^2 - 3x + 1$.

(b) $f'(x) = (3x - 1)(x + 3)$ is negative if one of the factors is positive and the other is negative. Since $3x - 1 < 0$ and $x + 3 > 0$ if and only if $x < \frac{1}{3}$ and $x > -3$, we see that $f'(x) < 0$ if $-3 < x < \frac{1}{3}$. Therefore, f is decreasing on $[-3, \frac{1}{3}]$. (Since $3x - 1 > 0$ and $x + 3 < 0$ if and only if $x > \frac{1}{3}$ and $x < -3$, there are no real numbers which make the first factor positive and the second factor negative.)

PRACTICE PROBLEM For the function $f(x) = 2x^3 - 3x^2 - 6x$, $-10 \leq x \leq 10$, find where the function is increasing.

Answer: On the intervals $[-10, -1]$ and $[2, 10]$. ∎

Intervals where function is monotonic

EXAMPLE 5 For the function $f(x) = 1/x$, find where it is increasing and where it is decreasing.

Solution: Since $f'(x) = -1/x^2$ is negative for all $x \neq 0$, f never increases in any interval. Since $f'(x) < 0$ for $x > 0$, f decreases where $x > 0$; similarly, since $f'(x) < 0$ for $x < 0$, f decreases where $x < 0$. Note we cannot say f decreases in any interval containing $x = 0$, since the function is not defined there. (See Figure 3.6.)

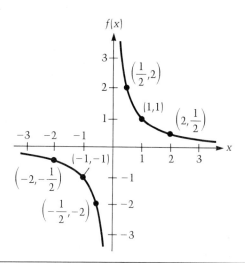

FIG. 3.6. $f(x) = 1/x$.

rate of change

For any differentiable function f, the derivative of f at x is defined to be the **rate of change** of f at x. Consider the two functions defined by

3.2 INCREASING AND DECREASING FUNCTIONS

$$F(x) = 2x^2 + 2x + 500$$

and

$$G(x) = x^2 + 4x.$$

Since $F'(x) = 4x + 2$ and $G'(x) = 2x + 4$, the rates of change of F and G at $x = 1$ are

$$F'(1) = 6 \quad \text{and} \quad G'(1) = 6.$$

This means, in both cases, that for a small change in x near 1

$$\frac{\text{corresponding change in } y}{\text{change in } x} = \frac{\Delta y}{\Delta x} \doteq 6.$$

Although this information is important, the *percentage rate of change of f at x* is more revealing; it is defined by

percentage rate of change

> **percentage rate of change of f at $x = \dfrac{f'(x)}{f(x)}(100)$.**

The percentage rate of change of $F(x) = 2x^2 + 2x + 500$ at 1 is

$$\frac{F'(1)}{F(1)}(100) = \frac{6}{504}(100) = 1.19\%,$$

and the percentage rate of change of $G(x) = x^2 + 4x$ at 1 is

$$\frac{G'(1)}{G(1)}(100) = \frac{6}{5}(100) = 120\%.$$

The percentage rate of change of F at 1 indicates that the rate of change of F at 1 is small compared to the function value at 1, while the percentage rate of change of G at 1 indicates that the rate of change of G at 1 is large compared to the function value at 1.

Percentage rate of change

EXAMPLE 6 For the daily production function $w(x)$ discussed earlier and defined by

$$w(x) = \frac{800x^2}{x^2 + 2352} + 200,$$

find the percentage rate of change of w at 10 and at 30.

Solution: We found that

$$w'(x) = \frac{(2352)(1600)x}{(x^2 + 2352)^2}.$$

Therefore, the percentage rate of change of w at 10 is

$$\frac{w'(10)}{w(10)} = \frac{6.259}{233}(100) \doteq 2.69\%,$$

and the percentage rate of change of w at 30 is

$$\frac{w'(30)}{w(30)} = \frac{10.675}{421}(100) \doteq 2.52\%.$$

Note: Although the rate of change is greater on the 30th day than on the 10th day, the percentage rate of change is less.

Psychology experiment

EXAMPLE 7 A psychologist finds that on the average a student in the first t hours of study of a foreign language can learn the meaning of

$$w(t) = \frac{100t^2}{t^2 + 27} + 10$$

new words provided $1 \leq t \leq 8$. (a) Use the derivative to show that $w(t)$ is an increasing function on $[1, 8]$. (b) During what time interval is the *rate of increase* increasing?

Solution

(a)

$$w'(t) = 100 \frac{(t^2 + 27)(2t) - t^2(2t)}{(t^2 + 27)^2}$$

$$= 100 \frac{54t}{(t^2 + 27)^2}.$$

3.2 INCREASING AND DECREASING FUNCTIONS

Since $w'(t) > 0$ for $1 < t < 8$, $w(t)$ is increasing on $[1, 8]$.

(b) Since $w'(t)$ is the rate of change of w with respect to t, we want to find the interval on which $w'(t)$ is an increasing function.

Differentiating $w'(t)$, we get

$$w''(t) = 5400 \frac{(t^2 + 27)^2 - t(2)(t^2 + 27)(2t)}{(t^2 + 27)^4}$$

$$= 5400 \frac{(t^2 + 27) - 4t^2}{(t^2 + 27)^3}$$

$$= 5400 \frac{27 - 3t^2}{(t^2 + 27)^3}.$$

We find that $w''(t) > 0$ for $1 < t < 3$ and $w''(t) < 0$ for $3 < t < 8$. Thus, the rate of increase in learning new words increases during the first three hours of study and then decreases.

Before turning to the exercises, let us make a final analysis of the daily production function $w(x)$. The derivative of $w(x)$ is

$$w'(x) = \frac{(2352)(1600)x}{(x^2 + 2352)^2},$$

and $w''(x)$ is found as follows.

$$w''(x) = (2352)(1600) \left[\frac{(x^2 + 2352)^2 - x(2)(x^2 + 2352)(2x)}{(x^2 + 2352)^4} \right]$$

$$= (2352)(1600) \left[\frac{x^2 + 2352 - 4x^2}{(x^2 + 2352)^3} \right]$$

$$= (2352)(1600) \left[\frac{2352 - 3x^2}{(x^2 + 2352)^3} \right].$$

Since $(x^2 + 2352)^3 > 0$, $w''(x) > 0$ provided

$$2352 - 3x^2 > 0$$

$$-3x^2 > -2352$$

$$x^2 < 784$$

$$x < 28 \quad \text{for} \quad x \geq 1.$$

This means that $w'(x)$ increases where $1 \leq x \leq 28$. Therefore, the number of units produced is not only increasing daily but it is also increasing at an increasing rate. We find that $w''(x) < 0$ for $x > 28$; thus, the rate of increase is decreasing if $x \geq 28$. Consequently, the worker's production level will increase by the greatest amount on the 28th day. (This answers Question 3 asked at the beginning of the section.) Since $w'(28) \doteq 11$, the greatest increase in one day is 11 units, which occurs on the 28th day.

EXERCISES

Set A–Increasing and Decreasing Functions

For each function in Exercises 1 through 24, use the derivative to determine the following, if they exist. (a) Intervals in the domain of the function where it is increasing. (b) Intervals in the domain of the function where it is decreasing.

1. $f(x) = x^4$, x any real
2. $f(x) = x^5$, x any real
3. $f(x) = (2x - 1)^3$, x any real
4. $g(x) = (3x - 8)^4$, x any real
5. $f(x) = x^{5/3}$, $-8 \leq x \leq 8$
6. $f(x) = x^{4/3}$, $-1 \leq x \leq 8$
7. $f(x) = 3x^2 - 12$, x any real
8. $f(x) = 2x^2 + 8x$, x any real
9. $f(x) = -4x^2 + 24x$, x any real
10. $f(x) = 6x^2 + 17$, x any real
11. $g(x) = 1/x^2$, $x \neq 0$
12. $f(x) = 1/x^3$, $x \neq 0$
13. $f(x) = x^3 - 3x^2$, $-\frac{1}{2} \leq x \leq 4$
14. $g(x) = x^3 + 6x^2$, $-8 \leq x \leq 1$
15. $f(x) = x^3 + 4x^2 - 3x + 1$, x any real number
16. $f(x) = 4x^3 - 7x^2$, x any real number
17. $f(x) = 2x^3 - 6x$, $-\frac{3}{2} \leq x \leq 4$
18. $g(x) = 2x^3 - 15x^2 + 36x + 3$, x any real number
19. $f(x) = x^3 + x^2 - 8x + 5$, x any real number
20. $f(x) = 4x^3 + 7x^2 - 10x - 7$, x any real number
21. $f(x) = \dfrac{x - 3}{x + 5}$
22. $f(x) = \dfrac{3x - 4}{6 - 2x}$

3.2 INCREASING AND DECREASING FUNCTIONS

23. $f(x) = x^2 + \dfrac{2}{x}$

24. $f(x) = x + \dfrac{4}{x}$

Set B—Miscellaneous Applications

Rate of change

25. The volume V of a sphere of radius r is given by
$$V(r) = \frac{4}{3}\pi r^3.$$
Find the rate of change of volume with respect to r.

Rate of change

26. A box with a square bottom x inches on a side has a height twice the length of the base. What is the rate of change of the surface area S with respect to x?

Percentage rate of change

27. (a) For the production function $w(x)$ discussed in this section, what is the percentage rate of change of w at $x = 5$?
    (b) What is the percentage rate of change of w at $x = 20$?

Level of production

28. (a) For the production function $w(x)$ discussed in this section, for what x is $w(x) = 999$? Explain the significance of the answer.
    (b) Explain why the point having 28 as its x-coordinate is often called the *point of diminishing returns*.

Production curve

29. A factory worker can produce a total of $f(t) = 60t + 9t^2 - t^3$ units of a product by the end of t hours on the job where $0 \leq t \leq 8$.
    (a) How many units does the worker produce during the first hour?
    (b) How many units does the worker produce during the second hour?
    (c) What is the average number of units per hour that the worker produces during an 8-hour period?
    (d) Why would you expect $f'(t)$ to be positive where $0 \leq t \leq 8$? Show that it is.
    (e) Prove that $f''(t) > 0$ where $t < 3$ and $f''(t) < 0$ where $t > 3$. Explain the significance in terms of production.

Learning curve

Use the following information in Exercises 30 through 35. Suppose it is found that a beginning typist, with only one hour practice each day, can type without error
$$f(x) = \frac{60x^2}{x^2 + 108} + 20$$
words per minute at the end of x weeks of typing where $x \geq 1$.

Proficiency level

30. How many words per minute could the typist type at the end of 10 weeks?

Change in proficiency

31. Use the derivative to show that the number of words per minute that can be typed increases each week (up to some bound).

Maximum proficiency

32. (a) Assuming the answer is an integer, what is the greatest number of words per minute such a typist can learn to type?
    (b) How many weeks does it take to reach this proficiency level?

Change in rate of change

33. (a) For what x is $f''(x) > 0$? For what x is $f''(x) < 0$?
    (b) What does $f'(x)$ represent where $f''(x) = 0$?

Percentage rate of change

34. (a) Find the rate of change of f at $x = 3$ and at $x = 10$.
    (b) Find the percentage rate of change of f at $x = 3$ and at $x = 10$.

Percentage rate of change

35. (a) Find the rate of change of f at $x = 6$ and at $x = 20$.
    (b) Find the percentage rate of change of f at $x = 6$ and at $x = 20$.

Elasticity of cost

36. If $C(x)$ is the total cost function of producing x units of a given product, then $\overline{C}(x) = C(x)/x$ is the average cost for x units and the **elasticity of cost** $E(x)$ is defined by

$$E(x) = \frac{C'(x)}{\overline{C}(x)}.$$

(a) Show that $E(x) = xC'(x)/C(x)$.
(b) Find $\overline{C}'(x)$ and show that the average cost is an increasing function provided $E(x) > 1$.

3.3 RELATIVE MAXIMA AND RELATIVE MINIMA

Consider the function whose graph is given in Figure 3.7. There are points on the graph of f which reveal some significant features about the behavior of the function. For example, the points with x_1 and x_6 as x-coordinates are *local high points* on the graph of f; the function values at these points are greater than at other points nearby. The points with x_3 and x_7 as x-coordinates are *local low points* on the graph of f. We also see that the highest point on the graph of f is at $x = x_8$; that is, $f(x_8)$ is the maximum function value for f. The points with x_2, x_4, and x_5 as x-coordinates are the points at which the curve changes from bending in one direction to bending in the other direction. Our task now is to make appropriate definitions for these important features so that we can use the tools of calculus to isolate them.

3.3 RELATIVE MAXIMA AND RELATIVE MINIMA

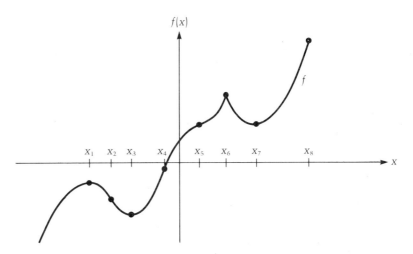

FIG. 3.7. Domain of f: $x \leq x_8$. Relative maxima: $f(x_1)$ and $f(x_6)$. Relative minima: $f(x_3)$ and $f(x_7)$. Absolute maximum: $f(x_8)$. Absolute minimum: none.

relative maximum

relative minimum

local extrema

Let f be a function and let t be a number in the domain of f. If there exists an *open interval in the domain of f containing t* such that $f(x) \leq f(t)$ for every x in the open interval, then $f(t)$ is called a **relative maximum,** or **local maximum, value** of f. Similarly, if there exists *an open interval in the domain of f containing t* such that $f(x) \geq f(t)$ for every x in the open interval, then $f(t)$ is called a **relative minimum,** or **local minimum, value** of f. Geometrically, if $f(t)$ is a local maximum (local minimum) value of f, then the point $(t, f(t))$ is a local high (low) point on the graph of f. Local maxima and minima of a function are called **local extrema** of the function. (See Figure 3.8.)

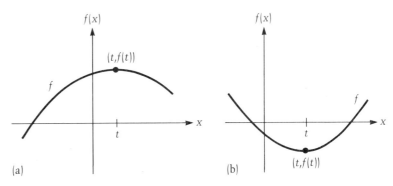

FIG. 3.8. (a) Relative maximum value: $f(t)$. (b) Relative minimum value: $f(t)$.

As shown in Figure 3.7, a relative maximum value for a function may be less than a relative minimum. However, if a function has both an absolute maximum and an absolute minimum, then the absolute maximum is greater than or equal to the absolute minimum. Furthermore, an absolute maximum (minimum) need not be a local maximum (minimum). (Observe in Figure 3.7 at $x = x_8$ that there is no open interval containing x_8 which is also totally contained *in the domain of f*; $f(x_8)$ is not a local maximum.) When the absolute maximum (minimum) is at an endpoint in the domain of a function, it is called an **endpoint extremum.**

<small>endpoint extremum</small>

Suppose t is in some open interval in the domain of a differentiable function f. It should be geometrically obvious from Figure 3.8 that if $f(t)$ is either a relative maximum or a relative minimum, then the tangent line at $(t, f(t))$ is parallel to the x-axis and $f'(t) = 0$. Therefore, for a differentiable function f, if $f'(t) \neq 0$, then $f(t)$ is neither a relative maximum nor a relative minimum.

In general, if t is in some open interval in the domain of f and if $f(t)$ is a relative maximum or relative minimum, then $f'(t) = 0$ or $f'(t)$ does not exist. These are *necessary conditions* for a relative maximum (minimum). Hence, if t is a number in some open interval in the domain of f such that $f'(t) = 0$ or $f'(t)$ is undefined, then $f(t)$ is a "candidate" for a relative maximum (or minimum) value. The following theorem gives *sufficient conditions* for local extrema.

THEOREM 2 FIRST DERIVATIVE TEST FOR RELATIVE MAXIMUM AND MINIMUM Assume that f is a continuous function at t and assume $f'(t) = 0$ or $f'(t)$ does not exist. Then we have the following.

(a) If there exists some open interval containing t such that $f'(x) > 0$ for $x < t$ and $f'(x) < 0$ for $x > t$, then $f(t)$ is a **relative maximum value** of f. (If the derivative "changes signs" from positive to negative as x increases through t, then $f(t)$ is a relative maximum. See Figure 3.9a.)

<small>relative maximum</small>

(b) If there exists some open interval containing t such that $f'(x) < 0$ for $x < t$ and $f'(x) > 0$ for $x > t$, then $f(t)$ is a **relative minimum value** of f. (If the derivative "changes signs" from negative to positive as x increases through t, then $f(t)$ is a relative minimum. See Figure 3.9b.)

<small>relative minimum</small>

3.3 RELATIVE MAXIMA AND RELATIVE MINIMA

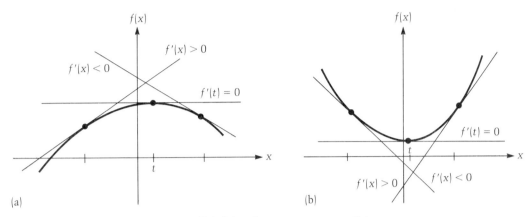

FIG. 3.9. (a) Relative maximum: $f(t)$. (b) Relative minimum: $f(t)$.

EXAMPLE 1 Consider the function f defined by $f(x) = x^2$, where $-2 \leq x \leq 2$. Since $f'(x) = 2x$, we see that $f'(x) = 0$ only at $x = 0$, and thus 0 is the only number between -2 and 2 for which the derivative is 0 or undefined. Therefore, $f(0) = 0$ is the only possible local extremum. Since

$$f'(x) = 2x < 0 \quad \text{for} \quad x < 0 \quad \text{and} \quad f'(x) = 2x > 0 \quad \text{for} \quad x > 0,$$

$f(0) = 0$ is a local minimum. Since $f(x) = x^2 \geq 0$ for any real number x, $f(0) = 0$ is also the absolute minimum value of the function. Since the function decreases on $[-2, 0]$ and increases on $[0, 2]$, the absolute maximum must be an endpoint extremum. Since $f(-2) = f(2) = 4$, the absolute maximum value of the function is 4; it is attained at both endpoints of the domain. (See Figure 3.10.)

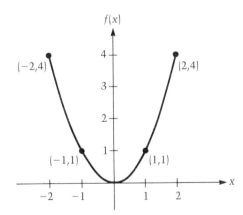

FIG. 3.10. $f(x) = x^2$, $-2 \leq x \leq 2$.

Extrema of a function

EXAMPLE 2 Consider the function f defined by $f(x) = x^3$, where $-1 \leq x \leq 2$. Since $f'(x) = 3x^2$, we see that 0 is the only number between -1 and 2 for which the derivative is 0 or undefined. Therefore, $f(0) = 0$ is the only possible *local extremum* for the function. However, since

$$f'(x) = 3x^2 > 0 \qquad \text{for both} \quad x < 0 \quad \text{and} \quad x > 0,$$

$f(0) = 0$ is not a local extremum. Since f is increasing in its domain, it should be clear that $f(-1) = -1$ is the absolute minimum of f and $f(2) = 8$ is the absolute maximum of f. (See Figure 3.11a.)

Extrema of a function

EXAMPLE 3 Consider the function f defined by $f(x) = x^{2/3}$ where $-3 \leq x \leq 3$. Since

$$f'(x) = \frac{2}{3} x^{-1/3} = \frac{2}{3x^{1/3}} \qquad \text{for} \quad x \neq 0,$$

we see that 0 is the only number between -3 and 3 for which the derivative is 0 or undefined. Hence, $f(0) = 0$ is the only possible local extremum for the function. Since $f'(x) < 0$ for $x < 0$ and $f'(x) > 0$ for $x > 0$, $f(0)$ is a local minimum value of f. Furthermore, we see that $f(0) = 0$ is the absolute minimum value of the function. (See Figure 3.11b.) The absolute maximum is $f(3) = f(-3) = (3)^{2/3}$. [*Warning:* Care should be taken to distinguish between *where* an extreme value for a function occurs (do-

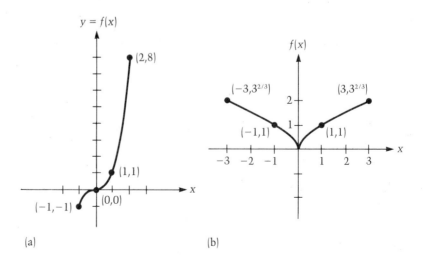

FIG. 3.11. (a) $f(x) = x^3$, $-1 \leq x \leq 2$. (b) $f(x) = x^{2/3}$, $-3 \leq x \leq 3$.

3.3 RELATIVE MAXIMA AND RELATIVE MINIMA

main value) and *what* the extreme value of the function is (range value).]

PRACTICE PROBLEM

(a) For $f(x) = x^{2/5} + 4$, $-1 \leq x \leq 32$, find any relative extrema.

(b) Find the absolute maximum.

(c) Find the absolute minimum.

Answer: (a) A relative minimum occurs at $x = 0$ and is 4. (b) $f(32) = 8$. (c) $f(0) = 4$. ∎

For a differentiable function, there is another set of sufficient conditions for local extrema called the *second derivative tests for relative maximum and minimum values.* They are quite useful when the second derivative exists and is not too difficult to obtain. The conditions are given in the following theorem.

THEOREM 3 SECOND DERIVATIVE TESTS FOR LOCAL EXTREMA Let f be a function and let t be a number in the domain of f such that $f'(t) = 0$. Then we have the following.

relative minimum

(a) If $f''(t)$ exists and is *positive*, then $f(t)$ is a **relative minimum value** of f. [If $f''(t) > 0$, then f' increases and since $f'(t) = 0$, the derivative must change sign from negative to positive, making $f(t)$ a relative minimum from Theorem 2. See Figure 3.12a.]

relative maximum

(b) If $f''(t)$ exists and is *negative*, then $f(t)$ is a **relative maximum value** of f. [If $f''(t) < 0$, then f' decreases and since $f'(t) = 0$, the derivative must change sign from positive to

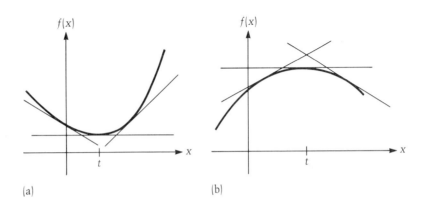

FIG. 3.12. (a) $f'(t) = 0$ and $f''(t) > 0$. (b) $f'(t) = 0$ and $f''(t) < 0$.

negative, making $f(t)$ a relative maximum from Theorem 2. See Figure 3.12b.]

Note: If $f''(t) = 0$, then $f(t)$ may or may not be a local extremum. In this case, the first derivative test can be used to decide.

Extrema of a function

EXAMPLE 4 Let h be the function defined by $h(x) = x + 4/x$. Then

$$h'(x) = 1 - \frac{4}{x^2} \quad \text{and} \quad h''(x) = \frac{8}{x^3}.$$

Although $h'(x)$ is undefined at $x = 0$, the function cannot have a maximum or minimum at zero since it is not in the domain of h. The only possible numbers for which the function can have a relative maximum or minimum are where $h'(x) = 0$, that is, at $x = 2$ and $x = -2$.

Since $h''(2) = 1 > 0$, we conclude from Theorem 3 that $h(2) = 4$ is a relative minimum. Since $h''(-2) = -1 < 0$, we conclude that $h(-2) = -4$ is a relative maximum. *Notes:* (1) This is an example of a function where a relative maximum is less than a relative minimum. (See Figure 3.13.) (2) For "large" x (in absolute

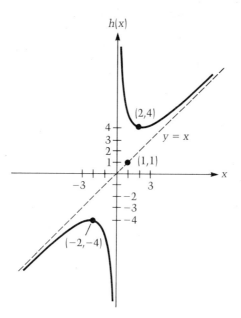

FIG. 3.13. Relative maximum: $h(-2) = 4$; relative minimum: $f(2) = 4$. $h(x) = x + \dfrac{4}{x}$.

value), $h(x) \doteq x$; therefore, the graph of h approaches $y = x$ and this line is an asymptote for the graph of h. (3) Similarly, $x = 0$ is also an asymptote for the graph of h.

concave upward

concave downward

inflection point

For a function f, if its derivative f' is an increasing function in some open interval, then the graph of f is said to be **concave upward** in the interval. Therefore, if $f''(x) > 0$ in some interval, then the slope $f'(x)$ is increasing and the graph of f bends upward. (See Figures 3.14 and 3.12a.) Similarly, if the derivative of f is a decreasing function in some open interval, then the graph of f is said to be **concave downward** in the interval. Therefore, if $f''(x) < 0$ in some interval, then the slope $f'(x)$ is decreasing and the graph of f bends downward. (See Figures 3.14 and 3.12b.) Finally, if $(t, f(t))$ is a point on the graph of a function f such that the concavity on one side of t is opposite from the concavity on the other side of t, then $(t, f(t))$ is called an **inflection point.** (See Figure 3.14.) Just as points in the domain where the first derivative is 0 or undefined are candidates for relative extrema, *points in the domain where the second derivative is 0 or undefined are candidates for inflection points.*

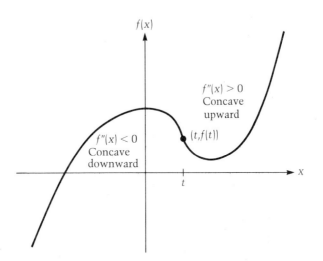

FIG. 3.14. $(t, f(t))$ is an inflection point.

Concavity

EXAMPLE 5 In Example 4, we found for the function

$$h(x) = x + \frac{4}{x} \quad \text{that} \quad h''(x) = \frac{8}{x^3}.$$

Since $h''(x) > 0$ for $x > 0$, the graph of h is concave upward for $x > 0$. (See Figure 3.13.) Since $h''(x) < 0$ for $x < 0$, the graph of h is concave downward for $x < 0$. Since 0 is not in the domain of h, there is no inflection point at $x = 0$.

Graphing

EXAMPLE 6 For $f(x) = x^4/4 - x^3 - 9x^2$, find relative maxima, relative minima, and inflection points, and sketch the graph of f.

Solution

$$f'(x) = x^3 - 3x^2 - 18x = x(x^2 - 3x - 18) = x(x - 6)(x + 3)$$

and

$$f''(x) = 3x^2 - 6x - 18 = 3(x^2 - 2x - 6) = 3(x - 4)(x + 2).$$

(a) Since $f'(6) = 0$ and $f''(6) > 0$, $f(6) = -216$ is a relative minimum. Since $f'(0) = 0$ and $f''(0) < 0$, $f(0) = 0$ is a relative maximum. Since $f'(-3) = 0$ and $f''(-3) > 0$, $f(-3) = -33.75$ is a relative minimum.

(b) Since $f''(x) > 0$ for $x < -2$, the graph is concave upward for $x < -2$. Since $f''(x) < 0$ for $-2 < x < 4$, the graph is concave downward in this interval. Since $f''(x) > 0$ for $x > 4$, the graph is concave upward for $x > 4$.

(c) The inflection points are $(-2, -24)$ and $(4, -144)$. (See Figure 3.15.)

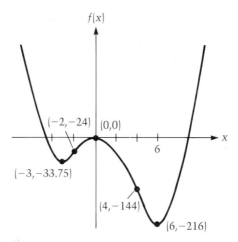

FIG. 3.15. Relative minimum points: $(6, -216)$ and $(-3, -33.75)$. Relative maximum points: $(0, 0)$. Inflection points: $(-2, -24)$ and $(4, -144)$.

3.3 RELATIVE MAXIMA AND RELATIVE MINIMA

Average cost

EXAMPLE 7 Assume $C(x)$ is a total cost function which is differentiable for $x > 0$ and assume there is an x in the domain which minimizes the average cost $\overline{C}(x) = C(x)/x$. Show that average cost is minimized at an x where average cost equals marginal cost.

Solution: We know that average cost is minimized at the point at which $\overline{C}'(x) = 0$. Differentiating $\overline{C}(x)$, we get

$$\overline{C}'(x) = \frac{xC'(x) - C(x)}{x^2};$$

$\overline{C}'(x) = 0$ if and only if $xC'(x) - C(x) = 0$, $xC'(x) = C(x)$, or

$$C'(x) = \frac{C(x)}{x}.$$

The preceding equation states that marginal cost equals average cost.

A geometric problem

EXAMPLE 8 Assume two sides and the included right angle of a rectangle are coincident with two sides and the included right angle of a right triangle whose two sides are 6 inches and 10 inches long. Find the maximum area of the inscribed rectangle as given in Figure 3.16.

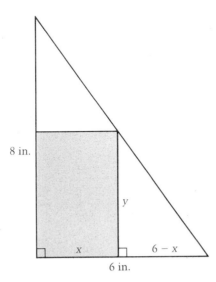

FIG. 3.16.

Solution: If x is the length of the rectangle and y is its height, we conclude from Figure 3.16 and the fact that the sides of similar triangles are proportional that

$$\frac{6}{8} = \frac{6-x}{y},$$

$$6y = 48 - 8x,$$

$$y = 8 - \frac{4}{3}x.$$

Therefore, the area of the rectangle is

$$A(x) = x\left(8 - \frac{4}{3}x\right) = 8x - \frac{4}{3}x^2.$$

Hence,

$$A'(x) = 8 - \frac{8}{3}x \quad \text{and} \quad A''(x) = -\frac{8}{3}.$$

Since $A'(x) = 0$ at $x = 3$ and since $A''(x)$ is negative for all x, $A(3) = 12$ in^2 is the maximum area of such a rectangle.

Any function which is continuous on a *closed* interval must have an absolute maximum and an absolute minimum value. (This important fact is not easy to prove.) If the domain of a function is not a closed interval, the function may or may not have an absolute extremum. Before we turn to the exercises, let us look at a simple (but nontrivial) example which clearly indicates the importance of giving proper attention to the domain of a function. The graph of the function

$$f(x) = x^3 - 3x^2 + 1, \quad x \text{ any real number,}$$

is given in Figure 3.17. Carefully consider each of the functions listed in the accompanying table with their corresponding ex-

$f(x) = x^3 - 3x^2 + 1$

Domain	Relative maximum point	Relative minimum point	Absolute maximum point	Absolute minimum point
All reals	(0, 1)	(2, −3)	None	None
$-1 \leq x \leq 0$	None	None	(0, 1)	(−1, −3)
$0 < x \leq 1$	None	None	None	(1, −1)
$1 \leq x \leq 3$	None	(2, −3)	(3, 1)	(2, −3)

3.3 RELATIVE MAXIMA AND RELATIVE MINIMA

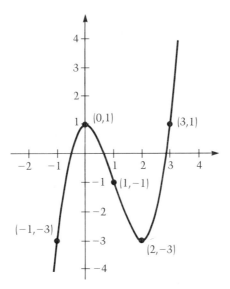

FIG. 3.17.
$f(x) = x^3 - 3x^2 + 1$.

treme points. (Each function has a graph which is part of the graph of the function defined above. The answers can be verified by looking at Figure 3.17.)

EXERCISES

Set A–Maximum and Minimum Points and Points of Inflection

For each function in Exercises 1 through 20, do the following. (a) Find the coordinates of any point where the function has a relative maximum. (b) Find the coordinates of any point where the function has a relative minimum. (c) Find the absolute maximum of f, if it exists. (d) Find the absolute minimum of f, if it exists. (e) Find any inflection points on the graph of f. (f) Discuss concavity for the graph of f. (g) Sketch the graph of f. *Note:* It should be helpful to observe that many of these functions are the same as those given in the exercises in Section 3.2.

1. $f(x) = x^4$, x any real
2. $f(x) = x^5$, x any real
3. $f(x) = (2x - 1)^3$, x any real
4. $g(x) = (3x - 8)^4$, x any real
5. $f(x) = x^{5/3}$, $-8 \le x \le 8$
6. $f(x) = x^{4/3}$, $-1 \le x \le 8$
7. $g(x) = \dfrac{1}{x^2}$, $x \neq 0$
8. $f(x) = \dfrac{1}{x^3}$, $x \neq 0$
9. $f(x) = x^3 - 3x^2$, $-\frac{1}{2} \le x \le 4$

10. $g(x) = x^3 + 6x^2$, $-8 \leq x \leq 1$
11. $f(x) = x^3 + 4x^2 - 3x + 1$, x any real number
12. $f(x) = 4x^3 - 7x^2$, x any real number
13. $f(x) = 2x^3 - 6x$, $-\frac{3}{2} \leq x \leq 4$
14. $g(x) = 2x^3 - 15x^2 - 36x + 3$, x any real number
15. $f(x) = x^3 + x^2 - 8x + 5$, x any real number
16. $f(x) = x^4 - 2x^2$, x any real number
17. $f(x) = x^4 - 4x^3 + 1$, x any real number
18. $f(x) = x^4 - 32x$, x any real number
19. $f(x) = x^2 + \dfrac{2}{x}$, x any real number different from zero
20. $f(x) = x + \dfrac{4}{x^2}$, x any real number different from zero

Set B—Miscellaneous Applications

Learning curve

21. A psychologist finds that after t days, $0 \leq t \leq 20$, a person is able to score $p(t) = 100t^2/(t^2 + 12)$ points on a performance test.
    (a) Does the person's score continue to increase throughout the 20 days?
    (b) Approximate the greatest score increase in a single day.

Average cost

22. The total cost function in dollars for producing x units of a given product is $C(x) = 2x^2 + 4x + 450$. What is the minimum *average cost* per unit that is possible?

Economic law of profit

23. Assume the cost function $C(x)$ and the revenue function $R(x)$ for a particular product are differentiable where $x > 0$, and assume there is an x in the domain which maximizes the profit. Show that the profit is maximized when marginal cost equals marginal revenue.

Spread of an epidemic

24. Assume that the rate at which an epidemic spreads through a community is jointly proportional to the number of persons who have the disease and the number who do not. Prove that the epidemic spreads most rapidly when half the individuals have caught the disease. [Hint: $f(x) = kx(P - x)$. Explain what the constants k and P represent. What is the domain of this function?]

Production costs

25. The cost in dollars for producing x thousand pencils is given by $C(x) = x^3 + 6x + 54$ where $0 \leq x \leq 5$.
    (a) What is the average cost function?
    (b) What production would minimize the average cost?
    (c) What is the minimum average cost per pencil obtainable?

A geometric problem

26. Consider the right triangle formed by the x-axis, the y-axis, and the tangent to the graph of $f(x) = 27 - x^2$. What is the minimum area for such a right triangle? Could there be such a right triangle with maximum area? [*Hint:* If $(a, 27 - a^2)$ is the point of tangency, first find the x- and y-intercepts of the tangent line.]

3.4 MORE APPLIED MAXIMUM AND MINIMUM PROBLEMS

As we have already discovered, the derivative is useful not only in graphing functions but also in solving practical problems. This section is devoted to looking further into the use of the derivative for solving applied maxima and minima problems.

In general, in an optimization problem, some quantity (dependent variable) is to be optimized as a result of its relationship to some other quantity (independent variable). In fact, as we shall find later, the quantity to be optimized can depend on several independent variables, but, for now, we stick to the simpler case. If, for example, the level of sales of a product depends upon the price of the product, we could let S represent sales and let p represent price. Then a model (function) could be constructed from the information given to represent the relationship between S and p. Finally, techniques of calculus are applied to find what value of p optimizes S.

The following examples should help to clarify all the important techniques involved.

Optimal dimensions for playground

EXAMPLE 1 A school decides to construct a rectangular playground, using an existing wall and 300 meters of fence. (See Figure 3.18.) What dimensions will maximize the area of the playground?

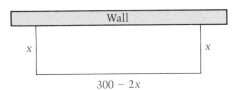

FIG. 3.18.

Solution: Let x meters be the width of each end of the playground. Since $2x$ meters of fence are necessary for the ends, $300 - 2x$ meters of fence are left for the side opposite the wall.

The area function A is given by

$$A(x) = x(300 - 2x)$$
$$= -2x^2 + 300x$$

where $0 < x < 300$. Finding the first and second derivatives, we get

$$A'(x) = -4x + 300,$$

and

$$A''(x) = -4.$$

Since $A'(x) = 0$ at $x = 75$ and since $A''(75) < 0$, $A(75)$ is a relative maximum. Since $A''(x) < 0$ for all x in the domain, the graph is always concave downward; from this we conclude that $A(75)$ is also the absolute maximum of the function. Hence, the dimensions should be 75 meters by 150 meters.

PRACTICE PROBLEM For the playground problem in Example 1, suppose 400 meters of fence were used.

(a) What dimensions will maximize the area?

(b) What is the maximum area obtainable?

Answer: (a) 100 meters by 200 meters.   (b) 20,000 m². ∎

Maximizing profits

EXAMPLE 2 The cost, in thousands of dollars, of producing x thousand units of a particular product is given by

$$C(x) = 0.08x^2 - 1.2x + 12, \quad x \geq 2,$$

and the revenue function, in thousands of dollars, for selling x thousand units is given by

$$R(x) = -0.08x^2 + 2.8x - 4, \quad x \geq 2.$$

(a) What is the cost of producing 3000 units?

(b) What is the revenue received from selling 3000 units?

(c) What is the profit (loss) from the sale of 3000 units?

(d) What is (are) the break-even point(s)?

(e) How many units will maximize profits?

(f) What is the maximum profit that can be obtained from this product?

(g) Graph the cost and revenue functions on the same coordinate system and label significant points.

(h) Graph the profit function.

Solution

(a) Since $C(3) = (0.08)(3)^2 - (1.2)(3) + 12 = 9.12$, the cost of producing 3000 units is $9120.

(b) Since $R(3) = (-0.08)(3)^2 + (2.8)(3) - 4 = 3.68$, the revenue from selling 3000 units is $3680.

(c) Since $P(3) = R(3) - C(3) = 3.68 - 9.12 = -5.44$, the loss incurred from producing and selling 3000 units is $5440.

(d) The break-even points are at $C(x) = R(x)$. That is,

$$0.08x^2 - 1.2x + 12 = -0.08x^2 + 2.8x - 4,$$
$$0.16x^2 - 4.0x + 16 = 0,$$
$$16x^2 - 400x + 1600 = 0,$$
$$x^2 - 25x + 100 = 0,$$
$$(x - 5)(x - 20) = 0.$$

The break-even points occur where $x = 5$ and $x = 20$. Thus, the break-even points occur when 5000 or 20,000 units are produced and sold.

(e) Since $P(x) = R(x) - C(x)$, the profit function is

$$P(x) = (-0.08x^2 + 2.8x - 4) - (0.08x^2 - 1.2x + 12)$$
$$= -0.16x^2 + 4.0x - 16.$$

The first and second derivatives of $P(x)$ are

$$P'(x) = -0.32x + 4.0 \quad \text{and} \quad P''(x) = -0.32.$$

Since $P'(x) = 0$ at $x = 12.5$ and since $P''(x) < 0$ for all x in the domain of $P(x)$, we conclude that producing and selling 12,500 units would maximize profits.

(f) Since $P(12.5) = -0.16(12.5)^2 + 4(12.5) - 16 = 9$, the maximum profit is $9000.

(g) See Figure 3.19a.

(h) See Figure 3.19b.

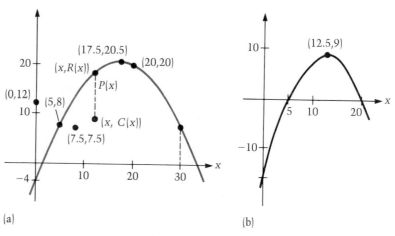

(a) (b)

FIG. 3.19. (a) $C(x) = 0.08x^2 - 1.2x + 12$, $R(x) = -0.08x^2 + 2.8x - 4$. (b) $P(x) = -0.16x^2 + 4.0x - 16$.

Maximum profit

EXAMPLE 3 Suppose a manufacturer determines that if a particular product is priced at x dollars per unit, $20 \leq x \leq 50$, then

$$1000(50 - x) \text{ units}$$

will be sold. That is, 30,000 units will be sold at \$20.00, and the consumer will not buy any units if the price is \$50.00 per unit. If the product costs \$18.00 per unit to make, then

$$x - 18 \text{ dollars}$$

is the profit per unit if $x > 18$ and the loss per unit if $x < 18$. The profit (loss) function in terms of selling price is given by

Profit = (number of units sold)(profit per unit).

Therefore, the profit function is defined by

$$P(x) = 1000(50 - x)(x - 18)$$
$$= -1000x^2 + 68{,}000x - 900{,}000 \text{ dollars}.$$

Finding the first and second derivatives, we get

$$P'(x) = -2000x + 68{,}000,$$

3.4 MORE APPLIED MAXIMUM AND MINIMUM PROBLEMS

and
$$P''(x) = -2000.$$

Since $P'(x) = 0$ at $x = 34$ and since $P''(34) < 0$, $P(34)$ is a relative maximum for P. In fact, since the graph of P is concave downward for $20 \leq x \leq 50$, we conclude that $P(34)$ is the absolute maximum of P. Consequently, profit is maximized when the price is set at $34 per unit and the maximum profit is

$$P(34) = 1000(16)(16) = \$256,000.$$

PRACTICE PROBLEM A manufacturer determines that if a particular product is priced at x dollars per unit, $35 \leq x \leq 80$, then $1000(80 - x)$ units will be sold.

(a) If the product costs $52 per unit to produce, what is the profit function?

(b) What price will maximize profits?

(c) What is the maximum profit?

(d) How many units maximize the profits?

Answer: (a) $P(x) = -1000(x^2 - 132x + 4160)$ (b) $66.00
(c) $196,000.   (d) 14,000 units. ■

Maximizing excursion boat income

EXAMPLE 4 An excursion boat with a capacity of 500 people runs a charter trip for a group of 275 people at $8.50 per person. The company offers a reduction in fare to all the passengers of 2 cents per person for each additional person over 275 taking the cruise.

(a) How many people must be on the cruise to maximize the income?

(b) What is the cost per person taking the cruise for this number?

(c) What is the total revenue that the company receives for such a cruise?

(d) What is the minimum income for such a cruise?

Solution

(a) Let x be the number *over* 275 taking the cruise. Then $0 \leq x \leq 225$ and the fare per person is

$$8.50 - 0.02x \text{ dollars.}$$

Since the number of people taking the cruise is 275 + x, the total revenue is

$$R(x) = (275 + x)(8.50 - 0.02x)$$
$$= -0.02x^2 + 3x + 2337.50 \text{ dollars.}$$

Finding the first and second derivatives, we get

$$R'(x) = -0.04x + 3 \quad \text{and} \quad R''(x) = -0.04.$$

Since $R'(x) = 0$ at $x = 75$, and since $R''(75) < 0$, $R(75)$ is a relative maximum. In fact, since $R''(x) < 0$, the graph is concave downward in the domain and $R(75)$ is the absolute maximum of the function. Therefore, $275 + 75 = 350$ people will maximize the revenue.

(b) The cost per person is $8.50 - 0.02(75) = \$7.00$.

(c) The maximum revenue is $R(75) = (350)(7.00) = \$2450$.

(d) The minimum revenue must be an endpoint extremum, so we evaluate $R(x)$ at $x = 0$ and $x = 225$.

$$R(0) = \$2337.50 \quad \text{and} \quad R(225) = \$2000.$$

Hence, the minimum revenue is $2,000 with $275 + 225 = 500$ people taking the cruise.

Transportation costs

EXAMPLE 5 At a speed of x miles per hour, $30 \le x \le 60$, a truck averages $360/x$ miles per gallon of gas. Suppose the driver of the truck is paid $10 per hour and suppose the price of gas is $1.44 per gallon.

(a) If the other costs for a 300-mile trip are assumed to be independent of the speed at which the truck is driven, what constant speed would minimize the cost of the trip?

(b) What would be the minimum cost for gas and wages?

(c) What would be the maximum cost for gas and wages if $30 \le x \le 60$?

Solution

(a) The number of gallons of gas needed for the trip is

$$300 \text{ miles} \div \frac{360}{x} \text{ miles per gallon} = \frac{5}{6}x \text{ gallons.}$$

3.4 MORE APPLIED MAXIMUM AND MINIMUM PROBLEMS

Therefore, the cost of the gas is

$$\left(\frac{5}{6}x\right)(\$1.44) = 1.20x \text{ dollars.}$$

Since distance = rate × time, the time required for the trip is

$$t = \frac{d}{r} = \frac{300}{x} \text{ hours.}$$

Therefore, the trucker's wages would be

$$\frac{300}{x}(\$10) = \frac{3000}{x} \text{ dollars,}$$

and the cost of the trip is

$$C(x) = 1.20x + \frac{3000}{x} \text{ dollars.}$$

Since the derivative of $C(x)$ is

$$C'(x) = 1.20 - \frac{3000}{x^2},$$

$C'(x) = 0$, where

$$1.20 = \frac{3000}{x^2},$$
$$x^2 = 2500,$$
$$x = \pm 50.$$

The only tenable solution to the practical problem is $x = 50$. Since $C''(x) = 6000/x^3 > 0$ if $x > 0$, the cost is minimized at $x = 50$ miles per hour.

(b) The minimum cost for gas and wages is $C(50) = \$120$.

(c) The maximum must be an endpoint extremum. Since $C(30) = \$136$ and $C(60) = \$122$, the maximum would be $\$136$.

Optimizing farm income

EXAMPLE 6 A farm manager finds for a given grain crop that the yield per acre is

$$\frac{40x}{x+9} + 60$$

bushels per acre when x bags per acre of a given fertilizer are used. If the grain can be sold for $10 a bushel and the fertilizer costs $16 per bag, how many bags of fertilizer per acre should be used to maximize the cash return? What is the maximum return?

Solution: The cash from the sale of the grain is

$$\left(\frac{40x}{x+9} + 60\right)10 = \frac{400x}{x+9} + 600$$

dollars per acre. The cost of the fertilizer per acre is $16x$ dollars. Therefore, the cash return is

$$C(x) = \frac{400x}{x+9} + 600 - 16x.$$

The derivative is

$$C'(x) = \frac{(x+9)400 - 400x}{(x+9)^2} - 16$$

$$= \frac{3600}{(x+9)^2} - 16.$$

We see that $C'(x) = 0$ if

$$\frac{3600}{(x+9)^2} = 16,$$

$$\frac{3600}{16} = (x+9)^2,$$

$$(x+9)^2 = 225,$$

$$x + 9 = \pm 15,$$

$$x = 6 \quad \text{or} \quad x = -24.$$

The only possible solution is $x = 6$. It is easy to verify that $C''(6) < 0$; hence $x = 6$ bags per acre maximizes the cash return.

3.4 MORE APPLIED MAXIMUM AND MINIMUM PROBLEMS

The maximum cash return is

$$C(6) = \frac{2400}{15} + 600 - 96 = 664$$

dollars per acre.

Machine costs

EXAMPLE 7 A construction firm plans to buy a machine whose initial cost is $10,000. Assume that the operating and maintenance costs of the machine for the first year will be $500 and assume these costs will increase at $250 each year. If the machine will have no resale value, how many years should it be kept in order to minimize the average annual cost?

Solution: The total cost for x years including operating and maintenance costs is

$$10{,}000 + [500 + 750 + 1000 + \cdots + (500 + 250(x-1))].$$

Using the fact that the sum of an arithmetic sequence is $n(a+b)/2$, where a is the first term, b is the last term, and n is the number of terms, we find the preceding sum to be

$$10{,}000 + x(500 + 500 + 250(x-1))/2 = 10{,}000 + 375x + 125x^2.$$

To find the average annual cost $\overline{C}(x)$ for x years we divide the total cost by x and obtain

$$\overline{C}(x) = \frac{10{,}000}{x} + 375 + 125x.$$

The domain of this function is, of course, the set of positive integers. As usual, in order to employ the tools of calculus, we assume that the domain is extended to the set of positive real numbers. Differentiating the function, we get

$$\overline{C}'(x) = -\frac{10{,}000}{x^2} + 125.$$

We find that $\overline{C}'(x) = 0$ provided

$$125x^2 = 10{,}000$$
$$x^2 = 80.$$

The only positive x for which $\overline{C}'(x) = 0$ is $x = \sqrt{80} \doteq 9$. Since $C''(x) = 20{,}000/x^3 > 0$ for any $x > 0$, $x = \sqrt{80}$ minimizes $\overline{C}(x)$. Since $\sqrt{80}$ is close to 9, it would appear that keeping the machine for 9 years would minimize the average cost. Since the solution is between 8 and 9, we evaluate $\overline{C}(8)$ and $\overline{C}(9)$ and find that $\overline{C}(8) = \$2625$ and $\overline{C}(9) = \$2611$ (to the nearest dollar). We conclude, as expected, that the machine should be kept 9 years to minimize average cost.

Optimal dimensions for a box

EXAMPLE 8 A box manufacturer wishes to construct a rectangular box with top and bottom whose length is twice its width. If the surface area of the box must be 60 in^2, what should be its dimensions in order to maximize the volume of the box? (See Figure 3.20.)

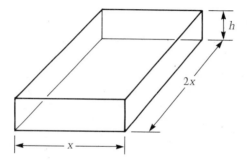

FIG. 3.20.
$V(x) = 20x - \dfrac{4}{3}x^3$.

Solution: Let h be the height of the box and let x be the width of the top and bottom. Since the length is twice the width, the length of the top and bottom is $2x$. Since the volume of the box is $V = 2x^2h$, our first task is to find V in terms of either x or h in order to have V as a function of one variable. Notice that the total surface area of the top and bottom is $4x^2$. If h is the height of the box, then $2xh$ is the total surface area of two of the sides and $4xh$ is the total surface area of the other two sides. Consequently

$$2xh + 4xh + 4x^2 = 60.$$

We solve the preceding equation for h in terms of x as follows:

$$6xh = 60 - 4x^2,$$

$$h = \frac{30 - 2x^2}{3x}.$$

3.4 MORE APPLIED MAXIMUM AND MINIMUM PROBLEMS

(*Note:* We could have solved the surface area equation for x in terms of h and obtained V in terms of h, but this approach would have required the use of the quadratic formula. We have here an example illustrating that a judicious choice in procedure makes the solution easier to obtain.)

Since the volume of the box is

$$\text{length} \times \text{width} \times \text{height},$$

we have

$$V(x) = 2x^2 \left(\frac{30 - 2x^2}{3x} \right)$$

$$= 20x - \frac{4}{3}x^3.$$

The derivative of V is

$$V'(x) = 20 - 4x^2.$$

Furthermore, $V'(x) = 0$, provided $4x^2 = 20$, $x^2 = 5$, $x = \pm\sqrt{5}$. Since the domain of V is the positive real numbers, $\sqrt{5}$ is the only candidate for a solution. Since $V''(x) = -8x$, $V''(x) < 0$ for any $x > 0$. Consequently, $V(\sqrt{5})$ is the absolute maximum of the volume function. The dimensions for the box with maximum volume are width = $\sqrt{5}$ inches, length = $2\sqrt{5}$ inches, and height = $4\sqrt{5}/3$ inches.

Minimizing the cost of a can

EXAMPLE 9 Suppose a large soup company plans to manufacture a can (right circular cylinder) that will contain K cubic units (a fixed volume) of soup. What should the ratio of the height to the radius be in order to minimize the can's surface area and thus its cost? (See Figure 3.21.)

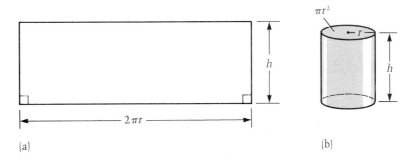

FIG. 3.21. Area: $2\pi rh$. Volume: $\pi r^2 h = K$.   (a)   (b)

Solution: The volume of the can is the area of the circular base times the altitude h. If r is the radius of the base, then $K = \pi r^2 h$. Therefore

$$h = \frac{K}{\pi r^2}.$$

The surface area of the can is the sum of the areas of the top, bottom, and side. The top and bottom areas are each πr^2. The area of the side is $2\pi r h$. (The area of the side could be derived by cutting the can vertically down the side and flattening it into a rectangular surface whose width is h and length is the circumference of the can.)

The total surface area is $2\pi r h + 2\pi r^2$. Since $h = K/\pi r^2$, we can determine the surface area S in terms of r by substituting into the preceding formula. The surface area is given by

$$S(r) = 2\pi r \left(\frac{K}{\pi r^2}\right) + 2\pi r^2$$

$$= \frac{2K}{r} + 2\pi r^2.$$

A sketch of the graph of this function or the physical aspects of the problem indicate that there exists a given radius which minimizes S. (See Figure 3.22 for the graphs of S using different

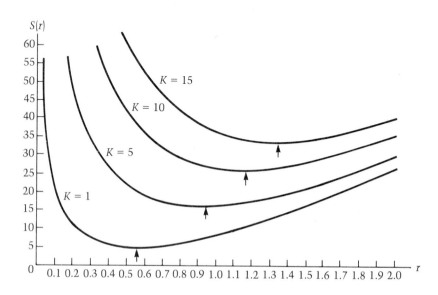

FIG. 3.22.

$S(r) = \dfrac{2K}{r} + 2\pi r^2.$

3.4 MORE APPLIED MAXIMUM AND MINIMUM PROBLEMS

numbers for K.) Hence, the minimum will be obtained where $S'(r) = 0$.

Differentiating the surface area function, we obtain

$$S'(r) = -\frac{2K}{r^2} + 4\pi r.$$

Now, $S'(r) = 0$ if and only if

$$4\pi r = \frac{2K}{r^2} \quad \text{or} \quad r = \frac{K}{2\pi r^2}.$$

Since $K = \pi r^2 h$, it follows that $S(r)$ will be minimized when

$$r = \frac{\pi r^2 h}{2\pi r^2} \quad \text{or} \quad r = \frac{h}{2}.$$

We conclude that the most economical ratio for the height to the radius is 2. In other words, the diameter of the can should equal its height. (It is easy to show that $S''(r) > 0$ for $r > 0$, so we could have concluded a minimum existed where $S'(r) = 0$ by the second derivative test.)

EXERCISES

Set A—Applied Maximum and Minimum Problems

Average cost

1. In Example 4, Section 1.6, we found that the average cost of producing q units of a certain product was given by

$$\overline{C}(q) = 2q + 4 + \frac{50}{q},$$

where $q \geq 1$. We estimated from the graph of the function that the minimum value was $\overline{C}(5) = 24$. Use the derivative to prove that this answer is correct.

Maximizing rental income

2. A manager of an apartment building with 100 apartments has each rented at $300 per month. If it is known that for each $5 increase in monthly rent, one additional vacancy will result, what should be the rent for each apartment in order to maximize the total rental income?

Maximizing rental income

3. A manager of an apartment building containing 180 apartments has each rented at $600 per month. If it is known that

for each $5 increase in monthly rent one additional vacancy will result, what should be the rent for each apartment in order to maximize the total rental income?

Maximizing excursion boat income

4. An excursion boat with a capacity of 500 people will run a charter trip for a group of 200 people at $10.00 per person. The company offers a reduction in fare to all the passengers of 2 cents per person for each additional person over 200 taking the cruise.
   (a) How many people will maximize the income?
   (b) What is the cost per person taking the cruise for this number?
   (c) What is the total revenue that the company receives for such a cruise?
   (d) What is the minimum income for such a cruise?

Maximizing theater ticket income

5. The management of a 300-seat movie theater will provide a special afternoon showing for organizations. The minimum reservation the theater will accept is for 100 people at $3.00 per person. However, the theater will reduce the price for all persons by 1 cent for each additional person attending.
   (a) What is the maximum income that can be obtained from such a showing?
   (b) What is the minimum income that can be obtained?

Maximizing profits

6. A manufacturer determines that if a particular product is priced at x dollars per unit for $x \geq 40$, then $1000(75 - x)$ units will be sold.
   (a) If the product costs $40 per unit to produce, what price will maximize profits?
   (b) How many units maximize the profits?
   (c) What is the maximum profit?

Minimizing costs

7. Suppose the cost, in thousands of dollars, for producing x thousand units of a particular product is given by

$$C(x) = 0.1x^2 - 3.7x + 63, \quad 2 \leq x \leq 40.$$

   (a) What is the cost of producing 5000 units?
   (b) How many units will minimize the cost?
   (c) What is the minimum cost for producing the product?

Maximizing revenue

8. Suppose the revenue, in thousands of dollars, from the sale of x thousand units of the product discussed in Exercise 7 is

$$R(x) = -0.1x^2 + 5.1x - 5, \quad 2 \leq x \leq 40.$$

   (a) What is the revenue from the sale of 5000 units?

3.4 MORE APPLIED MAXIMUM AND MINIMUM PROBLEMS

(b) How many units will maximize revenue?
(c) What is the maximum revenue?

9. (a) Use the cost and revenue functions in Exercises 7 and 8 to find the break-even point for the given product.
(b) What is the profit function?
(c) How many units will maximize profits?

Equipment usage

10. A company has an order for 12,500 units of a special product. It can rent machines to produce the units for $100 each and pay an individual $8 per hour to monitor all the machines used. If there are no additional costs and if each machine can produce 40 units per hour, how many machines should be rented to minimize the production cost?

Selling price

11. A company's present sales for a special type of light bulb selling for $8.60 each is 2000 bulbs per month. The company determines that for each 10 cent decrease in the price per bulb, an additional 50 bulbs will be sold each month. The bulbs cost $2.00 apiece to manufacture.
(a) What should be the selling price in order to maximize profits?
(b) What are the maximum monthly profits possible?

Delivery costs

12. On a 100-mile trip, a truck averages $600/x$ miles per gallon when it is driven at a constant speed of x miles per hour, $30 \leq x \leq 60$. If the driver earns $5 per hour and if gasoline costs $1.20 per gallon, at what constant speed should the driver drive in order to minimize the total cost of the trip? Would the answer be different if the trip was 300 miles?

Container dimensions

13. A container firm plans to use two different materials to make a rectangular box having a square bottom and a volume of 200 cm^3. If the cost of the material for the sides is $2\frac{1}{2}$ cents/cm^2 and if the cost of the material for the top and bottom is 4 cents/cm^2, what are the most economical dimensions for the box?

Container dimensions

14. A container firm plans to use two different materials to make a rectangular box having a square bottom and a volume of 648 in^3. If the cost of the material for the sides is 2 cents/in^2 and if the cost of the material for the top and bottom is 6 cents/in^2, what are the most economical dimensions for the box?

Textbook design

15. A book publisher plans to have a printed page contain 48 in^2 of print. If the top and bottom margins are each to be 1 inch wide and side margins are each to be $\frac{3}{4}$ inch wide, what page dimensions would require the minimum amount of paper? (See Figure 3.23.)

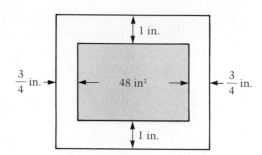

FIG. 3.23.

Dimensions of box

16. A box company plans to construct a box open at the top having a volume of 144 in³. If its base is to be a rectangle whose length is three times its width, find the dimensions of the box that will minimize the surface area.

Dimensions of box

17. Suppose a rectangular piece of tin with dimensions 12 inches by 17 inches is to be made into a box with an open top by cutting squares out of the corners and turning up the sides. What is the length of the sides of the squares that would give a box of maximum volume? (See Figure 3.24.)

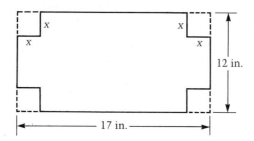

FIG. 3.24.

Selling price

18. The number of units sold of a given product depends upon the selling price p and is given by

$$x = 75 - \frac{3}{5}p, \quad 5 \leq p \leq 75.$$

(a) How many units would be sold if $p = 5$? $p = 20$? $p = 75$?
(b) Solve the equation in part (a) for p in terms of x to find an expression for $p(x)$, the selling price in terms of x.
(c) If the total cost of producing x units is given by

$$C(x) = x^2 + 13x + 500$$

3.4 MORE APPLIED MAXIMUM AND MINIMUM PROBLEMS

dollars, how many units would maximize profits? [Hint: $R(x) = xp(x)$ and $P(x) = R(x) - C(x)$.]

Selling price

19. A manufacturer of clock radios determines that if the selling price for each radio is p dollars, then the number of units sold in a week is given by $x = 800 - 16p$, $10 \leq p \leq 50$.
    (a) How many units would be sold in a week with a selling price of $10? Of $30? Of $50?
    (b) Solve the equation in part (a) for p in terms of x to find an expression for $p(x)$, the selling price in terms of x.
    (c) If the total cost of producing x units in one week is given by

    $$C(x) = 18x + 2400$$

    dollars, how many units would maximize profits? [Hint: $R(x) = xp(x)$ and $P(x) = R(x) - C(x)$.]
    (d) What should be the selling price to maximize profits?
    (e) What is the maximum profit possible?

Rental income

20. A manager of an apartment building with 100 apartments has each rented for $300 a month. If it is known that for each $5 increase in monthly rent, two additional vacancies will result, what rent should be charged for each apartment in order to maximize the total rental income?

Cruise fare

21. An excursion boat with a capacity of 500 people will run a charter trip for a group of 150 people at $8 per person. The company offers a reduction in fare to all passengers of 2 cents per person for each additional person over 150 taking the cruise.
    (a) What size group will maximize the income?
    (b) What is the cost per person taking the cruise for this number?
    (c) What is the total revenue that the company receives for such a cruise?

Playground dimensions

22. A county plans to fence part of a lot into a playground containing 16,200 ft². If a decorative fence which is to be used across the front of the playground is three times as expensive per lineal foot as the fence for the other three sides, what should be the dimensions of the playground if the goal is to minimize the cost of the fence?

Athletic field

23. An athletic field is to be composed of a rectangular plot and two semicircular plots on opposite ends of the rectangle, as in Figure 3.25a. A 440-yard track is to be constructed around the perimeter of the field. Determine the dimensions of the rectangular plot that will maximize its area.

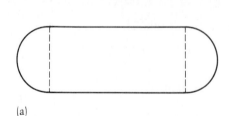 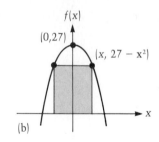

FIG. 3.25. (a) Athletic field. (b) $f(x) = 27 - x^2$.

A geometric problem

24. A rectangle with base on the x-axis is inscribed underneath the parabola $f(x) = 27 - x^2$ as in Figure 3.25b. What is the maximum area of such a rectangle?

Container dimensions

25. An engineer is to design a (right circular) cylinder for a company planning to market frozen biscuit dough. The ends of the cylinder are to be made from a metal costing three times as much as the cardboard for the side. What should be the ratio of the length h to the radius r in order to minimize the cost of the container?

Construction costs

26. The costs of land, excavation, architect fees, and foundation for an office building are $1,250,000 and are assumed to be independent of the height of the building. The cost of erecting the first floor is $500,000, the second $550,000, the third $600,000, etc. The minimum height is to be 5 stories and the maximum height 50 stories. If the net annual income (profit) from each floor is $75,000, how many stories should be constructed in order to maximize the annual rate of return on the investment in the building? (*Rate of return* is defined as *net income* divided by *total investment*.)

Textbook pricing

27. (a) Suppose a publisher finds that the cost $C(x)$ in dollars for producing $x \geq 10,000$ books is given by

$$C(x) = 175,000 + 10x.$$

What is the average cost of the first 10,000 books? What is the cost to produce each additional copy?

(b) The publisher determines that the number x of texts sold at $10 or above depends upon the price p of each book and is given by

$$x = 10,000 + 60,000\left(1 - \frac{p}{30}\right).$$

How many books would be sold if $p = 10$? How many books would be sold if $p = 30$? Solve the preceding equation for p in terms of x to find an expression for $p(x)$, the selling price in terms of x.

3.4 MORE APPLIED MAXIMUM AND MINIMUM PROBLEMS

(c) Since profit on the books is given by $P(x) = xp(x) - C(x)$, use this expression to find how many copies would maximize profits. What selling price will maximize profits on the book?

Textbook pricing

28. (a) Suppose a publisher finds that the cost $C(x)$ in dollars for producing $x \geq 10{,}000$ books is given by

$$C(x) = 50{,}000 + 9x.$$

What is the average cost of the first 10,000 books? What is the cost to produce each additional copy?

(b) The publisher determines that the number x of texts sold at \$10 or above depends upon the price p of each book and is given by

$$x = 50{,}000 + 70{,}000\left(1 - \frac{2p}{35}\right).$$

How many books would be sold if $p = 10$? How many books would be sold if $p = 30$? Solve the preceding equation for p in terms of x to find an expression for $p(x)$, the selling price in terms of x.

(c) Since profit on the books is given by $P(x) = xp(x) - C(x)$, use this equation to find how many copies would have to be sold to maximize profits. What should be the selling price to maximize profits on the book?

Trough dimensions

29. A sheet of tin 28 feet long and 16 inches wide is to be made into a trough by turning up equal sides at right angles to the bottom. Find the dimensions which give the maximum volume if the ends are to be closed with some other material. What is the maximum volume? (See Figure 3.26.)

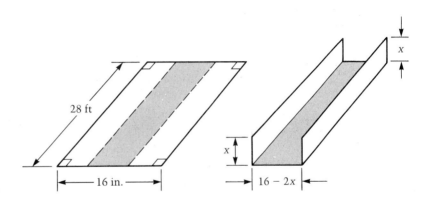

FIG. 3.26.

Utility problem

30. An electric company plans to erect a 20-foot pole and a 30-foot pole 50 feet apart on level ground. If the poles are to be anchored by a common guy wire from the top of each pole to a point on the ground between them, where should the wire be anchored to minimize the length of the wire needed? (See Figure 3.27.)

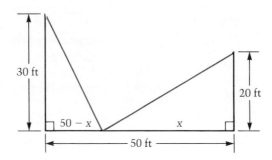

FIG. 3.27.

3.5 RELATED RATES (OPTIONAL)

Let us now look at another very important type of rate of change problem. Suppose the length of each edge of a cube is increasing at a rate of 4 cm/sec. If t is the time (independent) variable, then $s'(t) = 4$, or

$$\frac{ds}{dt} = 4.$$

Note that the volume V of the cube is also a function of time and $V(t) = [s(t)]^3$, where $s(t)$ is the length of one side at time t. The rate of change of volume with respect to time is $V'(t)$, or

$$\frac{dV}{dt}.$$

Therefore the rate of change of volume is

$$V'(t) = 3[s(t)]^2 s'(t). \tag{1}$$

The rate of change of volume when $s(t) = 2$ and $s'(t) = 4$ is

$$3[2]^2(4) = 48 \text{ cm}^3/\text{sec}.$$

3.5 RELATED RATES (OPTIONAL)

To express the derivative of $V(t)$ using the Leibniz notation, note that if $V = s^3$, then

$$\frac{dV}{dt} = \frac{dV}{ds}\frac{ds}{dt}$$

$$= 3s^2 \frac{ds}{dt}.$$

The preceding equation, of course, is equivalent to Eq. (1).

Suppose one side x of a right triangle is *increasing* at a rate of 2 cm/sec and the other side y is *decreasing* at a rate of 6 cm/sec. How fast is the area changing when x is 4 centimeters and y is 7 centimeters? In this case, both x and y are functions of t, and we are given that $x'(t) = 2$ and $y'(t) = -6$. [$y'(t)$ is negative since $y(t)$ is decreasing.] The area, which also depends upon t, is given by

$$A(t) = \frac{1}{2}[x(t)y(t)].$$

Using the product rule for differentiation, we find $A'(t)$ as follows:

$$A(t) = \frac{1}{2}[x(t)y'(t) + y(t)x'(t)].$$

Substituting in the values for $x(t)$, $y'(t)$, $y(t)$, and $x'(t)$, we get

$$A'(t) = \frac{1}{2}[(4)(-6) + (7)(2)]$$

$$= -5.$$

Therefore the area is *decreasing* at a rate of 5 cm²/sec.

related rate problems

The problems in this section are often called **related rate problems**. Related rate problems involve two or more related quantities which change with respect to time.

PRACTICE PROBLEM Suppose the radius of a right circular cylinder is increasing at a rate of 4 cm/min and the altitude is decreasing at a rate of 4 cm/min. How fast is the volume changing when the radius is 5 centimeters and the altitude is 6 centimeters? [*Hint:* $V(t) = \pi[r(t)]^2 h(t)$. Use the product rule and power rule for differentiation.]

Answer: It is increasing at the rate of 140 cm³/min. ∎

Cistern problem

EXAMPLE 1 Consider a cistern in the shape of an inverted right circular cone where the altitude is 15 feet and the radius is 5 feet. If water is being pumped into the cistern at the rate of 12 ft³/min,

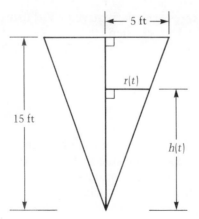

FIG. 3.28.
$V(t) = \dfrac{\pi}{27}[h(t)]^3.$

at what rate is the depth h of the water increasing when the depth is 4 feet? (See Figure 3.28.)

Solution: The radius and depth of the cone of water are functions of time t. By similar triangles we conclude that

$$\frac{5}{15} = \frac{r(t)}{h(t)}.$$

Thus $r(t) = h(t)/3$. (*Note:* Since the additional information is about the depth of the cone of water and not the radius, we solved for r in terms of h in order to find V in terms of h. If the additional information had been about r, we would have solved for h in terms of r.) Since $V = (\pi/3)r^2 h$ is the formula for the volume of a cone with radius r and altitude h, the volume of the cone of water is given by

$$\begin{aligned}
V(t) &= \frac{\pi}{3}[r(t)]^2 h(t) \\
&= \frac{\pi}{3}\left[\frac{h(t)}{3}\right]^2 h(t) \\
&= \frac{\pi}{27}[h(t)]^3.
\end{aligned}$$

Therefore the rate of change of volume with respect to time t is

$$V'(t) = \frac{\pi}{9}[h(t)]^2 h'(t).$$

Since $V'(t)$ is given to be 12 ft^3/min at any time t,

$$12 = \frac{\pi}{9}[h(t)]^2 h'(t).$$

Consequently the rate of change of the depth is given by

$$h'(t) = \frac{108}{\pi[h(t)]^2}.$$

Therefore, when $h(t) = 4$, we conclude that

$$h'(t) = \frac{108}{\pi(4)^2} = \frac{27}{4\pi},$$

and the depth is increasing at a rate of $27/4\pi \doteq 2.15$ ft/min.

Remark: To use the Leibniz notation, let $V = (\pi/27)h^3$. Since

$$\frac{dV}{dt} = \frac{dV}{dh}\frac{dh}{dt},$$

it follows that

$$\frac{dV}{dt} = \frac{\pi}{9}h^2 \frac{dh}{dt},$$

which is equivalent to our previous derivation for $V'(t)$.

PRACTICE PROBLEM Consider a cistern in the shape of an inverted right circular cone where the altitude is 10 feet and the radius is 5 feet. If the water level is rising at 2 ft/min when the depth is 4 feet, at what rate is water entering the cistern?

Answer: 8π ft^3/min. ∎

EXERCISES

Set A–Related Rate Problems

1. Each edge of a cube is decreasing at the rate of 0.3 cm/min. How fast is the volume of the cube changing when the length of each edge is 4 centimeters?

2. Each edge of a cube is increasing at 2 in/min. How fast is the

volume of the cube increasing when the length of each edge is 7 inches?

3. The volume of a cube is increasing at 150 cm³/sec. How fast is an edge increasing when its length is 12 centimeters?

4. If the volume of a sphere is changing at the rate of 8 ft³/min, how fast is the radius changing when the radius is 2 feet?

5. If the base of a triangle is increasing at 8 in/min and if the altitude is increasing at 3 in/min, what is the rate of change of the area when the base is 10 inches and the altitude is 8 inches?

6. If the base of a triangle is increasing at 5 in/min and if the altitude is decreasing at 2 in/min, what is the rate of change of the area when the base is 10 inches and the altitude is 8 inches?

7. The volume of a cube is increasing at 8 cm³/min. How fast is the total surface area increasing when the surface area of the cube is 96 cm²?

8. The volume of a sphere is increasing at 5 cm³/min. How fast is its surface area increasing when the surface area is 25π cm²? [*Hint:* Surface area is given by $S(r) = 4\pi r^2$.]

9. A cistern in the shape of an inverted right circular cone (tip down) has an altitude of 16 feet and a radius of 4 feet. If water is being pumped into the cistern at a rate of 13 ft³/min, at what rate is the depth h of the water increasing when the depth is 6 feet? [*Hint:* Surface area is given by $S(r) = 4\pi r^2$.]

10. Suppose the cistern in Exercise 9 rests on its base. If water is being pumped into the cistern at 13 ft³/min, at what rate is the depth h of the water increasing when the depth is 6 feet?

11. Air is escaping from a spherical balloon at the rate of 4 ft³/min. How fast is the surface area decreasing when the radius is 3 feet?

12. Water is being pumped into a hemispherical tank of radius 12 feet at the rate of 10 ft³/min. How fast is the water level rising when the water is 4 feet deep at the center? [*Hint:* The volume of a spherical segment is $V = \pi h^2(3R - h)/3$, where h is the altitude of the segment and R is the radius of the sphere.]

13. The altitude of an inverted right circular cone is 10 feet and the radius of its base is 15 feet. If water is being pumped into the cone at 15 ft³/sec and if the depth is increasing at 0.05 ft/sec when the depth is 4 feet, how fast is the water draining out through a hole in the bottom of the tank?

3.6 SUMMARY AND REVIEW EXERCISES

1. **INCREASING AND DECREASING FUNCTIONS**

   (a) A function f is **increasing** on $[a, b]$ if for each u and v in $[a, b]$ for which $u < v$ it follows that $f(u) < f(v)$. For a continuous function f on $[a, b]$, if $f'(x) > 0$ for $a < x < b$, then f is *increasing* on $[a, b]$.

   (b) A function f is **decreasing** on $[a, b]$ if for each u and v in $[a, b]$ for which $u < v$ it follows that $f(u) > f(v)$. For a continuous function f on $[a, b]$, if $f'(x) < 0$ for $a < x < b$, then f is *decreasing* on $[a, b]$.

2. **RELATIVE MAXIMA AND MINIMA**

   (a) If there exists an open interval containing t and in the domain of a function f such that $f(x) \leq f(t)$ for every x in the open interval, then $f(t)$ is called a **relative maximum,** or **local maximum, value** of f.

   (b) If there exists an open interval containing t and in the domain of a function f such that $f(x) \geq f(t)$ for every x in the open interval, then $f(t)$ is called a **relative minimum,** or **local minimum, value** of f.

   (c) Assume that f is a continuous function at t and assume $f'(t) = 0$ or $f'(t)$ does not exist. If there exists some open interval containing t such that $f'(x) > 0$ for $x < t$ and $f'(x) < 0$ for $x > t$, then $f(t)$ is a **relative maximum value** of f. If there exists some open interval containing t such that $f'(x) < 0$, $x < t$, and $f'(x) > 0$, $x > t$, then $f(t)$ is a **relative minimum value** of f. (This is the first derivative test for relative extrema.)

   (d) Let f be a function and let t be a number in the domain of f such that $f'(t) = 0$. If $f''(t)$ exists and is positive, then $f(t)$ is a **relative minimum value** of f. If $f''(t)$ is negative, then $f(t)$ is a **relative maximum value** of f. (This is the second derivative test for relative extrema.)

3. **CONCAVITY AND INFLECTION POINTS**

   (a) If $f''(t) > 0$ for a function f, then the graph of f is said to be **concave upward** at t. If $f''(t) < 0$, then the graph of f is said to be **concave downward** at t.

   (b) If $(t, f(t))$ is a point on the graph of a function f such that the concavity on one side of t is opposite from the concavity on the other side of t, then $(t, f(t))$ is called an **inflection point.**

4. OTHER APPLICATIONS OF THE DERIVATIVE

(a) For a differentiable function f, $f'(x)$ is the **rate of change** of f at x.

(b) The **percentage rate of change** of f at x is

$$\frac{f'(x)}{f(x)}(100).$$

5. REVIEW EXERCISES

1. (a) Determine where $f(x) = 2x^3 - 2x^2 - 32x + 1$ is increasing.
   (b) Determine where $f(x) = 2x^3 - 2x^2 - 32x + 1$ is decreasing.

2. (a) Determine where $f(x) = x/(x^2 + 1)$ is increasing.
   (b) Determine where $f(x) = x/(x^2 + 1)$ is decreasing.

3. On what part of the domain of $f(x) = x^2 + 1$ is the percentage rate of change of f at x an increasing function?

4. On what part of the domain is the slope of the tangent to the graph of $f(x) = x/(x^2 + 3)$ an increasing function?

For each function in Exercises 5 through 10, do the following. (a) Find the coordinates of any point where the function has a relative maximum. (b) Find the coordinates of any point where the function has a relative minimum. (c) Find the absolute maximum of f, if it exists. (d) Find the absolute minimum of f, if it exists. (e) Find any inflection points on the graph of f. (f) Discuss concavity for the graph of f. (g) Sketch the graph of f.

5. $f(x) = x^3 - 6x$
6. $f(x) = 9x - x^3$
7. $f(x) = \frac{1}{4}(x^4 - 8x^2)$
8. $f(x) = x^2 + 8/x$
9. $f(x) = x^4 - 8x^2$, $-4 \le x \le 3$
10. $f(x) = -x^3 + 6x^2$, $-1 \le x \le 6$

11. Suppose a rectangular piece of tin with dimensions 12 inches by 18 inches is to be made into a box with an open top by cutting squares out of the corners and turning up the sides. What is the length of the sides of the squares that would maximize the volume of such a box?

12. In the construction of storage boxes with square bottoms and open tops, a manufacturer plans to spend $36 for material on each box. If the material for the bottom of each box costs $3/m² and if the material for the sides costs $2/m², what should be the dimensions of the box to maximize the volume? What would be the maximum volume for such a box?

3.6 SUMMARY AND REVIEW EXERCISES

13. The publisher of a national magazine plans to have each page contain 96 in² of print. If each of the top and bottom margins of each page is to be ¾ inch wide and if each side margin is to be ½ inch wide, what are the most economical dimensions for the page?

14. A company has an order for 181,500 units of a given product. It can rent machines to produce the units for $200 each and pay an individual $12 an hour to monitor all the machines used. There are no additional costs and each machine can produce 90 units per hour.
    (a) How many machines should be rented to minimize the cost of production?
    (b) How much is the minimum cost per unit?

15. A company's unit cost for a hardware item is $24.50. At a selling price of $38.75 apiece, the company sells 1560 units per month. The company estimates that for each 25-cent decrease in the price per unit an additional 40 units will be sold each month.
    (a) What selling price would maximize profits?
    (b) What would be the maximum monthly profits possible?

16. A vertical wall 6 feet high stands on level ground 3 feet from a building. What is the shortest ladder resting on the ground that can reach over the wall and lean against the building? (See Figure 3.29.)

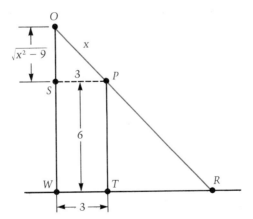

FIG. 3.29.

17. A manager of an apartment building containing 200 apartments has rented each at $600 per month. If it is known that for each $5 increase in monthly rent one additional vacancy

will result, what should be the rent charged for each apartment in order to maximize the total rental income?

18. A man on an island 3 miles from a straight shoreline wishes to reach, as soon as possible, a point on shore 4 miles from the closest point on shore. If he can average 2 miles per hour rowing and 4 miles per hour walking, what route should he take?

19. A company finds that it costs $C(x) = 300x + 1,400$ dollars each month to produce x units of a given product. It estimates that if p is the selling price in dollars for each unit, then $x = (10 - p/300)^2$ units will be sold each month provided $300 \leq p \leq 1500$.
   (a) What selling price will maximize profits? [*Hint:* Find p in terms of x. Then recall that $R(x) = xp(x)$.]
   (b) What is the maximum profit possible each month?

20. Two farmers who live on the same side of a straight river want to build an irrigation pumping station serving both their farms. The distances from the two farms to the river are 300 meters and 800 meters, respectively, and the perpendiculars from the farms to the river are 500 meters apart. Find the location of the pumping station at the river's edge that minimizes the total pumping distance to both farms. (See Figure 3.30.)

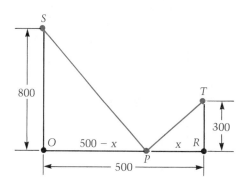

FIG. 3.30.

Gottfried Wilhelm Leibniz (1646–1716)

The Bettmann Archive

Gottfried Wilhelm Leibniz was born in Leipzig, Germany, on July 1, 1646. He entered the University of Leipzig as a student of law at the age of 15, and in 1666 he obtained his doctorate in law. It was not until 1672, at the age of 26, that he developed his interest in mathematics. This interest was initiated and encouraged by the famous physicist and mathematician Christiaan Huygens (1629–1695).

Leibniz developed the calculus between 1673 and 1676, some time after Newton had developed the subject but considerably before Newton published any of his results. Unfortunately, a battle raged for years over whether Newton or Leibniz should be credited with the creation of the calculus. During the fray both Newton and Leibniz were accused of plagiarism. But, in fact, as has been the occasion before and since in mathematics, the time was right for two fertile minds to make independently similar mathematical discoveries. Leibniz introduced the integral sign and many other calculus notations. It was in 1684 that Leibniz published a six-page paper that contained not only the calculus symbols dx and dy but also the basic differentiation rules.

Besides mathematics, Leibniz made important contributions to logic, mechanics, and nautical science. His interests were nearly always directed to what he considered to be relevant and practical problems. Eventually he was made a member of the Royal Society in England, and both he and Newton were among the first foreigners to be made members of the French Academy of Sciences. In 1669 Leibniz proposed the founding of the Berlin Academy of Sciences and became its first president. Truly a universal genius, he was a jack-of-all-trades. For many years he was in the diplomatic service and traveled widely to such places as Paris and London on diplomatic missions. As stated by Eric Temple Bell in *Men of Mathematics*, "As a diplomat and statesman Leibniz was as good as the cream of the best of them in any time or any place, and far brainier than all of them together."

CHAPTER

4

EXPONENTIAL AND LOGARITHMIC FUNCTIONS

4

- 4.1 INTRODUCTION
- 4.2 THE CHAIN RULE AND INVERSE FUNCTIONS
- 4.3 EXPONENTIAL AND LOGARITHMIC FUNCTIONS
- 4.4 DIFFERENTIATION OF THE EXPONENTIAL AND LOGARITHMIC FUNCTIONS
- 4.5 EXPONENTIAL GROWTH AND CONTINUOUS INTEREST
- 4.6 IMPLICIT AND LOGARITHMIC DIFFERENTIATION (OPTIONAL)
- 4.7 SUMMARY AND REVIEW EXERCISES

4.1 INTRODUCTION

In this chapter we investigate the exponential and logarithmic functions. We shall find that exponential functions, in particular, have many important applications. For example, learning curves, population growth, the spread of epidemics, growth of a country's GNP (gross national product), growth of bacteria in a culture, amount obtained from money deposited at a given interest rate with continuous compounding, and radioactive decay can all be described using exponential functions. Four of the problems we shall be able to solve before completing this chapter are:

1. (a) Find the amount resulting from a principal of $3000 deposited for 4 years at 7.5% annual interest rate with continuous compounding. (b) How long will it take to double the amount? (See Example 3, Section 4.5.)

2. A psychologist discovers that it takes the average person x hours to learn $f(x) = 126 - 126e^{-kx}$ meaningless words. (a) If it takes the average person 3 hours to learn 42 meaningless words, how long will it take him or her to learn 84 such words? (b) How long will it take to learn 120 meaningless words? (c) How long will it take to learn 125 meaningless words? (d) According to this model, what is the greatest number of meaningless words an average person can learn? (See Exercise 15, Section 4.5.)

3. An epidemiologist determines that the percentage of college students at a given college who will be infected x days from now with a flu virus is given by

$$f(x) = \frac{1}{1 + 9e^{-kx}}.$$

(a) What percentage of the students are infected with the disease at the present time? (b) If 50% of the students become infected in 20 days, what is the value of k? (c) How long will it take for 75% of the students to be infected? (d) Show that the rate of change of the spread of the epidemic is proportional to the product of the percentage infected and the percentage not infected. (See Exercise 1, Section 4.5.)

4. Suppose the number of bacteria in a culture increases at a rate proportional to the number present at time x. If 100,000 bacteria are present at a given time and 250,000 are present one hour later, how many bacteria are present in 3 hours? (See Example 1, Section 4.5.)

Before we turn our attention to such applications, we consider the differentiation theorem for the composition of two functions. This theorem is not only essential in finding the derivative of many functions but it is also useful in finding the derivative of the inverse of a function whose derivative is known. (This should become clear when we find the derivative of the logarithmic function.) ∎

4.2 THE CHAIN RULE AND INVERSE FUNCTIONS

Let us now look carefully at an important operation on functions called *composition*. In order to provide a better understanding of the composition of functions, let us consider first the following hypothetical situation. Assume that 50 men, each of whom is wearing a topcoat, decide to go to a restaurant. Also assume that when they arrive, they check their coats individually and no two coats are hung on the same hanger. If we let S be the set of men and T be the set of coats, then the set of ordered pairs (x, y) where each coat y in T is paired with its owner x in S is a function. We call this function g. If we let W be the set of coat hangers on which the coats are hung, then the set of ordered pairs (y, z) where y is in T, z is in W, and "y hangs on z" is a function. We call this function f. Now, each man x is paired with a hanger z, the one on which his coat hangs. (The man is probably informed of this pairing by being given a ticket with the number of his hanger.) Thus a new function with ordered pairs (x, z) is defined. The function is called the *composite f of g* (or the *composite of f with g*) and is denoted by $f \circ g$, or $f(g)$, (read "f of g"). Since $y = g(x)$ and $z = f(y)$, then

$$z = f(g(x)).$$

composite function

In general, if f and g are functions with domains D_f and D_g, respectively, then $F = f \circ g$ is called the **composite function** f *of* g and is defined by

$$F(x) = f(g(x)),$$

where x is in D_g and $g(x)$ is in D_f.

4.2 THE CHAIN RULE AND INVERSE FUNCTIONS

Suppose $f(x) = 2x - 6$ and $g(x) = \sqrt{x - 4}$. We see that $g(5) = 1$ and $f(1) = -4$. Therefore, $f(g(5)) = -4$, and -4 is the number in the range of $F(x) = f(g(x))$ paired with 5 in its domain. The domain of the composite function f of g contains all real numbers x in the domain of g such that $g(x)$ is in the domain of f. (This always makes the domain of $F(x) = f(g(x))$ a subset of the domain of g.) For $f(g(x))$ to make sense, we must have x in the domain of g to obtain $g(x)$, and we must have $g(x)$ in the domain of f to obtain $f(g(x))$. Since any real number is in the domain of $f(x) = 2x - 6$, the domain of F in this example is the same as the domain of g; it contains all $x \geq 4$. That is,

$$F(x) = 2\sqrt{x - 4} - 6 \quad \text{where} \quad x \geq 4.$$

For $f(x) = 2x - 6$ and $g(x) = \sqrt{x - 4}$, let us consider the composite function G defined by $G(x) = g(f(x))$. The domain of G contains only those real numbers x such that $f(x) \geq 4$, that is,

$$2x - 6 \geq 4, \quad 2x \geq 10, \quad \text{or} \quad x \geq 5.$$

Therefore,

$$\begin{aligned} G(x) &= g(f(x)) \\ &= \sqrt{(2x - 6) - 4} \\ &= \sqrt{2x - 10} \quad \text{where} \quad x \geq 5. \end{aligned}$$

Composition of functions

EXAMPLE 1 Let f and g be defined by $f(x) = x^{1/2}$ and $g(x) = 3x + 5$. If $h(x) = f(g(x))$, then

$$h(x) = (3x + 5)^{1/2}.$$

The domain of h consists of all x such that $3x + 5 \geq 0$, that is, $x \geq -\frac{5}{3}$. If H is the function defined by $H(x) = g(f(x))$, then

$$H(x) = 3x^{1/2} + 5.$$

The domain of H contains all x such that $x \geq 0$.

PRACTICE PROBLEM

(a) Let $f(x) = 7x - 2$ and let $g(x) = x^{3/2} + 4$. Find an equation defining $F(x) = f(g(x))$ and state the domain.

(b) Find an equation defining $H(x) = g(f(x))$ and state the domain.

Answer: (a) $F(x) = 7x^{3/2} + 26$, $x \geq 0$. (b) $H(x) = (7x - 2)^{3/2} + 4$, $x \geq \frac{2}{7}$. ∎

We have learned how to differentiate the sum, difference, product, and quotient of two functions. Let us now consider the differentiation theorem providing the rule for differentiating the composition of two functions. It is often called the **chain rule** and is one of the most used (and misused) theorems in calculus. Let f and g be two functions and let F be the composite function f of g defined by

$$F(x) = f(g(x)).$$

If g is differentiable at x, and f is differentiable at $g(x)$, then it can be proved that

chain rule

$$F'(x) = f'(g(x)) \cdot g'(x).$$

(The derivative of f of g is the derivative of f evaluated at $g(x)$ times the derivative of g at x.)
 If $f(x) = x^r$ and if $F(x) = f(g(x))$, then

$$F(x) = [g(x)]^r.$$

Since $f'(x) = rx^{r-1}$, it follows that $f'(g(x)) = r[g(x)]^{r-1}$. We conclude from the chain rule that

$$F'(x) = f'(g(x))g'(x) = r[g(x)]^{r-1}g'(x).$$

Consequently, we see that Theorem 4, Section 2.5, is a special case of the chain rule where the function f is defined by $f(x) = x^r$.

For the composition of functions, it is sometimes convenient to introduce an "intermediate" variable not only to define the function but also to find its derivative. For example, if we let $y = f(u)$ and $u = g(x)$, then

$$y(x) = f(g(x)).$$

With the Leibniz notation, since

$$\frac{dy}{dx} = y'(x), \qquad \frac{dy}{du} = f'(u), \quad \text{and} \quad \frac{du}{dx} = g'(x),$$

4.2 THE CHAIN RULE AND INVERSE FUNCTIONS

the chain rule

$$y'(x) = f'(g(x))g'(x)$$

can also be stated in the Leibniz notation as follows:

chain rule (Leibniz notation)

$$\frac{dy}{dx} = \frac{dy}{du} \cdot \frac{du}{dx}.$$

For example, to find the derivative of $y = (2x^3 - 6)^{5/2}$, we could let $u = 2x^3 - 6$ and $y = u^{5/2}$. Then

$$\frac{dy}{dx} = \frac{dy}{du}\frac{du}{dx}$$

$$= \frac{5}{2}u^{3/2}(6x^2)$$

$$= \frac{5}{2}(2x^3 - 6)^{3/2}(6x^2)$$

$$= 15x^2(2x^3 - 6)^{3/2}.$$

Chain rule

EXAMPLE 2 Suppose $y = u^{3/2} + u$ and $u = 4x + 1$. Find dy/dx in terms of x.

Solution

$$\frac{dy}{du} = \left(\frac{3}{2}\right)u^{1/2} + 1 \quad \text{and} \quad \frac{du}{dx} = 4.$$

Therefore,

$$\frac{dy}{dx} = \frac{dy}{du}\frac{du}{dx}$$

$$= \left(\frac{3}{2}u^{1/2} + 1\right)(4)$$

$$= 6u^{1/2} + 4$$

$$= 6(4x + 1)^{1/2} + 4.$$

Now we turn our attention to one more important differentiation theorem. By using the chain rule, we shall discover a reasonably easy method for finding the derivative of the inverse of a function in terms of the derivative of the function. First we define what is meant by the inverse of a function.

Let f be a function. The set defined by

$$g = \{(b, a) \mid (a, b) \text{ is in } f\}$$

inverse relation

is called the **inverse relation** of f. If no two different ordered pairs of f have the same second coordinates, then the inverse relation g is a function. This function g is called the **inverse function** of f and is usually denoted by f^{-1}. [*Note:* "−1" is not an exponent, and, in general, the inverse of f and the reciprocal of f are not equal; that is, $f^{-1}(x) \neq 1/f(x)$.]

inverse function

Let f be a function with inverse function f^{-1}. Geometrically, the graphs of f and f^{-1} are mirror images "reflected" in the graph of $H(x) = x$, the **identity function** on the set of real numbers. (See Figure 4.1.) If (a, b) is in f, then $f(a) = b$. Furthermore, (b, a) is in f^{-1} and $f^{-1}(b) = a$. For inverse functions f and f^{-1}, it is important to remember that $f(a) = b$ if and only if $f^{-1}(b) = a$. Therefore,

identity function

$$f(f^{-1}(b)) = f(a) = b \quad \text{and} \quad f^{-1}(f(a)) = f^{-1}(b) = a,$$

and the *composition* f of f^{-1} (or f^{-1} of f) is the *identity function*.

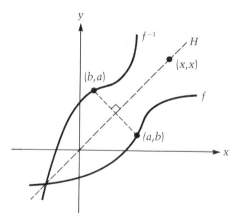

FIG. 4.1. Inverse functions: $f(a) = b$ and $f^{-1}(b) = a$.

The inverse relation of a function need not be a function. If F is defined by $F(x) = x^2$, then $(2, 4)$ and $(-2, 4)$ are two different elements in F with the same second coordinates. Thus the inverse relation of F is not a function. However, if we restrict the domain of F and define a different function f by $f(x) = x^2$, where $x \geq 0$, then the inverse of f is a function. (See Figure 4.2.) We call the

4.2 THE CHAIN RULE AND INVERSE FUNCTIONS

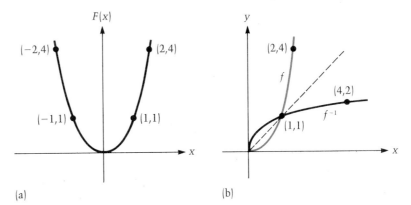

FIG. 4.2. (a) $F(x) = x^2$.
(b) $f(x) = x^2$, $x > 0$;
$f^{-1}(x) = \sqrt{x}$.

inverse of f the **square root function** and let $\sqrt{x}$ represent the number in the range of f^{-1} paired with x in the domain; thus $f^{-1}(x) = \sqrt{x}$. The domain and the range of f^{-1} are the set of nonnegative real numbers.

Let g be the function defined by $g(x) = 3x + 7$. We shall first prove that the inverse relation of g is a function. If $g(w) = g(z)$, then $3w + 7 = 3z + 7$; thus we conclude that $3w = 3z$ and $w = z$. In other words, if $w \neq z$, then $g(w) \neq g(z)$. Since the function g cannot have two different ordered pairs with the same coordinates, the inverse of g is a function. Now we would like to obtain a formula, if one exists, giving $g^{-1}(x)$ for each x in the domain of g^{-1}. In this and many similar cases such a formula can be found as follows. Substitute x for $g(x)$ and $g^{-1}(x)$ for x in the equation $g(x) = 3x + 7$ and then solve the resulting equation for $g^{-1}(x)$. The validity of this technique is based on the fact that $g(g^{-1}(x)) = x$. Hence

$$3g^{-1}(x) + 7 = x,$$
$$3g^{-1}(x) = x - 7,$$
$$g^{-1}(x) = \frac{1}{3}(x - 7).$$

Note, for example, that $g(2) = 13$ and $g^{-1}(13) = 2$. In general, if $g(a) = b$, then $g^{-1}(b) = a$.

PRACTICE PROBLEM Let f be the function defined by

$$f(x) = \frac{x}{2x + 1}, \quad x \neq -\frac{1}{2}.$$

Define the inverse function g by giving a formula for $g(x)$.

Answer: $g(x) = x/(1 - 2x)$, $x \neq 1/2$. ■

In the preceding practice problem, since $f(g(x)) = x$, the solution could be (was) found by solving

$$x = \frac{g(x)}{2g(x) + 1}$$

for $g(x)$. However, if $f(x) = x^3 + 5x + 2$ and if g is its inverse, it would be quite difficult to solve the equation

$$f(g(x)) = x, \quad \text{that is,} \quad [g(x)]^3 + 5g(x) + 2 = x$$

for $g(x)$ since it involves solving a cubic equation. Although we might not be able to obtain a formula for $g(x)$, we can find $g'(x)$ using the chain rule. Let us consider, in general, how this is done.

Let f and g be inverse functions. Assume g is differentiable at x, assume f is differentiable at $g(x)$, and assume $f'(g(x)) \neq 0$. Then, differentiating each side of the equation

$$f(g(x)) = x$$

using the chain rule, we obtain

$$f'(g(x))g'(x) = 1.$$

From this we arrive at the following important formula which gives g' in terms of f'.

inverse function theorem for derivatives

Differentiation Theorem for Inverse Functions f and g

$$g'(x) = \frac{1}{f'(g(x))}.$$

Chain rule and inverse functions

EXAMPLE 3 Let $f(x) = x^3$ and let g be its inverse function. Since $f'(x) = 3x^2$ and since

$$g'(x) = \frac{1}{f'(g(x))}$$

it follows that

$$g'(x) = \frac{1}{3[g(x)]^2}.$$

Observe that the inverse of the cube function $f(x) = x^3$ is the *cube root function* $g(x) = x^{1/3}$. Consequently

$$g'(x) = \frac{1}{3[x^{1/3}]^2} = \frac{1}{3}x^{-2/3}.$$

Note: This example verifies Theorem 1, Section 2.4, for $r = \frac{1}{3}$.

EXERCISES

Set A–Composition of Functions

For each of the functions in Exercises 1 through 6, find (a) $f(g(1))$ and (b) $g(f(2))$.

1. $f(x) = 4x + 3$, $g(x) = 5x + 1$
2. $f(x) = 2x - 1$, $g(x) = 7x + 8$
3. $f(x) = 3x + 1$, $g(x) = x^2$
4. $f(x) = x^2 + 1$, $g(x) = 2x + 3$
5. $f(x) = x + 1$, $g(x) = x^3$
6. $f(x) = x^2 - 4$, $g(x) = (x + 1)^3$

For the two functions f and g given in Exercises 7 through 16, find an equation defining (a) $F(x) = f(g(x))$ and (b) $H(x) = g(f(x))$.

7. $f(x) = 3x + 2$, $g(x) = 2x - 5$
8. $f(x) = 4x + 3$, $g(x) = 8x - 2$
9. $f(x) = x^2$, $g(x) = 3x - 1$
10. $f(x) = 2x + 7$, $g(x) = x^2 - 9$
11. $f(x) = x^{1/2}$, $g(x) = x^3$
12. $f(x) = x^3$, $g(x) = x^{1/2} + 4$
13. $f(x) = 2x - 1$, $g(x) = x^3 + 1$
14. $f(x) = x^2 - 4$, $g(x) = 4x - 12$
15. $f(x) = x^{1/2}$, $g(x) = x^3 + 4$
16. $f(x) = x^{2/3}$, $g(x) = x^3 + 8$
17. (a) For $f(x) = x^2$ and $g(x) = 3x - 1$, find $f'(g(x))$ and $g'(x)$.
    (b) For $F(x)$ found in Exercise 9 find $F'(x)$.
    (c) Verify that $F'(x) = f'(g(x))g'(x)$.
18. (a) For $f(x) = x^3$ and $g(x) = x^{1/2} + 4$, find $f'(g(x))$ and $g'(x)$.
    (b) For $F(x)$ found in Exercise 12, find $F'(x)$.
    (c) Verify that $F'(x) = f'(g(x))g'(x)$.

204   CHAPTER 4 EXPONENTIAL AND LOGARITHMIC FUNCTIONS

19. (a) For $f(x) = 2x - 1$ and $g(x) = x^3 + 1$, find $g'(f(x))$ and $f'(x)$.
    (b) For $H(x)$ found in Exercise 13, find $H'(x)$.
    (c) Verify that $H'(x) = g'(f(x))f'(x)$.

20. (a) For $f(x) = x^{2/3}$ and $g(x) = x^3 + 8$, find $g'(f(x))$ and $f'(x)$.
    (b) For $H(x)$ found in Exercise 16, find $H'(x)$.
    (c) Verify that $H'(x) = g'(f(x))f'(x)$.

Set B–Finding Derivatives of Inverse Functions

Let g be the inverse function of f in Exercises 21 through 26 and do the following. (a) Find $g'(5)$ by using the derivative of f and the inverse function theorem for derivatives. (b) Find $g'(5)$ by obtaining an equation defining g, finding its derivative by the basic differentiation theorems, and then evaluating it at 5.

21. $f(x) = 3x + 6$ 22. $f(x) = 6x + 17$
23. $f(x) = x^2 + 1, \ x \geq 0$ 24. $f(x) = x^3 - 3$
25. $f(x) = \sqrt{4x - 3}$ 26. $f(x) = \sqrt{3x + 1}$

4.3   EXPONENTIAL AND LOGARITHMIC FUNCTIONS

Let f be defined by $f(x) = b^x$, $b > 0$ and $b \neq 1$. It can be proved that if $b > 1$, then the function f is *continuous and increasing* on the set of real numbers. If $0 < b < 1$, then f is *continuous and decreasing* on the set of real numbers. Figure 4.3 shows sketches of the graphs of the functions defined by $f(x) = b^x$ with $b = \frac{1}{2}$, $b = 2$, and $b = 10$. The function $f(x) = b^x$ is called the **exponential function** *with base b*. For example, the exponential function with base 10 is defined by $f(x) = 10^x$.

exponential function

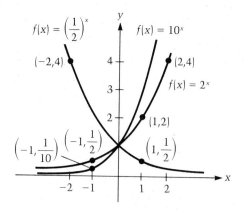

FIG. 4.3. Exponential functions.

4.3 EXPONENTIAL AND LOGARITHMIC FUNCTIONS

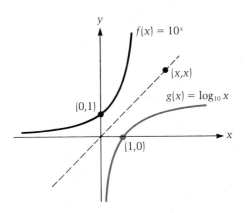

FIG. 4.4. Inverse functions: $f(x) = 10^x$ and $g(x) = \log_{10} x$.

If f is defined by $f(x) = 10^x$, then $f(-1) = 0.1$, $f(0) = 1$, $f(1) = 10$, and $f(2) = 100$; the corresponding ordered pairs in f (on the graph of f) are $(-1, 0.1)$, $(0, 1)$, $(1, 10)$, and $(2, 100)$. Since f is increasing and continuous, its inverse g is an increasing and continuous function. (See Figure 4.4.) The standard notation for the (range values of the) inverse function of f is

$$\log_{10} x,$$

read: "log x, base 10." From the fact that $f(-1) = 0.1$, it follows that $g(0.1) = \log_{10} 0.1 = -1$. Furthermore, $\log_{10} 1 = 0$, $\log_{10} 10 = 1$, and $\log_{10} 100 = 2$.

In general, if $f(x) = b^x$, $b > 0$, and $b \neq 1$, then its inverse function is $g(x) = \log_b x$. Since the range of the exponential function with base b is the set of positive real numbers, the domain of the logarithmic function with base b is the set of positive real numbers. Furthermore, since for any two inverse functions f and g

$$g(f(x)) = x \quad \text{and} \quad f(g(x)) = x,$$

the following two important identities for the logarithmic and exponential functions are obtained.

$$\log_b b^x = x \tag{1}$$

and

$$b^{(\log_b x)} = x \tag{2}$$

We assume that the student is familiar with the following three basic properties for logarithms; they are immediate consequences of the properties for exponents. If $b > 0$ and $b \neq 1$, then for any positive real numbers x and y we have the following.

Property 1 $\log_b xy = \log_b x + \log_b y$.
Property 2 $\log_b x^r = r(\log_b x)$, r a real number.
Property 3 $\log_b \dfrac{x}{y} = \log_b x - \log_b y$.

To show *Property 1* is valid, for example, note that

$$\text{if } u = \log_b x, \text{ then } x = b^u,$$

and

$$\text{if } v = \log_b y, \text{ then } y = b^v.$$

Hence, $xy = b^u b^v = b^{u+v}$ and

$$\log_b xy = u + v = \log_b x + \log_b y.$$

Logarithmic functions

EXAMPLE 1 In each of the following find the number represented by the variable which makes the equality a true statement.
(a) $\log_2 2^3 = x$. (b) $4^{\log_4 x} = 7$. (c) $\log_{10} x = -3$. (d) $\log_2 8 = x$.

Solution

(a) $x = 3$. (See Eq. 1.)   (b) $x = 7$. (See Eq. 2.)
(c) $10^{-3} = x$ and $x = 0.001$.   (d) $2^x = 8$ and $x = 3$.

Logarithmic functions

EXAMPLE 2 In each of the following, find the number represented by the variable which makes the equality a true statement.
(a) $\log_x 25 = -\tfrac{1}{2}$. (b) $\log_{10} x^{5/2} - \log_{10} x^{1/2} = 6$.

Solution

(a) $x^{-1/2} = 25,$ $x^{1/2} = \dfrac{1}{25},$ and $x = \dfrac{1}{625}.$

4.3 EXPONENTIAL AND LOGARITHMIC FUNCTIONS

(b) From *Property 3* we get $\log_{10} x^2 = 6$, and *Property 2* yields $2(\log_{10} x) = 6$. Therefore, $\log_{10} x = 3$ and $x = 10^3 = 1000$.

Logarithms can be used to solve exponential equations such as $3(2^x) = 15$. We proceed as follows. If $3(2^x) = 15$, then

$$2^x = 5.$$

Therefore (by taking the logarithm of each side), we obtain

$$\log_b 2^x = \log_b 5,$$
$$x(\log_b 2) = \log_b 5,$$
$$x = \frac{\log_b 5}{\log_b 2}.$$

Note that the base of the logarithmic function does not affect the technique for finding the answer. Using base 10 (on a calculator), we find that

$$x \doteq \frac{0.69897}{0.30103} \doteq 2.3219.$$

As we shall discover in Section 4.4, there are distinct advantages in using $e \doteq 2.71828$ as the base for the exponential and logarithmic functions. If

$$f(x) = e^x,$$

then its inverse function is

$$g(x) = \log_e x.$$

natural logarithm
The logarithmic function with e as base is called the **natural logarithm** and is usually denoted by

$$g(x) = \ln x.$$

Before turning to the exercises, let us consider three more examples. As we shall discover in Section 4.5, the functions discussed in these examples have considerable importance in applied problems.

Exponential functions

EXAMPLE 3 Let $f(x) = ce^{kx}$.

(a) Find $f(3)$ if $c = 100$ and $k = 0.06$.

(b) Find c if $k = 0.08$ and $f(2) = 469.40$.

(c) Find k if $c = 500$ and $f(4) = 610.70$.

Solution

(a) If $f(x) = ce^{kx}$, $c = 100$, and $k = 0.06$, then

$$f(3) = 100e^{0.06(3)} = 100e^{0.18}.$$

Using Table 3 in Appendix A (the "*e*-key" or "inv ℓn" on a calculator) we find

$$f(3) = 100(1.1972) = 119.72.$$

(b) If $f(x) = ce^{kx}$, $k = 0.08$ and $f(2) = 469.40$, then

$$469.40 = ce^{0.08(2)}$$
$$= ce^{0.16}$$
$$= 1.1735c$$
$$c = \frac{469.40}{1.1735} = 400.$$

(c) If $f(x) = ce^{kx}$, $c = 500$, and $f(4) = 610.70$, then

$$610.70 = 500e^{4k}$$
$$e^{4k} = \frac{610.70}{500} = 1.2214.$$

Let us solve for k by two different methods. First taking the natural logarithm of each side we obtain

$$\ell n\, e^{4k} = \ell n(1.2214),$$
$$4k \doteq 0.2,$$
$$k \doteq 0.05.$$

Using Table 3 in Appendix A, we find that

$$e^x = 1.2214 \quad \text{if} \quad x = 0.20.$$

4.3 EXPONENTIAL AND LOGARITHMIC FUNCTIONS

Therefore, $4k = 0.2$ and $k = 0.05$. (Generally, we shall not continue to use $\doteq$ except as a point of emphasis.)

Exponential function

EXAMPLE 4 Let $A(x) = 100e^{rx}$. If $A(5) = 134.98$, what is $A(10)$?

Solution: First we know that $134.98 = 100e^{5r}$; thus

$$e^{5r} = 1.3498.$$

At this point, we can solve for r as in Example 3. However, in problems such as this, we can often avoid that step as follows:

$$A(10) = 100e^{10r} = 100(e^{5r})^2 = 100(1.3498)^2 = 182.20.$$

Exponential functions

EXAMPLE 5 Let $f(x) = 100 - Ae^{-kx}$.
(a) Find $f(5)$ if $A = 50$ and $k = 0.2$.
(b) Find $f(10)$ if $f(0) = 40$ and $f(5) = 70$.

Solution

(a) If $f(x) = 100 - Ae^{-kx}$, $A = 50$, and $k = 0.2$, then

$$f(5) = 100 - 50e^{-0.2(5)}$$
$$= 100 - 50e^{-0.1}$$
$$= 100 - 50(0.9048)$$
$$= 54.76.$$

(b) If $f(x) = 100 - Ae^{-kx}$ and $f(0) = 40$, then

$$40 = 100 - A \quad \text{and} \quad A = 60.$$

Now if $f(x) = 100 - 60e^{-kx}$ and $f(5) = 70$, then

$$70 = 100 - 60e^{-5k},$$
$$60e^{-5k} = 30,$$
$$e^{-5k} = 0.5,$$
$$(e^{-k})^5 = 0.5,$$
$$e^{-k} = (0.5)^{1/5}.$$

Therefore,

$$f(10) = 100 - 60(e^{-k})^{10}$$
$$= 100 - 60(0.5)^2$$
$$= 100 - 60(0.25)$$
$$= 85.$$

EXERCISES

Set A—Exponential and Logarithmic Functions

In each of Exercises 1 through 18 find the number represented by the variable which makes the equality a true statement.

1. $\log_{10} 1000 = y$
2. $\log_2 8 = y$
3. $\log_{10} x = -3$
4. $\log_{10} x = \frac{1}{2}$
5. $\log_b 16 = 4$
6. $\text{lob}_b 64 = 3$
7. $\log_{10} 10^r = 7$
8. $\log_9 3^r = 2.5$
9. $\log_{10} 0.01 = x$
10. $\log_{10} 0.0001 = t$
11. $\log_b 25 = 2$
12. $\log_b \frac{1}{8} = 3$
13. $\log_b 0.0625 = -4$
14. $\ln e^2 = x$
15. $\ln x = 3$
16. $\ln x = -1$
17. $\ln x^3 = 6$
18. $\ln x^{1/2} = 2$
19. If $f(x) = ce^{kx}$, find $f(4)$ if $c = 200$ and $k = 0.08$.
20. If $f(x) = ce^{kx}$, find $f(10)$ if $c = 500$ and $k = 0.10$.
21. If $f(x) = ce^{kx}$, find c if $k = 0.06$ and $f(5) = 404.96$.
22. If $f(x) = ce^{kx}$, find c if $k = 0.08$ and $f(2) = 528.08$.
23. If $f(x) = ce^{kx}$, find k if $c = 100$ and $f(2) = 119.72$.
24. If $f(x) = ce^{kx}$, find k if $c = 150$ and $f(8) = 223.77$.
25. If $A(x) = 800e^{rx}$ and $A(3) = 1079.92$, find r.
26. If $A(x) = 1200e^{rx}$ and $A(2) = 1422.37$, find r.
27. If $f(x) = 200 - Ae^{-kx}$, $A = 120$ and $k = 0.05$, find $f(4)$.
28. If $f(x) = 300 - Ae^{-kx}$, $A = 200$, and $k = 0.08$, find $f(6)$.
29. If $f(x) = 400 - Ae^{-kx}$, $f(0) = 150$, and $f(2) = 250$, find $f(6)$.
30. If $f(x) = 300 - Ae^{-kx}$, $f(0) = 100$, and $f(4) = 150$, find $f(2)$.

4.4 DIFFERENTIATION OF THE EXPONENTIAL AND LOGARITHMIC FUNCTIONS

Let us turn our attention to the task of finding the derivative of $f(x) = e^x$ and the derivative of its inverse function $g(x) = \ell n\, x$. Recall that the derivative is defined as follows:

$$f'(x) = \lim_{\Delta x \to 0} \frac{f(x + \Delta x) - f(x)}{\Delta x}.$$

Hence, for $f(x) = e^x$,

$$f'(x) = \lim_{\Delta x \to 0} \frac{e^{x+\Delta x} - e^x}{\Delta x}$$

$$= \lim_{\Delta x \to 0} \frac{(e^x)(e^{\Delta x}) - e^x}{\Delta x}$$

$$= \lim_{\Delta x \to 0} \frac{e^x(e^{\Delta x} - 1)}{\Delta x}$$

$$= e^x \left(\lim_{\Delta x \to 0} \frac{e^{\Delta x} - 1}{\Delta x} \right).$$

Therefore, the derivative of $f(x) = e^x$ is e^x times some limit, if it exists. To pursue this matter, let us first find $e^{\Delta x}$ for values of Δx near zero. On a calculator we find that

$$e^{0.01} = 1.0100502,$$
$$e^{0.001} = 1.0010005,$$
$$e^{-0.001} = 0.9990005.$$

Therefore, the approximate value of

$$\frac{e^{\Delta x} - 1}{\Delta x}$$

at $\Delta x = 0.01$ is   1.005,
at $\Delta x = 0.001$ is   1.0005,
at $\Delta x = -0.001$ is   0.9995.

It seems reasonable (and it is a fact) that

$$\lim_{\Delta x \to 0} \frac{e^{\Delta x} - 1}{\Delta x} = 1.$$

Consequently,

> if $f(x) = e^x$ then $f'(x) = e^x$.

Geometrically, the slope of the line tangent to the graph of $f(x) = e^x$ at any point in its domain is the function value at that point. Except for the function $f(x) = 0$ for all x, $f(x) = ke^x$ is the *only* function such that $f'(x) = f(x)$—a remarkable fact which is also of importance in applications.

Exponential function

EXAMPLE 1 Find the derivative of $F(x) = e^{4x}$.

First Solution: Let $f(x) = e^x$ and $g(x) = 4x$. Then $F(x) = f(g(x))$. Since $f'(x) = e^x$ and $g'(x) = 4$,

$$F'(x) = f'(g(x))g'(x) = e^{4x}(4) = 4e^{4x}.$$

Second Solution: Let $y = e^u$ and $u = 4x$. Then

$$\frac{dy}{dx} = \frac{dy}{du}\frac{du}{dx}$$
$$= e^u(4)$$
$$= 4e^{4x}.$$

Using the chain rule, we obtain the following general differentiation theorem for the exponential function with e as base.

> If $f(x) = e^{g(x)}$ then $f'(x) = e^{g(x)}g'(x)$.

Derivative of exponential function

EXAMPLE 2 Find the derivative of $f(x) = e^{x^2}$.

Solution

$$f'(x) = e^{x^2}(2x) = 2xe^{x^2}.$$

4.4 DIFFERENTIATION OF THE EXPONENTIAL AND LOGARITHMIC FUNCTIONS

Derivative of exponential function

EXAMPLE 3 Find the derivative of $y = e^{5x} + xe^x$.

Solution

$$\frac{dy}{dx} = e^{5x}(5) + xe^x + e^x = 5e^{5x} + xe^x + e^x.$$

PRACTICE PROBLEM Find the derivative of $f(x) = x^2 e^{3x}$.

Answer: $f'(x) = 3x^2 e^{3x} + 2xe^{3x} = xe^{3x}(3x + 2)$. ∎

Geometric problem

EXAMPLE 4

(a) For the function $f(x) = xe^{-x} + 2$, find an equation of the line tangent to the graph of f at $x = 0$.

(b) For what x is the function increasing? For what x is the function decreasing?

(c) For what x is the graph concave upward? For what x is the graph concave downward?

(d) What is the inflection point on the graph of f?

Solution

(a) $f'(x) = xe^{-x}(-1) + e^{-x} = e^{-x}(1 - x)$. Since $f(0) = 2$ and since $f'(0) = 1$, the tangent contains the point $(0, 2)$ and has slope 1. Its equation is $y = x + 2$.

(b) Since $f'(x) = e^{-x}(1 - x) < 0$ for $x > 1$, f decreases on the set where $x \geq 1$. Since $f'(x) = e^{-x}(1 - x) > 0$ for $x < 1$, f increases on the set where $x \leq 1$. (Note: e^{-x} is *positive* for all x.)

(c) $f''(x) = e^{-x}(-1) - e^{-x}(1 - x) = e^{-x}(x - 2)$. Since $f''(x) > 0$ for $x > 2$, the graph of f is concave upward for $x > 2$. Since $f''(x) < 0$ for $x < 2$, the graph of f is concave downward for $x < 2$.

(d) From part (c) we see that there is an inflection point at $x = 2$. Since $f(2) = 2/e^2 + 2$, it is $(2, 2/e^2 + 2)$.

Now that we have found the derivative of $f(x) = e^x$ it is quite easy to find the derivative of $g(x) = \ln x$, using the theorem discussed in Section 4.2. Since

$$g'(x) = \frac{1}{f'(g(x))}$$

for inverse functions f and g, it follows that

$$g'(x) = \frac{1}{e^{\ln x}}.$$

Recall that $e^{\ln x} = x$. Hence

> if $g(x) = \ln x$, then $g'(x) = \dfrac{1}{x}$.

Observe that the natural logarithm function has the interesting property that the slope of the line tangent to the graph at any number in its domain is the reciprocal of that number.

Logarithmic function

EXAMPLE 5 Find the first and second derivatives of $f(x) = x^3(\ln x)$.

Solution: The first derivative is

$$f'(x) = x^3\left(\frac{1}{x}\right) + 3x^2(\ln x)$$
$$= x^2 + 3x^2(\ln x).$$

The second derivative is

$$f''(x) = 2x + (3x^2)\left(\frac{1}{x}\right) + 6x(\ln x)$$
$$= 2x + 3x + 6x(\ln x)$$
$$= 5x + 6x(\ln x).$$

PRACTICE PROBLEM Find the first and second derivatives of $f(x) = x(\ln x)$.

Answer: $f'(x) = 1 + \ln x$ and $f''(x) = 1/x$. ∎

To find the derivative of $F(x) = \ln g(x)$, let $f(x) = \ln x$. Then $F(x) = f(g(x))$ and, using the chain rule, we obtain the follow-

4.4 DIFFERENTIATION OF THE EXPONENTIAL AND LOGARITHMIC FUNCTIONS

ing general differentiation theorem for the natural logarithm function.

$$\text{If} \quad F(x) = \ln g(x) \quad \text{then} \quad F'(x) = \frac{g'(x)}{g(x)}.$$

Derivative of logarithmic function

EXAMPLE 6 Find the derivative of $f(x) = \ln(x^3 + 1)$.

Solution

$$f'(x) = \frac{3x^2}{x^3 + 1}$$

PRACTICE PROBLEM Find the derivative of $f(x) = \ln(x^2 + 3x)$.

Answer: $f'(x) = (2x + 3)/(x^2 + 3x)$. ∎

Note that if $g(x) = b^x$, $b > 0$ and $b \neq 1$, then

$$\ln g(x) = \ln b^x = (\ln b)x.$$

Differentiating each side, we get

$$\frac{g'(x)}{g(x)} = \ln b.$$

Therefore, $g'(x) = g(x)(\ln b) = b^x(\ln b)$. This gives us the following formula.

$$\text{If} \quad g(x) = b^x, \quad \text{then} \quad g'(x) = b^x(\ln b).$$

Choosing e as the base for the exponential and logarithmic functions has many advantages and is standard practice in mathematical analysis. It is always possible to change from base e to any other base by a simple formula which we now derive.

$$\text{If} \quad y = \log_b x, \quad \text{then} \quad b^y = x.$$

Taking the natural logarithm of each side of the second equation yields

$$\ln (b^y) = \ln x,$$
$$y(\ln b) = \ln x,$$
$$y = \frac{\ln x}{\ln b}.$$

Consequently,

$$\log_b x = \frac{\ln x}{\ln b}.$$

From the preceding formula, we obtain immediately that

$$\text{if } f(x) = \log_b x, \quad \text{then } f'(x) = \frac{1}{(\ln b)x}.$$

Derivative of logarithmic function

EXAMPLE 1 If $F(x) = \log_{10} (x^2 + 1)$, find $F'(x)$.

Solution: Let $g(x) = x^2 + 1$ and let $f(x) = \log_{10} x$. Then F is defined by $F(x) = f(g(x))$. Using the chain rule and the differentiation theorem for the logarithmic function with base 10, we get

$$F'(x) = \frac{1}{(\ln 10)(x^2 + 1)} (2x) = \frac{2x}{x^2 + 1} \log_{10} e.$$

EXERCISES

Set A–Derivatives of Exponential and Logarithmic Functions

In Exercises 1 through 20 find the first and second derivatives of each function.

1. $f(x) = e^{5x}$
2. $f(x) = \ln 3x$

3. $f(x) = e^{x^3}$
4. $f(x) = e^{-2x}$
5. $f(x) = xe^{-3x}$
6. $f(x) = x^2 e^x$
7. $f(x) = x[\ln(x+1)]$
8. $f(x) = \ln(e^x + 1)$
9. $f(x) = 3^x$
10. $f(x) = x^2 + e^{x/2} + 2^x$
11. $f(x) = \ln(x + e^x)$
12. $f(x) = \ln(x^2 + 2x)$
13. $f(x) = x(\ln x)$
14. $f(x) = \ln(\ln x)$
15. $f(x) = x - x^2(\ln x)$
16. $f(x) = e^x(\ln x)$
17. $f(x) = (\ln x)^2$
18. $f(x) = x^2 e^{1/x}$
19. $f(x) = \dfrac{x}{\ln x}$
20. $f(x) = \dfrac{\ln x}{x}$

Set B—Geometric Applications for the Exponential and Logarithmic Functions

21. For $f(x) = x(\ln x)$, where $x > 0$, determine maxima, minima, concavity, and inflection points—if they exist. Determine where the function is increasing and where it is decreasing. Sketch a graph of the function.

22. For $f(x) = x^2(\ln x)$, where $x > 0$, determine maxima, minima, concavity, and inflection points—if they exist. Determine where the function is increasing and where it is decreasing. Sketch a graph of the function.

23. For $f(x) = xe^x$, determine maxima, minima, concavity, and inflection points—if they exist. Determine where the function is increasing and where it is decreasing. Sketch a graph of the function.

24. For $f(x) = xe^{2x}$, determine maxima, minima, concavity, and inflection points—if they exist. Determine where the function is increasing and where it is decreasing. Sketch a graph of the function.

25. Find an equation of the line tangent to the graph of $f(x) = e^{3x} - 1$ at the point $x = 0$.

26. Find an equation of the line tangent to the graph of $f(x) = x^2(\ln x) + 3$ at $x = 1$.

4.5 EXPONENTIAL GROWTH AND CONTINUOUS INTEREST

Let us now look at some important applications of exponential functions. First consider the function f defined by

$$f(x) = ce^{kx},$$

where c and k are constants. Since $f'(x)$ is defined to be the *instantaneous rate of change of f at x* and since

$$f'(x) = ce^{kx}(k) = k(ce^{kx}) = kf(x),$$

we see that the rate of change of f at x is directly proportional to the value of f at x. Conversely, it will be shown in Section 5.3 that if f is a differentiable function (not identically zero) such that $f'(x) = kf(x)$, then $f(x) = ce^{kx}$, where c is an arbitrary constant Physical laws concerning growth and decay lead directly to exponential functions. For example, if $f(x)$ is the amount of radioactive substance present at time x, then the rate of decay (rate of change of f with respect to x) is directly proportional to the amount present at time x; that is, $f'(x) = kf(x)$. Also, if $g(x)$ is the number of bacteria in a culture at time x, then the rate of growth of the number is directly proportional to $g(x)$, the amount present at time x. Therefore, $g'(x) = kg(x)$ and $g(x) = ce^{kx}$. Note if $g(x) = ce^{kx}$, then $g(0) = c$, and c represents the amount at time $x = 0$, the initial amount.

Growth of bacteria

EXAMPLE 1 Suppose the number of bacteria in a culture increases at a rate proportional to the number present at time x. If 100,000 bacteria are present at a given time and 250,000 are present 1 hour later, how many bacteria are present in 3 hours?

Solution: Let $f(x)$ be the number of bacteria present at time x. Since $f'(x) = kf(x)$, it follows that $f(x) = ce^{kx}$, where c and k are constants to be determined. We let $f(0)$ be the initial amount present, that is, the amount present at time 0. Therefore, since $f(0) = c$, we conclude that $c = 100,000$. Consequently

$$f(x) = 100,000e^{kx}.$$

Since $f(1) = 100,000e^k = 250,000$, it follows that $e^k = 2.5$. Therefore, at any time x, we have

$$f(x) = 100,000(2.5)^x.$$

For $x = 3$, $f(3) = 100,000(2.5)^3 = 1,562,500$.

Radioactive decay

EXAMPLE 2 If the half-life of radium is approximately 1600 years, how long will it take three fourths of a given amount of radium to disintegrate? ("Half-life" means one-half of a given

4.5 EXPONENTIAL GROWTH AND CONTINUOUS INTEREST

amount will be gone after 1600 years. It is not hard to show that half-life is independent of the original amount present.)

Solution: Let $g(x)$ be the amount present at time x and assume $g'(x) = kg(x)$. If we let $g(0) = 1$ (that is, we let the initial amount be unity, for example, 1 gram), then we need to find x such that $g(x) = \frac{1}{4}$. If $g'(x) = kg(x)$, then $g(x) = ce^{kx}$ and $g(0) = c$. Thus, $c = 1$, and the general formula is

$$g(x) = e^{kx}.$$

Therefore, $g(1600) = e^{1600k} = \frac{1}{2}$. We can solve for k by taking the natural logarithm of each side of $e^{1600k} = \frac{1}{2}$. For this problem, it is probably easier to note that $e^{1600k} = (e^k)^{1600}$ and obtain

$$e^k = (0.5)^{1/1600}$$

and

$$g(x) = (0.5)^{x/1600}.$$

To find how many years are required for the amount present to be one fourth of the original quantity, we solve the equation

$$(0.5)^{x/1600} = \frac{1}{4}.$$

First we find that

$$(0.5)^x = (0.25)^{1600}.$$

Now using logarithms (or a calculator), we find that

$$x(\log_{10} 0.5) = 1600(\log_{10} 0.25)$$

$$x = \frac{(1600)(-0.60206)}{-0.30103}$$

$$= 3200 \text{ years.}$$

---

continuous compounding

In Chapter 8 we will show that if a principal P is deposited for t years at an annual interest rate r with **continuous compounding,** then the amount A on deposit at time t is given by

$$A(t) = Pe^{rt}.$$

Note that
$$A'(t) = P(e^{rt})r = rA(t).$$
The rate of change of the amount at time t is directly proportional to the amount on deposit at time t where r, the annual rate, is the constant of proportionality. (Occasionally, continuous interest is defined this way. Then the formula $A(t) = Pe^{rt}$ is "recovered" as the solution to the derivative (differential) equation
$$A'(t) = rA(t).$$
A discussion of differential equations is given in Section 5.7.)

Continuous compounding

EXAMPLE 3

(a) Find the amount resulting from a principal of $3000 deposited for 4 years at a 7.5% annual interest rate with continuous compounding.

(b) How long would it take to double the amount?

Solution

(a) Since $A(t) = Pe^{rt}$, it follows that
$$A(4) = 3000e^{(0.075)(4)}$$
$$= 3000e^{0.3}$$
$$= 3000(1.3499)$$
$$= 4049.70.$$

The amount would be $4049.70. *Note:* The interest earned is $1049.70.

(b) We want to find t such that $A(t) = 6000$. It is the solution to the following equation.
$$6000 = 3000e^{0.075t},$$
$$e^{0.075t} = 2,$$
$$0.075t = \ln 2$$
$$0.075t = 0.693\ 15,$$
$$t \doteq 9.241\ 96.$$

Since 0.242 of a year is approximately 2.9 months, the amount would be doubled in approximately 9 years and 3 months. (For an extensive discussion of continuous compounding of interest and other types of interest problems, see Chapter 8.)

4.5 EXPONENTIAL GROWTH AND CONTINUOUS INTEREST

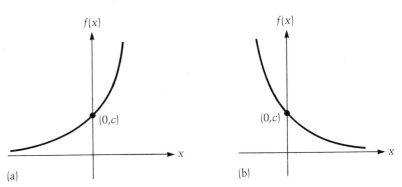

FIG. 4.5. Unlimited growth/decay. (a) $f(x) = ce^{kx}$, $c > 0$ and $k > 0$. (b) $f(x) = ce^{kx}$, $c > 0$ and $k < 0$.

unlimited growth/decay function

We have already discussed the function $f(x) = ce^{kx}$ called the **unlimited growth/decay function.** Graphs of this function are given in Figure 4.5.

Consider the function defined by

$$f(x) = a - Ae^{-kx}, \quad x \geq 0,$$

limited growth function

where $A > 0$ and $k > 0$. This is called the **limited growth function.** (See Figure 4.6.) The first and second derivatives of the limited growth function are as follows:

$$f'(x) = Ake^{-kx}, \tag{1}$$

$$f''(x) = -Ak^2 e^{-kx}. \tag{2}$$

Equation (1) tells us that f is an increasing function and Eq. (2) tells us that the *rate of increase* is *decreasing*. In other words, the graph of $f(x)$ always goes up and is concave downward. Since

$$f(x) = a - Ae^{-kx} = a - \frac{A}{e^{kx}},$$

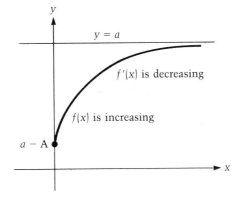

FIG. 4.6. Limited growth equation: $f(x) = a - Ae^{-kx}$.

and since A/e^{kx} is positive but approaches 0 as x increases, the range of the limited growth function approaches a but does not attain it. Note further that

$$f'(x) = k(Ae^{-kx}) = k[a - f(x)].$$

This states that the rate of change of the function at x is directly proportional to the difference between (its upper limit) a and $f(x)$. Psychologists have found that this function describes many learning situations; it is sometimes called a *learning curve*. Sociologists have also found that it describes many "diffusion processes," such as the spread of information or propaganda; hence it is sometimes called a *diffusion curve*.

Draft announcement

EXAMPLE 4 Suppose the government decides to announce that it will reinstitute the draft and suppose it is assumed that the fractional part of the population which will have heard the announcement in x days is given by

$$f(x) = 1 - Ae^{-kx}.$$

If 60% of the population hears the announcement in 10 days, approximately how long will it take for 90% to hear the announcement?

Solution: Since $f(0) = 0$, it follows that $0 = 1 - A$ and $A = 1$. Since $f(10) = 0.60$,

$$0.60 = 1 - e^{-10k}$$
$$e^{-10k} = 0.40,$$
$$e^{-k} = (0.40)^{1/10}.$$

Now we need to solve $f(x) = 0.90$ for x:

$$0.90 = 1 - (e^{-k})^x,$$
$$(e^{-k})^x = 0.10,$$
$$(0.40)^{x/10} = 0.10.$$

Taking the natural logarithm of each side of the preceding equation yields

4.5 EXPONENTIAL GROWTH AND CONTINUOUS INTEREST

$$\frac{x}{10}(\ln 0.4) = \ln 0.1$$

$$x = \frac{10(\ln 0.1)}{\ln 0.4}$$

$$\doteq 25.13 \text{ days}.$$

Another very important growth equation involving the exponential function arises from the following situation. Suppose that among a population of fixed size a, $f(0)$ individuals are infected with a virus at the present time. Let $f(x)$ be the number infected x days from now and assume that the rate of change, $f'(x)$, is proportional to the product of the number infected, $f(x)$, and the number not infected, $a - f(x)$. That is, assume that

$$f'(x) = kf(x)[a - f(x)].$$

logistic equation

We can show that the following function, called the **logistic equation**, satisfies the preceding conditions. It is

$$f(x) = \frac{a}{1 + ce^{-akx}}, \quad c > 0.$$

(In Section 5.7 we shall show how this function can be *derived* from the conditions. We leave for the exercises the verification that it does satisfy the conditions.) The graph of the logistic equation is given in Figure 4.7. (Verifying the features of its graph will

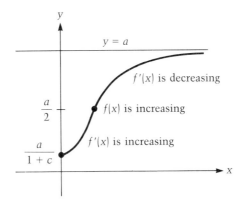

FIG. 4.7.
Logistic equation:
$f(x) = \dfrac{a}{1 + ce^{-akx}}$.

be left for the exercises.) The logistic equation is also important in problems containing a predetermined upper bound on growth.

Logistic equation

EXAMPLE 5 Suppose that 200 animals of a given species are placed in a game preserve and assume that their number will increase according to the logistic equation. If their number increases to 400 in 3 years and if it is assumed that the preserve cannot support more than 1000 such animals, how long will it take for the number on the preserve to increase to 800?

Solution: In the logistic equation $a = 1000$, $f(0) = 200$, $f(3) = 400$, and we need to find x such that $f(x) = 800$. From

$$f(x) = \frac{a}{1 + ce^{-akx}},$$

$a = 1000$, and $f(0) = 200$, we get

$$200 = \frac{1000}{1 + c},$$
$$1 + c = 5,$$
$$c = 4.$$

Now since $f(3) = 400$, we have

$$400 = \frac{1000}{1 + 4e^{-3000k}},$$
$$1 + 4e^{-3000k} = 2.5,$$
$$4e^{-3000k} = 1.5,$$
$$e^{-3000k} = 0.375.$$

Taking the natural logarithm of each side of the preceding equation yields

$$-3000k = \ln 0.375,$$
$$k = \frac{-0.980\,829}{-3000}$$
$$= 0.000\,326\,9.$$

Therefore, the general formula is

$$f(x) = \frac{1000}{1 + 4e^{-0.3269x}}.$$

Now we proceed to find x such that $f(x) = 800$.

$$800 = \frac{1000}{1 + 4e^{-0.3269x}}.$$

$$8 + 32e^{-0.3269x} = 10,$$

$$e^{-0.3269x} = \frac{1}{16},$$

$$-0.3269x = \ell n \, 0.0625,$$

$$x = \frac{-2.772\,589}{-0.3269}$$

$$\doteq 8.48 \text{ years.}$$

EXERCISES

Set A–Applications of Exponential Functions

Spread of disease

1. An epidemiologist determines that the percentage of college students at a given college who will be infected x days from now with a flu virus is given by

$$f(x) = \frac{100}{1 + 9e^{-kx}}.$$

(a) What percentage of the students are infected with the disease at the present time?
(b) If 50% of the students become infected in 20 days, what is the value of k?
(c) How long will it take for 75% of the students to be infected?
(d) Show that the rate of change of the spread of the epidemic is proportional to the product of the percentage infected and the percentage not infected.

Doubling a deposit at continuous interest

2. For a given amount deposited at an annual interest rate of $r\%$ with continuous compounding, prove that it takes approximately
$$\frac{0.693}{r}$$
years to double the amount.

USA growth in population

3. (a) In 1940 the population of the USA was 132,165,000 and in 1950 it was 151,326,000. The growth in population is assumed to follow the unlimited growth function $f(x) = ce^{kx}$. Determine the given function.
   (b) Use part (a) to approximate the 1980 population of the USA.
   (c) Use the fact that the 1980 population of the USA was 226,505,000 to show that the error in the approximation found in part (b) is less than 0.3%.
   (d) Use your function to estimate the population in the USA in the year 2000.

Continuous compounding

4. Find the amount resulting from a principal of $6000 deposited for four years at 8% compounded continuously.

Continuous compounding

5. Find the amount resulting from a principal of $10,000 deposited for 30 months at 12% compounded continuously.

Continuous compounding

6. How much would an individual need to deposit at 6% compounded continuously in order to have $10,000 on deposit 4 years from now?

Continuous compounding

7. How much would an individual need to deposit at 10% compounded continuously in order to have $5000 on deposit 6 years from now?

Continuous compounding

8. Suppose the interest in 1 year on $3000 earning continuous interest is $233.65. What is the annual interest rate?

Continuous compounding

9. Suppose the interest accrued in 2 years on $4000 earning continuous interest is $717.57. What is the annual interest rate?

Growth of bacteria

10. If the number of bacteria in a certain culture at a given time is 1000 and if the number present at the end of 5 hours is 10,000, what is the number of bacteria present at the end of 1 hour? (Assume unlimited exponential growth.)

Growth of bacteria

11. If the number of bacteria in a certain culture at a given time is 1000 and if the number present at the end of 5 hours is 5000, what is the number of bacteria present at the end of 30 minutes?

Radioactive decay 12. What percentage of a given amount of radium disintegrates in 100 years? (See Example 2.)

Radioactive decay 13. How long will it take for a given amount of radium to disintegrate to 75% of the initial amount? (See Example 2.)

Diffusion curve 14. (a) In Example 4, how many days would it take for 80% of the population to hear the announcement?
(b) In Example 4, how many days would it take for 95% of the population to hear the announcement?

Learning curve 15. (a) A psychologist discovers that an average person can learn a maximum number of 126 meaningless words. If it takes the average person 3 hours to learn $\frac{1}{3}$ this number of words, how long will it take the person to learn $\frac{2}{3}$ this number? Assume that the learning curve is given by $f(x) = a - Ae^{-kx}$, where $a = 126$.
(b) How long will it take for the person to learn 120 such words?

Logistic equation 16. In Example 5, how long will it take the given species of animals on the preserve to increase to 900?

Logistic equation 17. For the logistic function (equation)

$$f(x) = \frac{a}{1 + ce^{-akx}},$$

prove that $f'(x) = kf(x)[a - f(x)]$.

Logistic equation 18. (a) For the logistic function, find $f(0)$.
(b) Prove for the logistic function that $f(x) < a$ for all $x > 0$.
(c) Prove that the logistic function is increasing for $x > 0$.

Logistic equation 19. (a) Prove for the logistic function that the rate of change $f'(x)$ is increasing when $f(x) < a/2$.
(b) Prove for the logistic function that the rate of change $f'(x)$ is decreasing when $f(x) > a/2$.
(c) For the logistic function, for what x is $f'(x)$ a maximum?

4.6 IMPLICIT AND LOGARITHMIC DIFFERENTIATION (OPTIONAL)

The graph of the equation $x^2 + y^2 = 25$ is a circle and is not the graph of a function since, for example, $(3, 4)$ and $(3, -4)$ are two points on the graph with the same first coordinates. (See Figure

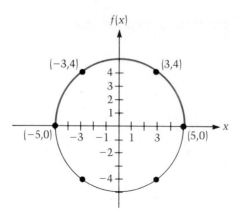

FIG. 4.8. $f(x) = \sqrt{25 - x^2}$.

implicit differentiation

4.8.) However, with appropriate restrictions on the range values the equation will define a function, and thus the techniques of calculus can be used to study graphs of such relations. It should be geometrically clear that many different "pieces" of the graph of the circle could be selected which would represent functions. If we restrict y to the set of nonnegative real numbers, for example, then the set of ordered pairs (x, y) satisfying $x^2 + y^2 = 25$ is a function. Since $y^2 = 25 - x^2$ and since $y \geq 0$, the graph of this function is the semicircle defined by $y = \sqrt{25 - x^2}$, or $f(x) = \sqrt{25 - x^2}$. The equation $x^2 + y^2 = 25$ with $y \geq 0$ is said to define the function **implicitly**, whereas $f(x) = \sqrt{25 - x^2}$ with $-5 \leq x \leq 5$ is said to define the function **explicitly**. If y were restricted to less than or equal to 0 and x to the interval $0 \leq x \leq 5$, the graph would be the part of the circle in the fourth quadrant defined by $y = -\sqrt{25 - x^2}$, $0 \leq x \leq 5$. Let us now discuss a method for differentiating a function defined implicitly; the process is called **implicit differentiation**.

Assume some restriction is made (such as $y \geq 0$) so that the equation $x^2 + y^2 = 25$ defines a function f. Then for each x there is a unique $f(x)$ such that

$$x^2 + [f(x)]^2 = 25.$$

Differentiating each side of the equation and using the chain rule to find the derivative of $[f(x)]^2$ we get

$$2x + 2f(x)f'(x) = 0.$$

Assuming $f(x) \neq 0$ and solving the preceding equation for $f'(x)$, we obtain

4.6 IMPLICIT AND LOGARITHMIC DIFFERENTIATION (OPTIONAL)

$$f'(x) = \frac{-x}{f(x)}.$$

If $f(x) > 0$, then

$$f(x) = \sqrt{25 - x^2} \quad \text{and} \quad f'(x) = \frac{-x}{\sqrt{25 - x^2}}.$$

This result is the same as would have been obtained by differentiating $f(x) = \sqrt{25 - x^2}$ directly (explicitly).

Implicit differentiation

EXAMPLE 1 Assume $y^2 + 2xy + x^3 = 25$ defines a function implicitly. For each x in the domain, let $f(x)$ be the unique range value corresponding to x. Then for each x

$$[f(x)]^2 + 2xf(x) + x^3 = 25.$$

Differentiating each side of the equation, we get

$$2f(x)f'(x) + 2xf'(x) + 2f(x) + 3x^2 = 0.$$

Solving the preceding equation for $f'(x)$, we find that

$$f'(x) = \frac{-3x^2 - 2f(x)}{2(x + f(x))}.$$

PRACTICE PROBLEM Assume $y^3 + 3xy + x^4 = 2$ defines a function implicitly. Use implicit differentiation to find $f'(x)$ in terms of x and $f(x)$.

Answer

$$f'(x) = -\frac{4x^3 + 3f(x)}{3[f(x)]^2 + 3x}. \blacksquare$$

As seen in the preceding examples, implicit differentiation usually gives $f'(x)$ in terms of x and $f(x)$ instead of just x. This should be expected since if there are two points with the same x-coordinate satisfying the original equation, then the y-coordinate is needed to identify the point at which the derivative is to be found.

Tangent to curve

EXAMPLE 2 Find the slope of the line tangent to the curve

$$x^3 - 2x^2y^3 + 3y = -x - 8$$

at the point $(1, 2)$.

Solution: If $f(x)$ is the number in the range paired with x in the domain of a function f defined implicitly by the equation, then

$$x^3 - 2x^2[f(x)]^3 + 3f(x) = -x - 8.$$

Differentiating each side of the preceding equation, we obtain

$$3x^2 - 2x^2(3)[f(x)]^2 f'(x) - 4x[f(x)]^3 + 3f'(x) = -1.$$

Simplifying and solving for $f'(x)$ yields

$$3f'(x) - 6x^2[(f(x)]^2 f'(x) = 4x[f(x)]^3 - 3x^2 - 1,$$

$$f'(x) = \frac{4x[f(x)]^3 - 3x^2 - 1}{3 - 6x^2[f(x)]^2}.$$

Since $f(1) = 2$, we conclude that

$$f'(1) = \frac{4(2)^3 - 3 - 1}{3 - 6(4)} = -\frac{28}{21} = -\frac{4}{3}.$$

In each of the preceding examples we could have replaced $f'(x)$ by dy/dx, and this is often done. We could have also replaced $f(x)$ by y and $f'(x)$ by $y'(x)$ or by just y'. The latter notation y' has the advantage of being the least cumbersome and it is often used for this reason. One disadvantage of the y' notation is that the independent variable is not indicated.

Implicit differentiation

EXAMPLE 3 If $x^3 + 3xy + y^2 = 7$ defines a function f implicitly, find y''.

Solution: Differentiating each side of the equation, we get

$$3x^2 + 3xy' + 3y + 2yy' = 0. \tag{1}$$

Differentiating each side of the preceding equation yields

$$6x + 3xy'' + 3y' + 3y' + 2yy'' + 2(y')^2 = 0.$$

4.6 IMPLICIT AND LOGARITHMIC DIFFERENTIATION (OPTIONAL)

Solving the preceding equation for y'', we find that

$$y'' = -\frac{6x + 6y' + 2(y')^2}{3x + 2y}. \tag{2}$$

Note: The second derivative could be expressed in terms of x and y by solving Eq. (1) for y' and then substituting the result in Eq. (2).

PRACTICE PROBLEM (a) Use implicit differentiation to find y' in terms of x and y for the function defined implicitly by $x^2 + 2xy + y = 5$. (b) Find y' after solving the equation for y in terms of x. (c) Verify that the same solution is obtained in parts (a) and (b).

Answer: (a) $y' = -(2x + 2y)/(2x + 1)$. (b) $y = (5 - x^2)/(2x + 1)$; $y' = -(2x^2 + 2x + 10)/(2x + 1)^2$. ∎

We should emphasize that one definite advantage of implicit differentiation is that it can be used when explicit differentiation may be very difficult or even impossible. For example, solving

$$x^3 - 2x^2y^3 + 3y = -x - 8$$

for y and then differentiating explicitly would be a problem of considerable difficulty. We found in Example 2 that implicit differentiation for this equation was quite straightforward.

Slope of tangent

EXAMPLE 4

(a) Use implicit differentiation to find $y'(1)$ at the point $(1, 2)$ on the curve $xy^2 - 2xy + x = 1$.

(b) Solve the equation in part (a) for y, differentiate (explicitly), and verify your answer.

Solution

(a) Differentiating $xy^2 - 2xy + x = 1$ implicitly, we obtain

$$x(2yy') + y^2 - 2xy' - 2y + 1 = 0,$$
$$(2xy - 2x)y' = -(y^2 - 2y + 1),$$
$$y' = \frac{y^2 - 2y + 1}{2x - 2xy}.$$

Since $y = 2$ where $x = 1$,

$$y'(1) = \frac{4 - 4 + 1}{2 - 4} = -\frac{1}{2}.$$

(b) Using the quadratic formula to solve

$$xy^2 - 2xy + (x - 1) = 0$$

for y, we obtain

$$y = \frac{2x \pm \sqrt{4x^2 - 4x(x - 1)}}{2x}$$

$$= \frac{2x \pm \sqrt{4x}}{2x}$$

$$= 1 \pm x^{-1/2}.$$

We know which square root to take because we know that $y = 2$ at $x = 1$. Since $1 + x^{-1/2} = 2$ at $x = 1$, we take the positive square root and conclude

$$y = 1 + x^{-1/2}.$$

Differentiating, we obtain

$$y' = -\frac{1}{2}x^{-3/2}.$$

Consequently, $y'(1) = -\frac{1}{2}$, the solution obtained in part (a).

Although we have formulas for the derivatives of $f(x) = x^r$ where r is a real number, $g(x) = e^x$, and $F(x) = 2^x$, we have not found the derivatives of such functions as $g(x) = x^x$ for $x > 0$. One technique to find the derivative of such a function is called **logarithmic differentiation**; it is done as follows.

logarithmic differentiation

To find the derivative of $g(x) = x^x$, we first *take the natural logarithm of each side of the equation* and obtain

$$\ell n\, g(x) = \ell n\, x^x$$

$$\ell n\, g(x) = x(\ell n\, x).$$

Then, differentiating each side of the preceding equation yields

4.6 IMPLICIT AND LOGARITHMIC DIFFERENTIATION (OPTIONAL)

$$\frac{g'(x)}{g(x)} = x\left(\frac{1}{x}\right) + \ell n\ x$$

$$g'(x) = g(x)[1 + \ell n\ x]$$

$$g'(x) = x^x[1 + \ell n\ x]$$

Boat problem

EXAMPLE 5 Suppose the deck of a boat is 30 feet above the harbor bed and suppose it is anchored with 90 feet of rope out, which is kept taut and straight by the current. If the rope is hauled in at 5 ft/min, how fast is the boat moving through the water when there is 50 feet of rope out? Find the acceleration of the boat at this time.

First Solution (See Figure 4.9): We are given that $s'(t) = -5$ ft/min and we wish to find $y'(t)$ where $s(t) = 50$. (*Note:* Since $s(t)$ is decreasing, $s'(t)$ is negative.) By the Pythagorean theorem,

$$[s(t)]^2 = [y(t)]^2 + 900.$$

Using *implicit differentiation*, we obtain

$$2s(t)s'(t) = 2y(t)y'(t).$$

Therefore,

$$y'(t) = \frac{s(t)s'(t)}{y(t)}. \tag{1}$$

If $s(t) = 50$, then by the Pythagorean theorem, $y(t) = 40$. Hence

$$y'(t) = \frac{50(-5)}{40} = -\frac{25}{4},$$

and $y'(t) = -6.25$ ft/min. The *speed* (absolute value of velocity) of the boat through the water with 50 feet of rope out is 6.25 ft/min.

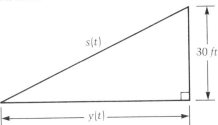

FIG. 4.9.
$y(t) = \sqrt{s^2(t) - 900}.$

Since $s'(t) = -5$ ft/min at any time, from Eq. (1) we have

$$y'(t) = \frac{-5s(t)}{y(t)}.$$

Using the quotient theorem for differentiation, we obtain the second derivative. It is

$$y''(t) = \frac{-5[y(t)s'(t) - s(t)y'(t)]}{[y(t)]^2}.$$

Since the second derivative is the acceleration of the boat, using the fact that if $s(t) = 50$, then $y(t) = 40$ and $y'(t) = -\frac{25}{4}$, we get

$$y''(t) = \frac{-5[(40)(-5) - (50)(-\frac{25}{4})]}{(40)^2}$$

$$= -\frac{45}{128} \doteq -0.3516.$$

Since $y''(t)$ is negative, $y'(t)$ is a decreasing function. But since $y'(t)$ is negative, this means that $y'(t)$ is becoming "more negative." Therefore, the speed of the boat increases; the acceleration is approximately 0.3516 ft/min.^{2}.

Second Solution: We could have first found a formula for $s(t)$. If $s(0) = 90$, then at time t, $s(t) = 90 - 5t$. Now $s(t) = 50$ when $90 - 5t = 50$, or $t = 8$. Since

$$y(t) = \sqrt{s^2(t) - 900}$$
$$= \sqrt{(90 - 5t)^2 - 900}$$
$$= 5(t^2 - 36t + 288)^{1/2},$$

we find that

$$y'(t) = \frac{5(t - 18)}{(t^2 - 36t + 288)^{1/2}}.$$

Therefore,

$$y'(8) = \frac{5(-10)}{(64)^{1/2}} = -\frac{25}{4}.$$

EXERCISES

Set A—Implicit Differentiation

Assume that y' and y'' exist for the functions defined implicitly by the equations in Exercises 1 through 12. (a) Find an equation defining y' in terms of x and y. (b) Find an equation defining y'' in terms of x, y, and y'.

1. $y^2 = 4x$
2. $x^2 + 3xy + y = 11$
3. $x^3 - y^2 = 7$
4. $xy^3 + x = 5$
5. $xy + y^2 = 4 - x$
6. $x^2 y^2 = 9$
7. $2xy + y^2 = 10$
8. $y^2 - 3xy = 7x$
9. $x^3 + y^3 = 1$
10. $x + y^3 = 8$
11. $xy + 3x + y = 11$
12. $5xy - 2x + 3y = 17$

Assume y' exists for the functions defined implicitly by the equations in Exercises 13 through 16. Find an equation defining y' in terms of x and y.

13. $x^2 e^y + \ln(x^2 + y) = 10$
14. $ye^y + \ln(x + 3y) = x^2$
15. $e^{xy} + x(\ln y) = x^3 + 7$
16. $xe^{xy} + y(\ln x) = 2x + 5$

Set B—Logarithmic Differentiation

Use logarithmic differentiation to find $f'(x)$ for each function in Exercises 17 through 24.

17. $f(x) = x(2x + 1)^{3/2}$
18. $f(x) = x(4x + 3)^{1/2}$
19. $f(x) = \dfrac{3x + 7}{5x - 1}$
20. $f(x) = \dfrac{7x - 5}{4x + 9}$
21. $f(x) = x^2(5x - 1)^{7/5}$
22. $f(x) = x^3(x^2 + 1)^{3/2}$
23. $f(x) = (\ln x)^x$
24. $f(x) = x^{\ln x}$

Set C—Miscellaneous Exercises

Graphing   25. Assume that the equation $x^2 + 2xy + y^2 = 4$ defines a function f. Find $f'(x)$ for each x in the domain. Graph the equation.

Tangent line   26. For the curve $2x^2 - xy + y^2 - 6x + 5y - 8 = 0$, find an equation of the tangent line at the point $(1, 2)$.

Tangent line   27. Find an equation of the tangent to $x^2 + xy - y^2 + 2x = 3$ at the point $(1, 1)$.

Tangent line   28. Find an equation of the tangent to $y^3 + xy^2 - 3x + 3 = 0$ at the point $(2, 1)$.

Second derivative   29. Evaluate y'' at $(1, 1)$ on the curve $x^2 + xy - y^2 + 2x = 3$. [*Hint:* See Exercise 27.]

4.7 SUMMARY AND REVIEW EXERCISES

1. EXPONENTIAL AND LOGARITHMIC FUNCTIONS

(a) Let $b > 0$ and $b \neq 1$. The function $f(x) = b^x$ is called an **exponential function** with base b.

(b) The inverse of the exponential function with base b is the **logarithmic function** with base b, denoted by $\log_b x$.

(c) Since $f(g(x)) = g(f(x)) = x$ for inverse functions,
$$b^{\log_b x} = x \quad \text{and} \quad \log_b b^x = x.$$

(d) The basic **logarithm properties** are as follows:
$$\log_b xy = \log_b x + \log_b y,$$
$$\log_b x^r = r(\log_b x),$$
$$\log_b \frac{x}{y} = \log_b x - \log_b y.$$

(e) The **natural logarithm** is the logarithm function with e as base. The notation $\ln x$ is generally used instead of $\log_e x$.

2. DIFFERENTIATION OF THE LOGARITHMIC AND EXPONENTIAL FUNCTIONS

(a) If $f(x) = \log_b x$, then $f'(x) = \frac{1}{x}(\log_b e) = \frac{1}{x(\ln b)}$.

(b) If $f(x) = \ln x$, then $f'(x) = \frac{1}{x}$.

(c) If $f(x) = \ln g(x)$, then $f'(x) = \frac{g'(x)}{g(x)}$.

(d) If $f(x) = b^x$, then $f'(x) = b^x(\ln b)$.

(e) If $f(x) = e^x$, then $f'(x) = e^x$.

(f) If $f(x) = e^{g(x)}$, then $f'(x) = e^{g(x)}g'(x)$.

3. EXPONENTIAL GROWTH FUNCTIONS

(a) $f(x) = ce^{kx}$ is the **unlimited growth/decay function.** It has the property that $f'(x) = kf(x)$; this means that the rate of change of f at x is directly proportional to the value of f at x.

(b) $f(x) = a - Ae^{-kx}$ is the **limited growth function** where $k > 0$. It has the property that $f'(x) = k[a - f(x)]$; this means that the rate of change of f at x is directly proportional to the difference between the upper limit a of the function and the value of f at x.

(c) $f(x) = \dfrac{a}{1 + ce^{-akx}}$ is the **logistic equation**. It has the property that $f'(x) = kf(x)[a - f(x)]$; this means that the rate of change of f at x is jointly proportional to the value of f at x and the difference between the upper limit a of the function and the value of f at x.

(d) If a principal P is deposited for t years at an annual interest rate r with **continuous compounding** then the amount A on deposit at time t is given by $A(t) = Pe^{rt}$.

4. CHAIN RULE AND INVERSE FUNCTION THEOREM

(a) Let f and g be two functions and let F be the composite function f of g defined by

$$F(x) = f(g(x)).$$

If g is differentiable at x and f is differentiable at $g(x)$, then

$$F'(x) = f'(g(x))g'(x).$$

(b) Let f and g be inverse functions. Assume g is differentiable at x, assume f is differentiable at $g(x)$, and assume $f'(g(x)) \neq 0$. Then

$$g'(x) = \dfrac{1}{f'(g(x))}$$

5. REVIEW EXERCISES

1. Find the number represented by the variable which makes the equality a true statement.
   (a) $\log_b 32 = -5$ (b) $\ln x = -\tfrac{1}{2}$ (c) $e^{\ln 8} = x$

2. If $f(x) = 500 - Ae^{-kx}$, $f(0) = 200$, and $f(3) = 300$, find $f(5)$.

In Exercises 3 through 8 find the first and second derivatives of the given functions.

3. $f(x) = \ln(x^2 + 1)$ 　　　　4. $f(x) = x^3 e^{1/x}$
5. $f(x) = x^{-1} e^x$ 　　　　　6. $f(x) = x^3 (\ln x)$
7. $f(x) = e^{2x}(\ln x)$ 　　　　8. $f(x) = x^{1/2}(\ln x)$

In Exercises 9 through 12, let g be the inverse function of f and do the following. (a) Find $g'(8)$ by using the derivative of f and the theorem on derivatives of inverse functions. (b) Find $g'(8)$ by obtaining an equation defining g, finding its derivative by the basic differentiation theorems, and then evaluating it at 8.

9. $f(x) = 3x + 15$
10. $f(x) = -2x + 9$
11. $f(x) = x^3 + 7$
12. $f(x) = x^2 + 2x, x \geq 0$

13. For $f(x) = xe^{3x}$, determine maxima, minima, concavity, and inflection points—if they exist. Determine where the function is increasing and where it is decreasing.

14. Find an equation of the line tangent to the graph of $f(x) = x^3(\ln x) + 2$ at $x = 1$.

15. Find an equation of the line tangent to the graph of $f(x) = x^2 e^{3x} + 2x + 1$ at $x = 0$.

16. (a) Find the amount resulting from a principal of $5000 deposited for 4 years at an 8.25% annual interest rate with continuous compounding.
    (b) How long would it take to double the amount?

17. Suppose that in two years a principal of $4000 earning continuous interest yields a return of $647.33. What is the annual interest rate?

18. Suppose the number of bacteria in a culture increases at a rate proportional to the number present at time x. If 100,000 bacteria are present at a given time and 300,000 are present 2 hours later, how many bacteria are present in $3\frac{1}{2}$ hours?

19. Referring to Example 4 of Section 4.5 suppose it is found that it takes 15 days for 60% of the population to hear the announcement. How long will it take for 90% to hear the news?

20. Suppose that 100 animals of a given species are placed in a game preserve and assume that their number will increase according to the logistic equation. If their number increases to 400 in 6 years and if the preserve cannot support more than 1000 such animals, how long will it take for the number on the preserve to increase to 800?

Carl Friedrich Gauss (1777–1855)

The Bettmann Archive

Carl Friedrich Gauss was born in Brunswick, Germany, on April 20, 1777. His father was a hard-working, poorly educated man who had little money or personal enthusiasm to contribute to his son's academic pursuits. However, Gauss received encouragement from his mother and from Johann Bartels, a friend who helped Gauss in his early study of mathematics and who later helped him to obtain entrance into Caroline College in Brunswick. Even more important for Gauss's career, Bartels helped him obtain from the Duke of Brunswick the financial assistance needed for college, a financial assistance that continued during the Duke's lifetime.

By the time Gauss was 30 years old, he was known throughout Europe. Also at that age he became Director of the Göttingen Observatory, a position he held until his death. He had many students who later became famous; his last student was Richard Dedekind who was one of the first mathematicians to put the real number system on a sound logical foundation.

Gauss made extensive contributions to nearly every branch of mathematics. In 1799, at the age of 22, in his doctoral dissertation he gave the first satisfactory proof of what is called the "fundamental theorem of algebra" (that a polynomial equation with complex coefficients has at least one complex root). He contributed extensively to the fields of analysis, probability, and number theory. His penchant for number theory is clearly indicated by his statement that "Mathematics is the queen of the sciences and the theory of numbers is the queen of mathematics." In his later years he turned his efforts more to applied mathematics and made several very important contributions to both physics and astronomy. In 1809 his second great work entitled *Theoria motus corporum coelestium* (Theory of Motion of Heavenly Bodies) was published. He has been called the "prince of mathematicians," and he ranks along with Archimedes and Newton as one of the three greatest mathematicians of all time.

CHAPTER

THE ANTIDERIVATIVE AND ITS APPLICATIONS

5

- 5.1 INTRODUCTION
- 5.2 THE ANTIDERIVATIVE OF A FUNCTION
- 5.3 ANTIDERIVATIVES INVOLVING EXPONENTIAL AND LOGARITHMIC FUNCTIONS
- 5.4 APPLICATIONS OF THE ANTIDERIVATIVE
- 5.5 INTEGRATION BY SUBSTITUTION
- 5.6 INTEGRATION BY PARTS
- 5.7 DIFFERENTIAL EQUATIONS (OPTIONAL)
- 5.8 SUMMARY AND REVIEW EXERCISES

5.1 INTRODUCTION

In this chapter we first define what is called the antiderivative of a function. Thereafter, we consider the basic techniques for finding antiderivatives of functions and then learn how to solve such problems as the following.

1. Suppose the marginal cost of producing q units of a given product is $50 + 20q^{1/2}$ dollars. If the total cost (including fixed costs) of producing the first unit is $100, what is the cost of producing the first 16 units? (See Example 1, Section 5.4.)

2. Suppose the slope of the line tangent to the graph of a function f is $3x^2 + 2x$ at each x in the domain of the function. Given that the graph of f contains the point $(2, -1)$, find an equation defining the function. (See Exercise 1, Section 5.4.)

3. It is estimated that in t months from now the population in a metropolitan area will be changing at a rate of $50 + 30t^{1/2}$ people per month. If the population today is 100,000, what will the population be in 9 months from today? (See Exercise 17, Section 5.4.)

4. The marginal profit on the sale of x units of a given product is $92 - x/2$ dollars. If the profit from the sale of 10 units is $595, what is the profit (loss) from the sale of 4 units? How many units would need to be sold in order to maximize the profit and what would be the maximum profit possible? (See Example 4, Section 5.4.)

5. Suppose the rate of a person's heartbeat t minutes after completing a series of exercises is given by $130 - 4t$ beats per minute, $0 \le t \le 15$. How many times does the person's heart beat during the 15 minutes?

There are many other applications of the antiderivative, but the discussion of some of them will have to wait until we have mastered the techniques for finding antiderivatives. In Chapter 6 we shall discuss a truly remarkable application of the antiderivative which will allow us to solve an even larger variety of practical problems. ■

5.2 THE ANTIDERIVATIVE OF A FUNCTION

We know that one useful application of the derivative of a function is in finding the slope of the tangent to the graph of a function at each point in its domain. Suppose we know that the slope of the tangent at each point on the graph of a function is given by

$$m(x) = 2x^2 - 4x + 1.$$

Can we determine the function? As we shall find in Section 5.4, the answer is "yes," provided we are given one point on the graph of the function.

In general, the problem that confronts us is to find for a given function f a function g (if it exists) which has f as its derivative. For example, if f is the function defined by $f(x) = 2x$, then *one* function having f as its derivative is $g(x) = x^2$ since $g'(x) = 2x$. This leads to the following definition.

antiderivative (indefinite integral)

> **DEFINITION 5.1** Let g be a differentiable function. If $g'(x) = f(x)$ for each x in the domain of g, then g is an **antiderivative** of f. An antiderivative of a function f is also called an **indefinite integral** of f.

Since the derivative of g defined by $g(x) = x^3$ is the function f defined by $f(x) = 3x^2$, g is an antiderivative of f. Furthermore, since the derivative of G defined by $G(x) = x^3 + 6$ is also $f(x) = 3x^2$, G is another antiderivative of f. Therefore, we see that two antiderivatives of a given function can differ by *at least* a constant. It is also true, though more difficult to show, that any two antiderivatives of a given function f can differ by *at most* a constant. Consequently, since $g(x) = x^2$ is an antiderivative of $f(x) = 2x$, *any* antiderivative of f must be of the form $x^2 + C$ where C is a constant. We call $x^2 + C$ the (general) antiderivative of f. Symbolically, if $g'(x) = f(x)$, then we write

$$\int f(x)\, dx = g(x) + C.$$

constant of integration

If $\int f(x)\, dx = g(x) + C$, then $f(x)$ is called the **integrand,** $g(x) + C$ is called *the* **antiderivative** of $f(x)$, and C is called the **constant of**

5.2 THE ANTIDERIVATIVE OF A FUNCTION

integration. Using "dx" as part of the antiderivative notation is not only traditional but also useful in remembering some important formulas and theorems. Furthermore, it clearly stipulates the independent variable when other letters are used for constants and variables in an antidifferentiation problem.

ILLUSTRATION

$$\int 2x \, dx = x^2 + C; \qquad \int 2t \, dt = t^2 + C.$$

Obviously, if we have found the derivatives of one hundred different functions, then one hundred antiderivative "formulas" are immediately available to us. However, since it is difficult to remember such a large number of specific antiderivatives, it is useful to develop a few general theorems which enable us to find the antiderivatives of many functions without having to remember too many specific formulas. For example, we know that if

$$g(x) = \frac{x^{p+1}}{p+1} \quad \text{for any real number} \quad p \neq -1,$$

then $g'(x) = x^p$. This leads immediately to the following antiderivative formula (theorem).

power formula

FORMULA 5.1 $\quad \displaystyle\int x^p \, dx = \frac{x^{p+1}}{p+1} + C \quad$ provided $\quad p \neq -1$.

Antiderivative

EXAMPLE 1 $\quad \int x^6 \, dx = x^7/7 + C \quad$ and $\quad \int x^{-1/2} \, dx = 2x^{1/2} + C$.

PRACTICE PROBLEM Find $\int x^{-2/3} \, dx$.

Answer: $3x^{1/3} + C$. ∎

Corresponding to each general differentiation theorem is a general antidifferentiation theorem. An immediate consequence of the fact that the derivative of a constant times a function is the constant times the derivative of the function is the following.

FORMULA 5.2 $\displaystyle\int kf(x)\,dx = k\int f(x)\,dx,\quad k$ a constant.

(The antiderivative of a constant times a function is the constant times the antiderivative of the function.)

EXAMPLE 2 $\int 8x^3\,dx = 8\int x^3\,dx = 8(x^4/4 + C) = 2x^4 + C_1$.

Note: $C_1 = 8C$.

An immediate consequence of the fact that the derivative of the sum of two functions is the sum of their derivatives is the following.

FORMULA 5.3 $\displaystyle\int [f(x) + g(x)]\,dx = \int f(x)\,dx + \int g(x)\,dx$.

(The antiderivative of the sum of two functions is the sum of their antiderivatives.)

Notes

(1) Formulas 5.2 and 5.3 imply that

$$\int [f(x) - g(x)]\,dx = \int f(x)\,dx - \int g(x)\,dx.$$

(2) The antidifferentiation theorems resulting from the product theorem for differentiation and from the chain rule are discussed in Sections 5.5 and 5.6.

Antiderivative

EXAMPLE 3 Use Formulas 5.1, 5.2, and 5.3 to find

$$\int (12x^5 + 3x^2 - 8x)\,dx.$$

5.2 THE ANTIDERIVATIVE OF A FUNCTION

Solution

$$\int (12x^5 + 3x^2 - 8x)\, dx = \int 12x^5\, dx + \int 3x^2\, dx - \int 8x\, dx$$

$$= 12 \int x^5\, dx + 3 \int x^2\, dx - 8 \int x\, dx$$

$$= 2x^6 + x^3 - 4x^2 + C.$$

Note: Technically, a constant of integration is associated with each of the three integrals on the right-hand side of the equation in this example. But, as is standard practice, these constants of integration are combined into one constant C. For example,

$$\int (3x^2 + 6x)\, dx = \int 3x^2\, dx + \int 6x\, dx$$

$$= (x^3 + C_1) + (3x^2 + C_2).$$

Letting $C = C_1 + C_2$, we obtain

$$\int (3x^2 + 6x)\, dx = x^3 + 3x^2 + C.$$

PRACTICE PROBLEM Find $\int (4x^3 - 2x + 11)\, dx$.

Answer: $x^4 - x^2 + 11x + C$. ∎

If f is a differentiable function and if

$$F(x) = \frac{[f(x)]^{p+1}}{p+1} \quad \text{for a real number} \quad p \neq -1,$$

then we know from the chain rule that $F'(x) = [f(x)]^p f'(x)$. From this fact we obtain the following important formula.

general power formula

FORMULA 5.4 $\quad \int [f(x)]^p f'(x)\, dx = \dfrac{[f(x)]^{p+1}}{p+1} + C \quad \text{if} \quad p \neq -1.$

Note: If $f(x) = x$ then Formula 5.1 is a special case of Formula 5.4.

Antiderivative

EXAMPLE 4 Find $\int (x^3 + 1)^4 (3x^2)\, dx$.

Solution: If $f(x) = x^3 + 1$, then $f'(x) = 3x^2$ and Formula 5.4 is immediately applicable. We obtain

$$\int (x^3 + 1)^4 (3x^2)\, dx = \frac{(x^3 + 1)^5}{5} + C.$$

PRACTICE PROBLEM Find $\int (x^4 + 2)^5 (4x^3)\, dx$.

Answer: $\frac{1}{6}(x^4 + 2)^6 + C.$ ∎

Although the antiderivatives in Example 4 and the practice problem could be obtained by expressing the given product as a polynomial and then performing term-by-term integration, it is much easier to apply Formula 5.4. In the following example, term-by-term integration is not possible, so we need to apply Formula 5.4 to find the antiderivative.

Antiderivative

EXAMPLE 5 Find the antiderivative of $F(x) = x(x^2 + 3)^{1/3}$.

Solution: First we note that if $f(x) = x^2 + 3$, then $f'(x) = 2x$. Then we make Formula 5.4 applicable by an arithmetic operation and by using Formula 5.2.

$$\int x(x^2 + 3)^{1/3}\, dx = \int (x^2 + 3)^{1/3} \left(\frac{1}{2}\right)(2x)\, dx$$

$$= \frac{1}{2} \int (x^2 + 3)^{1/3} (2x)\, dx$$

$$= \frac{1}{2} \frac{(x^2 + 3)^{4/3}}{4/3} + C$$

$$= \frac{3}{8}(x^2 + 3)^{4/3} + C.$$

Antiderivative

EXAMPLE 6 Find $\int (x^3 - 3x)^{2/5}(x^2 - 1)\, dx$.

5.2 THE ANTIDERIVATIVE OF A FUNCTION

Solution

$$\int (x^3 - 3x)^{2/5}(x^2 - 1)\, dx = \int (x^3 - 3x)^{2/5}\left(\frac{1}{3}\right)(3x^2 - 3)\, dx$$

$$= \frac{1}{3}\int (x^3 - 3x)^{2/5}(3x^2 - 3)\, dx$$

$$= \frac{1}{3} \cdot \frac{(x^3 - 3x)^{7/5}}{7/5} + C$$

$$= \frac{5}{21}(x^3 - 3x)^{7/5} + C.$$

PRACTICE PROBLEM Find $\int (x^2 - 2x + 3)^{4/3}(x - 1)\, dx$.

Answer: $\frac{3}{14}(x^2 - 2x + 3)^{7/3} + C.$ ∎

EXERCISES

Set A—Antiderivatives of Functions

In each of Exercises 1 through 30 find the antiderivative of the given function and check your answer by differentiation.

1. $\int x^5\, dx$
2. $\int x^{-4}\, dx$
3. $\int x^{-2}\, dx$
4. $\int x^3\, dx$
5. $\int x^{2/3}\, dx$
6. $\int 3x^{7/5}\, dx$
7. $\int x^{-3/2}\, dx$
8. $\int x^{-3/4}\, dx$
9. $\int 4x^{7/3}\, dx$
10. $\int 7x^{2/5}\, dx$
11. $\int (3x^2 - 6x + 7)\, dx$
12. $\int (4x^3 + 5x - 15)\, dx$
13. $\int (8x^3 - 6x^2 + 21)\, dx$
14. $\int (3x^4 - 2x^2 + 7x)\, dx$
15. $\int (x^{2/3} - 2x^{-3/2})\, dx$
16. $\int (3x^{-1/2} - 7x^{-2})\, dx$

17. $\displaystyle\int 3(3x+5)^3\,dx$

18. $\displaystyle\int 2(2x+1)^5\,dx$

19. $\displaystyle\int (x+1)(3x-1)\,dx$

20. $\displaystyle\int (4x-1)(2x-5)\,dx$

21. $\displaystyle\int (5x-1)^5\,dx$

22. $\displaystyle\int (3x-2)^{7/2}\,dx$

23. $\displaystyle\int 2x(x^2+4)^{3/2}\,dx$

24. $\displaystyle\int 3x^2(x^3-2)^4\,dx$

25. $\displaystyle\int x^{1/2}(x^{3/2}-1)^{7/3}\,dx$

26. $\displaystyle\int x^{1/3}(x^{4/3}+8)^{-1/2}\,dx$

27. $\displaystyle\int x^2(6x^3+11)^{4/5}\,dx$

28. $\displaystyle\int (3x^2+4x)^{15}(3x+2)\,dx$

29. $\displaystyle\int \frac{x}{(x^2+1)^3}\,dx$

30. $\displaystyle\int \frac{2x+1}{(x^2+x+3)^{2/3}}\,dx$

5.3 ANTIDERIVATIVES INVOLVING EXPONENTIAL AND LOGARITHMIC FUNCTIONS

In this section we consider a few more antiderivative formulas and continue to learn how to find the antiderivatives of functions. Then, in the next four sections, we put this knowledge to work in many practical applications.

In Section 4.3 we found that if

$$f(x) = e^{g(x)}, \quad \text{then} \quad f'(x) = e^{g(x)}g'(x).$$

From this fact and the definition of the antiderivative, we obtain the following.

FORMULA 5.5 $\displaystyle\int e^{g(x)}g'(x)\,dx = e^{g(x)} + C.$

Note: In Formula 5.5, if $g(x) = x$, then $g'(x) = 1$ and we obtain

$$\int e^x\,dx = e^x + C.$$

Antiderivative

EXAMPLE 1 Find $\int e^{3x}\,dx$.

5.3 ANTIDERIVATIVES INVOLVING EXPONENTIAL AND LOGARITHMIC FUNCTIONS

Solution: If we let $g(x) = 3x$, then $g'(x) = 3$. To apply Formula 5.5, we proceed as follows.

$$\int e^{3x}\, dx = \frac{1}{3}\int e^{3x}(3)\, dx$$

$$= \frac{1}{3} e^{3x} + C.$$

PRACTICE PROBLEM Find $\int e^{7x}\, dx$.

Answer: $\frac{1}{7} e^{7x} + C.$ ∎

Antiderivative

EXAMPLE 2 Find $\int x e^{x^2}\, dx$.

Solution

$$\int x e^{x^2}\, dx = \frac{1}{2}\int e^{x^2}(2x)\, dx$$

$$= \frac{1}{2} e^{x^2} + C.$$

Also in Section 4.3, we found that if

$$f(x) = \ln g(x), \qquad \text{then} \quad f'(x) = \frac{g'(x)}{g(x)}.$$

From this fact we obtain the following.

FORMULA 5.6 $\displaystyle\int \frac{g'(x)}{g(x)}\, dx = \ln g(x) + C.$

Notes

(1) We assume $g(x) > 0$.

(2) In Formula 5.6, if $g(x) = x$, then $g'(x) = 1$ and we obtain the following important formula.

FORMULA 5.7 $\int \dfrac{1}{x}\,dx = \ell n\, x + C.$

Antiderivative

EXAMPLE 3 Find $\int \dfrac{1}{3x + 5}\,dx.$

Solution

$$\int \dfrac{1}{3x + 5}\,dx = \dfrac{1}{3}\int \dfrac{3}{3x + 5}\,dx$$

$$= \dfrac{1}{3}\ell n(3x + 5) + C$$

Antiderivative

EXAMPLE 4 Find $\int \dfrac{x}{x^2 + 1}\,dx.$

Solution

$$\int \dfrac{x}{x^2 + 1}\,dx = \dfrac{1}{2}\int \dfrac{2x}{x^2 + 1}\,dx$$

$$= \dfrac{1}{2}\ell n(x^2 + 1) + C.$$

PRACTICE PROBLEM Find $\int \dfrac{6x^2}{x^3 + 4}\,dx.$

Answer: $2[\ell n(x^3 + 4)] + C.$ ∎

Suppose we know that the rate of change of some differentiable function f at x is directly proportional to the value of f at x. That is, suppose

$$f'(x) = kf(x). \qquad (1)$$

Let us see how Formula 5.6 can be used to obtain an expression for f. First, we divide each side of Eq. (1) by $f(x)$ and obtain

$$\dfrac{f'(x)}{f(x)} = k.$$

5.3 ANTIDERIVATIVES INVOLVING EXPONENTIAL AND LOGARITHMIC FUNCTIONS

Since antiderivatives differ only by a constant, we obtain

$$\int \frac{f'(x)}{f(x)} \, dx = \int k \, dx + C_1.$$

Applying Formula 5.6 and using the fact that $\int k \, dx = kx + C_2$, we obtain

$$\ell n \, f(x) = kx + C \quad \text{where} \quad C = C_1 + C_2.$$

From the definition of the logarithm function it follows that

$$f(x) = e^{kx+C}$$
$$= e^{kx}e^C.$$

Letting $c = e^C$, we obtain

$$f(x) = ce^{kx}.$$

Recall that this is the exponential growth/decay function discussed in Section 4.5. Furthermore, note that $f(0) = c$.

Our last antiderivative formula results from the fact that for $b > 0$ and $b \neq 1$,

$$\text{if} \quad f(x) = b^x, \quad \text{then} \quad f'(x) = b^x(\ell n \, b).$$

We conclude that

$$\int b^x(\ell n \, b) \, dx = b^x + C_1,$$

or

FORMULA 5.8 $\displaystyle \int b^x \, dx = \frac{b^x}{\ell n \, b} + C.$

Antiderivative

EXAMPLE 5 Find $\int 2^x \, dx$.

Solution: $\displaystyle \int 2^x \, dx = \frac{2^x}{\ell n \, 2} + C.$

EXERCISES

Set A–Antiderivatives of Functions

In each of Exercises 1 through 22 find the antiderivative of the given function and check your answer by differentiation.

1. $\int e^{5x}\, dx$
2. $\int e^{6x}\, dx$
3. $\int e^{-2x}\, dx$
4. $\int e^{-3x}\, dx$
5. $\int (e^{3x} + x^{3/4})\, dx$
6. $\int (e^{-5x} + x^{1/5})\, dx$
7. $\int x^2 e^{(x^3)}\, dx$
8. $\int x^3 e^{(x^4)}\, dx$
9. $\int x^{-1/2} e^{\sqrt{x}}\, dx$
10. $\int x^{-2/3} e^{\sqrt[3]{x}}\, dx$
11. $\int \frac{4}{4x+5}\, dx$
12. $\int \frac{3}{3x-7}\, dx$
13. $\int \frac{1}{2x-5}\, dx$
14. $\int \frac{1}{5x-9}\, dx$
15. $\int \frac{x}{2x^2+1}\, dx$
16. $\int \frac{x}{3x^2+1}\, dx$
17. $\int \frac{x^2}{x^3+1}\, dx$
18. $\int \frac{x^3}{3x^4+1}\, dx$
19. $\int 3^x\, dx$
20. $\int 4^x\, dx$
21. $\int \frac{1}{x^2-1}\, dx$ $\left[\text{Hint: } \frac{1}{x^2-1} = \frac{1}{2}\left(\frac{1}{x-1} - \frac{1}{x+1}\right).\right]$
22. $\int \frac{1}{4x^2-1}\, dx$ $\left[\text{Hint: } \frac{1}{4x^2-1} = \frac{1}{2}\left(\frac{1}{2x-1} - \frac{1}{2x+1}\right).\right]$

5.4 APPLICATIONS OF THE ANTIDERIVATIVE

Most applications of the antiderivative which we look at in this section result from knowing the rate of change of some quantity (dependent variable) with respect to another quantity (independent variable). Although some of the examples represent idealized real-world problems, they do exhibit ideas and techniques that we

5.4 APPLICATIONS OF THE ANTIDERIVATIVE

must comprehend before we are able to pursue more complicated problems. At the end of the section, we analyze a real-world problem, decide what are reasonable assumptions to make in order to use the tools of calculus to obtain a solution, and then determine the resulting mathematical model for the problem.

Let us begin by looking at two problems, one given in the Introduction, Section 5.1, and one mentioned at the beginning of Section 5.2.

Marginal cost

EXAMPLE 1 Suppose the marginal cost of producing q units of a given product is $50 + 20/q^{1/2}$ dollars. If the total cost (including fixed costs) of producing the first unit is $100, what is the cost function and what is the cost of producing the first 16 units?

Solution: Since $C'(q) = 50 + 20q^{-1/2}$,

$$C(q) = \int (50 + 20q^{-1/2})\, dq$$

$$= 50q + 40q^{1/2} + C_1.$$

Since $C(1) = 100$, $100 = 50 + 40 + C_1$ and $C_1 = 10$. Therefore,

$$C(q) = 50q + 40q^{1/2} + 10$$

is the cost function, and

$$C(16) = 800 + 160 + 10 = \$970.$$

Note: $C(0) = 10 = C_1$ are the fixed costs.

PRACTICE PROBLEM Suppose the marginal cost for producing q units of a given product is $30 + 12q^{1/2}$ dollars. If the total cost (including fixed costs) of producing the first 4 units is $300, what is the cost function and what is the cost of producing 9 units?

Answer: $C(q) = 30q + 8q^{3/2} + 116$ and $C(9) = \$602$. ∎

Slope of tangent

EXAMPLE 2 Suppose the slope of the tangent line to the graph of a function f at each x in its domain is given by $m(x) = x^2 - 4x + 1$. If the graph of f contains the point $(3, 1)$, find an equation defining the function.

Solution: Since $f'(x) = m(x)$, $f(x) = \int (x^2 - 4x + 1)\,dx$ and

$$f(x) = \frac{x^3}{3} - 2x^2 + x + C.$$

Since the graph of f contains $(3, 1)$, it follows that $f(3) = 1$ and

$$1 = 9 - 18 + 3 + C.$$

Therefore, $C = 7$ and $f(x) = x^3/3 - 2x^2 + x + 7$.

Marginal revenue

EXAMPLE 3 The marginal revenue in dollars from selling x units each month of a product is given by $R'(x) = 20x - 0.06x^2$. Given that $R(0) = 0$, find the revenue function.

Solution: Since $R'(x) = 20x - 0.06x^2$,

$$R(x) = \int R'(x)\,dx$$

$$= \int (20x - 0.06x^2)\,dx$$

$$= 10x^2 - 0.02x^3 + C.$$

Furthermore, since $R(0) = 0$, $C = 0$ and

$$R(x) = 10x^2 - 0.02x^3.$$

Marginal profit

EXAMPLE 4 The marginal profit on the sale of x units of a given product is $92 - x/2$ dollars. If the profit from the sale of 10 units is $595, what is the profit (loss) from the sale of 4 units? How many units would need to be sold in order to maximize the profit? What would be the maximum profit possible?

Solution: Since $P'(x) = 92 - x/2$,

$$P(x) = \int \left(92 - \frac{x}{2}\right) dx$$

and

$$P(x) = 92x - \frac{x^2}{4} + C.$$

5.4 APPLICATIONS OF THE ANTIDERIVATIVE

Furthermore, since $P(10) = 595$,

$$595 = 920 - 25 + C$$

and

$$C = -300.$$

Hence, $P(x) = 92x - x^2/4 - 300$, $P(4) = 368 - 4 - 300 = 64$, and a profit of \$64 results from the sale of 4 units. Since $P'(x) = 0$ when

$$92 - \frac{x}{2} = 0$$

$$\frac{x}{2} = 92$$

$$x = 184$$

and since $P''(x) = -\frac{1}{2} < 0$ for all x, profit is maximized at $x = 184$. The maximum profit is $P(184) = 16{,}928 - 8464 - 300 = \$8164$.

Concavity and relative minimum

EXAMPLE 5 Suppose the rate of change in the slope of the tangent to the graph of a twice differentiable function f is 4 at each x in its domain. If the function has a relative minimum at $(-1, 1)$, what is an equation defining f?

Solution: Since $f''(x)$ is the rate of change of the slope of the tangent $f'(x)$, $f''(x) = 4$. Therefore,

$$f'(x) = \int 4 \, dx = 4x + C_1.$$

Since f has a relative minimum at $x = -1$,

$$f'(-1) = 0, \qquad 4(-1) + C_1 = 0, \qquad \text{and} \qquad C_1 = 4.$$

Now since $f'(x) = 4x + 4$,

$$f(x) = \int (4x + 4) \, dx = 2x^2 + 4x + C_2.$$

Since the graph of f contains the point $(-1, 1)$,

$$f(-1) = 1, \qquad 2 - 4 + C_2 = 1, \qquad \text{and} \qquad C_2 = 3.$$

Therefore, $f(x) = 2x^2 + 4x + 3$.

PRACTICE PROBLEM Find the solution to Example 5 if $f''(x) = 6$.

Answer: $f(x) = 3x^2 + 6x + 4$. ∎

Velocity of a car

EXAMPLE 6 The velocity of a car at t seconds after starting from rest is $v(t) = 11t$ ft/sec, $0 \leq t \leq 8$.

(a) How fast is the car going when $t = 8$ seconds?

(b) How far has the car traveled in 8 seconds?

Solution

(a) $v(8) = 88$ ft/sec. *Note:* Since there are 5280 feet in 1 mile and 3600 seconds in 1 hour, the velocity is

$$\frac{88 \times 3600}{5280} = 60 \text{ mph.}$$

(b) Since $v(t) = s'(t)$ where $s(t)$ is the distance function,

$$s(t) = \frac{11}{2}t^2 + C.$$

Since $s(0) = 0$, $s(t) = \left(\frac{11}{2}\right)t^2$. At $t = 8$,

$$s(8) = 352 \text{ feet} = 0.0666 \text{ mile.}$$

Let us consider a problem where certain assumptions are made to create a mathematical model for its solution. Suppose an underwater pipe carrying oil bursts and begins to leak oil at a constant rate of 1000 ft^3/hr. As the oil comes to the top of the water, it begins to create a circular oil slick having a constant thickness of 0.36 inch. The oil company responsible (as well as some environmentalists) wants to know how long it will take to create an oil slick with a surface area of 1,742,000 ft^2 ($\frac{1}{16}$ square mile)? To find out, let $r(t)$ be the radius of the oil slick at time t hours from the time the pipe bursts. Since the thickness of the oil is a constant 0.36 inch = 0.36/12 = 0.03 foot, the volume of the oil on the surface at any time t is that of a right circular cylinder. Since $V = \pi r^2 h$,

$$V(t) = 0.03\pi [r(t)]^2.$$

Since $V'(t) = 1000$, $V(t) = 1000t + C$. At $t = 0$, $V(0) = 0$ and we conclude that $V(t) = 1000t$ ft^3. Consequently

$$1000t = 0.03\pi[r(t)]^2.$$

Since the oil slick is circular, we seek the time t when

$$\pi[r(t)]^2 = 1{,}742{,}400.$$

It is when

$$1000t = 0.03(1{,}742{,}400)$$
$$t = 52.272 \text{ hours} \doteq 52 \text{ hours } 16 \text{ minutes.}$$

EXERCISES

Set A—Applications of the Antiderivative

Slope of tangent

1. Suppose the slope of the tangent line to the graph of a function f is $3x^2 + 2x$ at each x in the domain of the function. If the graph of f contains the point $(2, -1)$, find an equation defining the function.

Slope of tangent

2. Suppose the slope of the tangent line to the graph of a function f is $x^2 - x + 1$ at each x in the domain of the function. If the graph of f contains the point $(-1, 4)$, what is an equation defining the function?

Concavity and relative maximum

3. Suppose the rate of change in the slope of the tangent to the graph of a twice differentiable function f is -2 at each x in its domain. If the function has a relative maximum at $(2, 4)$, what is an equation defining f?

Concavity and relative minimum

4. Suppose the rate of change in the slope of the tangent to the graph of a twice differentiable function f is 6 at each x in its domain. If the function has a relative minimum at $(-2, 1)$, what is an equation defining f?

Marginal revenue

5. The marginal revenue in dollars from selling x units per month of a product is given by $R'(x) = 36x - 0.03x^2$. If $R(0) = 0$, what is the revenue function?

Marginal cost

6. Suppose the marginal cost for producing q units of a given product is $60 + 10/q^{1/2}$ dollars. If the total cost (including fixed costs) of producing the first four units is $340, what is the cost of producing the first 25 units?

Marginal profit

7. The marginal profit on the sale of x units of a given product is $84 - 2x$ dollars. If the profit from the sale of 6 units is $68, what is the profit (loss) from the sale of 4 units? How many units would need to be sold in order to maximize the profit and what would be the maximum profit possible?

Marginal cost

8. The marginal cost in dollars for producing x units of a given product each month is $C'(x) = 400 - x/4$. Suppose it costs $8900 to produce 20 units.
   (a) What is the monthly cost function?
   (b) What is the cost of producing 50 units?
   (c) What is the average cost of producing 50 units?
   (d) Use marginal analysis to approximate the cost of the 51st unit?

Marginal cost

9. Assume the marginal cost in dollars of producing x units per week of a given product is given by $C'(x) = 100 - 0.2x$ and assume the cost of producing 10 units is $1200.
   (a) Find the weekly cost function.
   (b) Find the cost of producing 50 units.
   (c) What is the average cost of the first 50 units?
   (d) Use marginal analysis to approximate the cost of the 51st unit.

Velocity of a car

10. The velocity of a car at t seconds after starting from rest is $v(t) = 16t$ ft/sec, $0 \leq t \leq 8$.
    (a) How fast is the car going at the end of 8 seconds?
    (b) How far has the car traveled in 6 seconds?

Machine depreciation

11. A machine is bought for $12,000 and its value V depreciates at a rate of $1000e^{-t/5}$ dollars per year in t years.
    (a) Find its value $V(t)$ at the end of t years. [*Hint:* $V'(t) = -1000e^{-t/5}$.]
    (b) What is its approximate value after 2 years?
    (c) What is its approximate value after 10 years?

Machine depreciation

12. A machine is bought for $3000 and its rate of change in value V is given by $V'(t) = \dfrac{-500}{(t+1)}$ dollars at t months.
    (a) What is its value $V(t)$ at the end of t months?
    (b) In approximately how many months would the machine be worthless?

Altitude of a bullet

13. The initial velocity of a bullet fired vertically upward is 800 ft/sec. Find its altitude h after t seconds. [*Hint:* Since we ignore air resistance, the bullet is decelerating at 32 ft/sec^2 due to gravity; that is, $h''(t) = -32$.]

Distance a car travels

14. A car starting from rest accelerates at 3.6 ft/sec^2 for 6 seconds. What is its velocity at the end of 6 seconds? How far does it travel in 6 seconds?

5.5 INTEGRATION BY SUBSTITUTION

Population growth

15. Suppose the rate of change of population P of a certain country during the next t years is given by $e^{0.03t}$ million per year. If the population today is 60 million, what will the population be in 20 years?

Human heartbeat

16. Suppose the rate of a person's heartbeat t minutes after completing a series of exercises is given by $130 - 4t$ beats per minute, $0 \le t \le 15$. How many times does the person's heartbeat during the 15 minutes?

Population growth

17. It is estimated that in t months from now the population in a metropolitan area will be changing at a rate of $50 + 30t^{1/2}$ people per month. If the population today is 100,000, what will the population be in 9 months from today?

5.5 INTEGRATION BY SUBSTITUTION

In this section we discover how the *chain rule* (differentiation theorem for composition of functions) can be used to develop a powerful technique for finding antiderivatives; it is called the *substitution technique*. From the chain rule we know that

substitution technique

$$[f(g(x))]' = f'(g(x))g'(x).$$

Therefore, from the definition of the antiderivative, we conclude that

$$\int f'(g(x))g'(x)\,dx = f(g(x)) + C.$$

To find the antiderivative $\int F(x)\,dx$ for some given function F, if we can find functions f and g such that

$$F(x) = f'(g(x))g'(x),$$

then

$$\int F(x)\,dx = f(g(x)) + C.$$

Finding f and g, even if they exist, could be a difficult task without some "rules" to help in this endeavor. In general, to find $\int F(x)\,dx$ by the substitution technique, we let $u = g(x)$ and *change the variable*. Note that if $u = g(x)$, then the differential is $du = g'(x)\,dx$. Futheremore,

change of variable

$$\int f'(g(x))g'(x)\,dx = \int f'(u)\,du = f(u) + C$$

where $u = g(x)$. The substitution (change of variable) technique, as generally used, is exhibited in the following examples.

Substitution technique

EXAMPLE 1 Find $\int x(x^2 + 1)^{3/2}\, dx$ by the substitution technique.

Solution: In this example there is a natural choice for $u = g(x)$ since $\int (x^2 + 1)^{3/2} x\, dx$ is practically in the form $\int [g(x)]^n g'(x)\, dx$. We let $u = x^2 + 1$. Then $du = 2x\, dx$ and $x\, dx = (\tfrac{1}{2})\, du$. By substituting into

$$\int (x^2 + 1)^{3/2} x\, dx,$$

we get

$$\int u^{3/2} \left(\frac{1}{2}\right) du = \frac{1}{2} \int u^{3/2}\, du$$

$$= \frac{1}{5} u^{5/2} + C.$$

By substituting $u = x^2 + 1$, we obtain

$$\int (x^2 + 1)^{3/2} x\, dx = \frac{1}{5}(x^2 + 1)^{5/2} + C.$$

PRACTICE PROBLEM Change the variable to find $\int x^2(x^3 - 1)^{4/3}\, dx$.

Answer: $(\tfrac{1}{7})(x^3 - 1)^{7/3} + C$. ∎

Substitution technique

EXAMPLE 2 Find $\int x(2x + 1)^{1/2}\, dx$ by the substitution technique.

Solution: Since this antiderivative is not in the form $\int [g(x)]^n g'(x)\, dx$, the choice for $u = g(x)$ is not so clear. However, as in this case, whenever an expression is raised to a power, a standard choice to try is to let u be the expression. That is, let $u = 2x + 1$. Therefore

$$du = 2\, dx,$$

and solving $u = 2x + 1$ for x, we get

$$x = \frac{1}{2}(u - 1).$$

5.5 INTEGRATION BY SUBSTITUTION

Substituting the preceding expressions into

$$\int x(2x + 1)^{1/2}\, dx,$$

we get

$$\int \frac{1}{2}(u - 1)u^{1/2}\left(\frac{1}{2}\right) du = \frac{1}{4}\int (u^{3/2} - u^{1/2})\, du$$

$$= \frac{1}{10}u^{5/2} - \frac{1}{6}u^{3/2} + C.$$

Therefore, we obtain

$$\int x(2x + 1)^{1/2}\, dx = \frac{1}{10}(2x + 1)^{5/2} - \frac{1}{6}(2x + 1)^{3/2} + C.$$

Note: We could check the solution by finding its derivative.

Substitution technique

EXAMPLE 3 Use the substitution technique to find $\int x(x - 4)^5\, dx$.

Solution: Let $u = x - 4$. Then $x = u + 4$ and $dx = du$. Substitution yields

$$\int (u + 4)u^5\, dx = \int (u^6 + 4u^5)\, du$$

$$= \frac{u^7}{7} + \frac{2u^6}{3} + C$$

$$= \frac{u^6}{21}(3u + 14) + C.$$

Hence, we find that

$$\int x(x - 4)^5\, dx = \frac{1}{21}(x - 4)^6(3x + 2) + C.$$

Note: We could have obtained the antiderivative by first expanding $(x - 4)^5$ and then integrating the product $x(x - 4)^5$ term by term.

EXERCISES

Set A—Integration by Substitution

Use the substitution technique to find each antiderivative.

1. $\int x(x^2 + 1)^{4/3}\, dx$
2. $\int x^2(x^3 + 4)^{3/2}\, dx$
3. $\int x^3(x^4 + 1)^{1/2}\, dx$
4. $\int x(x^2 + 2)^{2/5}\, dx$
5. $\int x(9x + 2)^{1/2}\, dx$
6. $\int x(2x + 13)^{1/2}\, dx$
7. $\int x(3x + 1)^{12}\, dx$
8. $\int x(2x + 3)^7\, dx$
9. $\int x^2(3x - 6)^{1/2}\, dx$
10. $\int x^2(3x + 1)^{2/3}\, dx$
11. $\int x(x + \sqrt{2x + 1})\, dx$
12. $\int x(1 + \sqrt{x + 1})\, dx$

Set B—Applications

13. If the marginal cost for x units of a product is given by $C'(x) = x(3x + 4)^{1/2}$ and if initial costs $C(0) = 200$ dollars, what is the cost function?

14. The slope of the tangent to the graph of a function f which contains the origin is $x(5x + 8)^{1/3}$ at each x in its domain. Find $f(x)$ in terms of x.

5.6 INTEGRATION BY PARTS

Let us now consider a general technique for finding antiderivatives of functions which is a consequence of the differentiation theorem for the product of two functions. Recall that the derivative of the product of two differentiable functions u and v is given by $(uv)' = uv' + u'v$. Since two antiderivatives of a function differ at most by a constant, it follows that

$$uv + C = \int (uv' + u'v)$$

where C is a constant. Since the antiderivative of a sum is the sum of the antiderivatives, except for a possible constant,

$$uv + C = \int u'v + \int u'v,$$

5.6 INTEGRATION BY PARTS

and by rearranging the terms, we obtain

$$\int uv' = uv - \int u'v + C.$$

The preceding equation is called the **integration by parts formula**. It is also expressed as follows:

Integration by parts

$$\int u(x)v'(x)\,dx = u(x)v(x) - \int u'(x)v(x)\,dx + C.$$

The following examples indicate how the integration by parts formula can be used in determining the antiderivatives of many functions.

Integration by parts

EXAMPLE 1 Find $\int x^2(\ell n\, x)\,dx$.

Solution: (When using the integration by parts formula, our first task is to pick $u(x)$ and $v'(x)$ so that their product is the integrand, in this case $x^2(\ell n\, x)$. After this example we give some rules that should make the following choices for $u(x)$ and $v'(x)$ seem less arbitrary.) Let $u(x) = \ell n\, x$ and let $v'(x) = x^2$ so that $u(x)v'(x) = x^2(\ell n\, x)$. Now obtain u' and v. They are

$$u'(x) = \frac{1}{x} \quad \text{and} \quad v(x) = \frac{x^3}{3}.$$

Since $\int u(x)v'(x)\,dx = u(x)v(x) - \int u'(x)v(x)\,dx + C$,

$$\int x^2(\ell n\, x)\,dx = \frac{x^3}{3}\ell n\, x - \int \frac{1}{x} \cdot \frac{x^3}{3}\,dx + C$$

$$= \frac{x^3}{3}\ell n\, x - \frac{1}{3}\int x^2\,dx + C$$

$$= \frac{x^3}{3}\ell n\, x - \frac{x^3}{9} + C.$$

[The answer can (should) be checked by differentiation.]

PRACTICE PROBLEM Find $\int x^3 (\ln x)\, dx$.

Answer: $(x^4/4)(\ln x) - x^4/16 + C$. ∎

We should make several things clear about the integration by parts technique.

1. To find $\int f(x)\, dx$ by this method we must choose u and v' so that $u(x)v'(x) = f(x)$.

2. Since it is necessary to determine an antiderivative of v', it is important to let v' be a function we can integrate. Thus we tend to choose v' first and let $u = f/v'$.

3. We should also try to pick v' so that the antiderivative $\int u'v$ will be easier to obtain than $\int uv'$.

4. As we shall discover, it may be necessary not only to make "unobvious" choices for u and v', but also to repeat the integration by parts process several times before we find the antiderivative of a given function.

5. Last, but not least, the technique may not "work," in the sense that it does not lead to an antiderivative we can calculate.

Since we are studying a technique, it is essential to consider carefully several examples and then work the exercises in order to develop some facility in finding antiderivatives by means of the integration by parts formula.

Integration by parts

EXAMPLE 2 Find $\int xe^x\, dx$.

Solution: There are many choices for u and v' that can be tried, but the "natural" ones to consider are either (a) letting $u(x) = e^x$ and $v'(x) = x$ or (b) letting $u(x) = x$ and $v'(x) = e^x$. If we let $u(x) = x$ and $v'(x) = e^x$, then $u'(x) = 1$ and $v(x) = e^x$. With this selection for u and v', u' is a "simpler" function to deal with than u. Furthermore, $\int u'v$ is easy to obtain. Using the integration by parts formula, we obtain

$$\int xe^x\, dx = xe^x - \int e^x\, dx + C$$

$$= xe^x - e^x + C.$$

5.6 INTEGRATION BY PARTS

Integration by parts

EXAMPLE 3 Find $\int x^2 e^x \, dx$.

Solution: Let $u(x) = x^2$ and $v'(x) = e^x$. Then $u'(x) = 2x$ and $v(x) = e^x$. Therefore, using the integration by parts formula, we obtain

$$\int x^2 e^x \, dx = x^2 e^x - 2 \int x e^x \, dx + C.$$

Using the antiderivative obtained in Example 2 and substituting in the preceding equality, we get

$$\int x^2 e^x \, dx = x^2 e^x - 2x e^x + 2e^x + C.$$

(Essentially, the integration by parts technique was applied twice to obtain this antiderivative.)

Integration by parts

EXAMPLE 4 Find $\int x(2x + 1)^{1/2} \, dx$.

Solution: Let $u(x) = x$ and let $v'(x) = (2x + 1)^{1/2}$. Then $u'(x) = 1$ and since

$$v'(x) = \frac{1}{2}(2x + 1)^{1/2}(2),$$

it follows from Formula 5.4 that

$$v(x) = \frac{1}{3}(2x + 1)^{3/2}.$$

Therefore

$$\int x(2x + 1)^{1/2} \, dx = \frac{x}{3}(2x + 1)^{3/2} - \frac{1}{3} \int (2x + 1)^{3/2} \, dx$$

$$= \frac{x}{3}(2x + 1)^{3/2} - \frac{1}{15}(2x + 1)^{5/2} + C$$

$$= \frac{1}{15}(2x + 1)^{3/2}(5x - 2x - 1) + C$$

$$= \frac{1}{15}(2x + 1)^{3/2}(3x - 1) + C.$$

Note: In Section 5.5 we used the substitution technique to find the antiderivatives of functions such as those in this example and the next.

Integration by parts

EXAMPLE 5 Find $\int x(x - 4)^7 \, dx$.

Solution: (The antiderivative could be found by expanding $(x - 4)^7$ and then integrating term by term, but we use integration by parts for reasons which should be obvious if the two approaches are compared.) Let $u(x) = x$ and $v'(x) = (x - 4)^7$. Then $u'(x) = 1$ and $v(x) = (x - 4)^8/8$. Therefore,

$$\int x(x - 4)^7 \, dx = \frac{x}{8}(x - 4)^8 - \frac{1}{8}\int (x - 4)^8 \, dx$$

$$= \frac{x}{8}(x - 4)^8 - \frac{1}{72}(x - 4)^9 + C$$

$$= (x - 4)^8\left(\frac{x}{8} - \frac{x}{72} + \frac{1}{18}\right) + C$$

$$= \frac{1}{18}(x - 4)^8(2x + 1) + C.$$

Integration by parts

EXAMPLE 6 Find $\int (\ln x) \, dx$.

Solution: Let $u(x) = \ln x$ and $v'(x) = 1$. Then $u'(x) = 1/x$ and $v(x) = x$. Consequently,

$$\int (\ln x) \, dx = x(\ln x) - \int \frac{1}{x} x \, dx + C$$

$$= x(\ln x) - x + C.$$

In practice, "u" and "v" are often used instead of "$u(x)$" and "$v(x)$" to simplify the notation. In this case, we let $u' = u'(x)$ and $v' = v'(x)$. These abridged notations are used in the following example.

Integration by parts

EXAMPLE 7 Find $\int x^3(x^2 + 3)^{1/2} \, dx$.

Solution: We find (after considerable thought and perhaps some false starts) that we should let $v' = x(x^2 + 3)^{1/2}$ in order to apply

Formula 5.4. Therefore, we let $u = x^2$ so that $uv' = x^3(x^2 + 3)^{1/2}$. Now, $u' = 2x$ and

$$v = \frac{1}{2} \int (x^2 + 3)^{1/2}(2x)\, dx$$

$$= \frac{1}{3}(x^2 + 3)^{3/2}.$$

Integration by parts yields

$$\int x^3(x^2 + 3)^{1/2}\, dx = \frac{x^2}{3}(x^2 + 3)^{3/2} - \frac{1}{3}\int (x^2 + 3)^{3/2}(2x)\, dx$$

$$= \frac{x^2}{3}(x^2 + 3)^{3/2} - \frac{2}{15}(x^2 + 3)^{5/2} + C.$$

EXERCISES

Set A—Integration by Parts

Use integration by parts to obtain the antiderivatives in Exercises 1 through 10.

1. $\int x(3x + 1)^{1/2}\, dx$
2. $\int x(5x + 2)^{-1/2}\, dx$
3. $\int x(x - 4)^5\, dx$
4. $\int x(2x - 7)^6\, dx$
5. $\int xe^{-x}\, dx$
6. $\int x^2 e^{-x}\, dx$
7. $\int x(2x + 1)^{1/3}\, dx$
8. $\int x(3x + 5)^4\, dx$
9. $\int x^3 e^x\, dx$
10. $\int x^4(\ln x)\, dx$

Set B—Applications

11. The slope of the tangent to the graph of a function f at each x in its domain is $x(3x + 8)^{2/3}$. If the graph contains the point $(0, \frac{3}{5})$, find an equation defining f.
12. The slope of the tangent to the graph of a function f at each x

in its domain is xe^{3x}. If the graph of f contains the point $(0, \frac{26}{9})$, what is an equation defining f?

13. In t seconds an object is moving at $te^{t/6}$ ft/sec where $0 \leq t \leq 18$. How far does it travel in 18 seconds?

5.7 DIFFERENTIAL EQUATIONS (OPTIONAL)

In the study of algebra and trigonometry we learn how to solve various types of equations such as

1. $3x^2 + 5x - 6 = 0$ (polynomial equation)
2. $\sqrt{3x + 1} + \sqrt{x - 1} = x + 1$ (algebraic equation)
3. $\log x^2 + \log x = 3$ (logarithmic equation)
4. $2 \sin^2 x + \sin x - 1 = 0$ (trigonometric equation)
5. $2^x = 5$ (exponential equation)

For each of these equations the solution set consists of exactly those numbers making the equality a true statement. This section is devoted to solving equations involving derivatives. These equations, which arise naturally in many applications of calculus, are called **differential equations.** We have already seen examples in Section 5.4, where, for example, we sought a function $C(x)$ such that $C'(x) = 50 + 20x^{-1/2}$.

Another example of a differential equation is

$$\frac{d^2y}{dx^2} - 2\frac{dy}{dx} = 3y \quad \text{or} \quad y''(x) - 2y'(x) = 3y(x).$$

The **function** defined by the equation $y = e^{3x}$ is a **solution** of the given differential equation; that is, it is a function such that when it and its first and second derivatives are substituted into the equation, the equality becomes a true statement. Note that

$$\text{if} \quad y = e^{3x}, \quad \text{then} \quad \frac{dy}{dx} = 3e^{3x}, \quad \frac{d^2y}{dx^2} = 9e^{3x},$$

and

$$\frac{d^2y}{dx^2} - 2\frac{dy}{dx} = 9e^{3x} - 2(3e^{3x}) = 3e^{3x} = 3y.$$

We see that substituting $y'' = 9e^{3x}$, $y' = 3e^{3x}$, and $y = e^{3x}$ into the given differential equation reduces it to an identity for each x in the domain of the function $y = e^{3x}$. The given differential equation is also often written as

$$y'' - 2y' = 3y.$$

Note that if $y = e^{-x}$, then $y' = -e^{-x}$, $y'' = e^{-x}$, and

$$y'' - 2y' = e^{-x} - 2(-e^{-x}) = 3e^{-x} = 3y.$$

Thus $y = e^{-x}$ is *another* solution to the given differential equation. It can be proved that all solutions to this differential equation are of the form

$$y = c_1 e^{3x} + c_2 e^{-x} \quad \text{where} \quad c_1 \text{ and } c_2 \text{ are constants.}$$

Whenever we have found an antiderivative (indefinite integral) of a function, a solution to the simplest type of differential equation was obtained. For example

$$\int 3x^2 \, dx = x^3 + C$$

is equivalent to stating that

$$y = x^3 + C$$

is the solution of the differential equation

$$y' = 3x^2 \quad \text{or} \quad \frac{dy}{dx} = 3x^2.$$

A solution of the differential equation

$$s''(t) = 32 \quad \text{or} \quad \frac{d^2 s}{dt^2} = 32$$

is a function whose second derivative is 32. We know that if $s''(t) = 32$, then, by finding the antiderivative of each side, we obtain

$$s'(t) = 32t + C_1,$$

and from the preceding equation we obtain

$$s(t) = 16t^2 + C_1 t + C_2$$

initial conditions

where the arbitrary constants are determined from certain **initial (or boundary) conditions.** For example, if $s(t) = 16t^2 + C_1 t + C_2$ is interpreted as a distance function giving the distance s an object travels in time t, then $s'(t)$ is the velocity at time t and $s''(t)$ is the acceleration. If we are told that the initial velocity $s'(0)$ is 96 ft/sec, then

$$s'(0) = C_1 = 96.$$

If it is also given that $s(0) = 0$, then $C_2 = 0$ and

$$s(t) = 16t^2 + 96t$$

particular solution

is the **particular solution** of the differential equation with the given initial conditions.

order

One of the most important characteristics of a differential equation is its order. The **order** of a differential equation is the order of the highest derivative in the equation. For example,

$$y'(x) = 3x^2$$

is a *first-order* differential equation and

$$s''(t) = 32$$

is a *second-order* differential equation. We shall confine our study to just two types of first-order differential equations, namely,

(1) those as encountered in Sections 5.2, 5.3, and 5.4 which can be solved directly by integration and

(2) those called "separable" which we now define and learn how to solve.

If a differential equation of the form

$$y' = f(x, y) \quad \text{or} \quad \frac{dy}{dx} = f(x, y)$$

can also be written in the form

$$\frac{dy}{dx} = \frac{u(x)}{v(y)},$$

5.7 DIFFERENTIAL EQUATIONS (OPTIONAL)

separable
general
solution

then it is called **separable.** If the variables can be "separated" in this manner, then the general solution takes the form

$$\int v(y)\, dy = \int u(x)\, dx + C.$$

For example, to solve

$$x^2 + e^x - y\frac{dy}{dx} = 0,$$

we can proceed as follows. First we solve the equation for dy/dx,

$$-y\frac{dy}{dx} = -(x^2 + e^x),$$

$$\frac{dy}{dx} = \frac{x^2 + e^x}{y}.$$

For this differential equation, $u(x) = x^2 + e^x$ and $v(y) = y$. Hence, the general solution is

$$\int y\, dy = \int (x^2 + e^x)\, dx + C,$$

$$\frac{y^2}{2} = \frac{x^3}{3} + e^x + C.$$

This (implicit) solution of the separable differential equation can be verified by implicit differentiation. We find that

$$\frac{2yy'}{2} = \frac{3x^2}{3} + e^x \quad \text{or} \quad yy' = x^2 + e^x.$$

In practice, to solve a separable differential equation, one "puts the x's on one side of the equation and the y's on the other" and then integrates. Consider the following examples.

CHAPTER 5 THE ANTIDERIVATIVE AND ITS APPLICATIONS

Separable differential equation

EXAMPLE 1 Find the general solution of $xy' + y = 0$.

Solution

$$x\frac{dy}{dx} = -y,$$

$$\frac{1}{y}\,dy = -\frac{1}{x}\,dx,$$

$$\int \frac{1}{y}\,dy = -\int \frac{1}{x}\,dx + C,$$

$$\ln y = -\ln x + C,$$

$$\ln y + \ln x = C,$$

$$\ln xy = C,$$

$$xy = e^C.$$

Letting $k = e^C$, we find that $y = k/x$. This general solution is easy to check by noting that

$$xy' + y = x\left(-\frac{k}{x^2}\right) + \frac{k}{x} = 0.$$

Unlimited growth equation

EXAMPLE 2 Solve the differential equation

$$\frac{dy}{dx} = ky$$

where k is a constant. (This equation is sometimes called the *unlimited growth/decay equation*. See Section 4.5.)

Solution

$$\frac{dy}{dx} = ky,$$

$$\frac{1}{y}\,dy = k\,dx,$$

$$\int \frac{1}{y}\,dy = k\int dx + C,$$

$$\ln y = kx + C,$$

$$y = e^{kx+C},$$

$$y = (e^{kx})(e^C).$$

5.7 DIFFERENTIAL EQUATIONS (OPTIONAL)

Letting $c = e^C$, we find that the general solution is

$$y = ce^{kx}.$$

Limited growth equation

EXAMPLE 3 Solve

$$\frac{dy}{dx} = k(a - y)$$

where k and a are constants. (This expression states that the rate of change of y with respect to x is directly proportional to the difference between the constant a and the dependent variable y. We assume $y < a$.)

Solution

$$\frac{dy}{dx} = k(a - y),$$

$$\frac{1}{a - y} dy = k\, dx,$$

$$-\int \frac{-1}{a - y} dy = k \int dx + C,$$

$$-\ln(a - y) = kx + C,$$

$$\ln(a - y) = -kx - C,$$

$$a - y = (e^{-kx})(e^{-C}).$$

Letting $A = e^{-C}$, we find the general solution (see Figure 5.1) is

$$y = a - Ae^{-kx}.$$

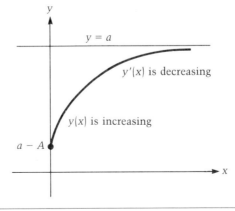

FIG. 5.1. Limited growth equation: $y = a - Ae^{-kx}$.

Separable differential equation

EXAMPLE 4 Solve $\dfrac{dy}{dx} = \dfrac{xy + y}{x + xy}$.

Solution

$$\frac{dy}{dx} = \frac{y(x+1)}{x(y+1)},$$

$$\frac{y+1}{y}\,dy = \frac{x+1}{x}\,dx,$$

$$\int \left(1 + \frac{1}{y}\right) dy = \int \left(1 + \frac{1}{x}\right) dx.$$

Thus $y + \ln y = x + \ln x + c$ is the general solution. The solution also can be expressed in another form in the following way:

$$\ln \frac{y}{x} = x - y + c,$$

$$\frac{y}{x} = e^{x-y+c} = e^c e^{x-y} = c_1 e^{x-y}$$

where $c_1 = e^c$. Thus, $y = c_1 x e^{x-y}$ is the general solution.

We encountered the solution to the differential equation given in Example 2 when we discussed exponential growth in Section 4.5. The equation expresses a general law of unrestricted growth/decay. It states that the *rate of change of y* (the dependent variable) with respect to x (the independent variable) is directly proportional to y. For example, if $y(x)$ represents the number of bacteria in a culture at time x, then

$$\frac{dy}{dx} = ky$$

expresses a law that the bacteria growth (rate of change in th amount) is directly proportional to the number y present at given time x. This equation, as we know, can also be used to describe such phenomena as radioactive decay and population growth. Since the range of the solution (growth) is unbounded and since physical factors place an upper limit on population growth, this equation cannot represent population growth over long periods of time.

A reasonable model (differential equation) describing population growth, spread of an epidemic throughout a population,

5.7 DIFFERENTIAL EQUATIONS (OPTIONAL)

spread of a rumor throughout a population, growth of bacteria in a culture, etc., where a represents the upper limit on the growth is given by

$$\frac{dy}{dx} = ky(a - y).$$

We can see for y near zero or for y near a that the rate of change (growth) is small. (This equation was discussed in Section 4.4.) Let us solve this separable differential equation.

$$\frac{dy}{dx} = ky(a - y)$$

$$\frac{1}{y(a - y)} dy = k\, dx. \tag{1}$$

It is easy to show that

$$\frac{1}{a}\left(\frac{1}{y} + \frac{1}{a - y}\right) = \frac{1}{y(a - y)}.$$

From the preceding equation and Eq. (1), we get

$$\frac{1}{a}\int\left(\frac{1}{y} + \frac{1}{a - y}\right) dy = k\int dx + C,$$

$$\frac{1}{a}[\ln y - \ln(a - y)] = kx + C,$$

$$\ln\frac{y}{a - y} = kax + Ca,$$

$$\frac{y}{a - y} = (e^{akx})(e^{Ca}).$$

Letting $1/c = e^{Ca}$, we get

$$\frac{y}{a - y} = \frac{e^{akx}}{c},$$

$$cy = ae^{akx} - ye^{akx},$$

$$ye^{akx} + cy = ae^{akx},$$

$$(e^{akx} + c)y = ae^{akx},$$

$$y = \frac{ae^{akx}}{e^{akx} + c},$$

$$y = \frac{a}{1 + ce^{-akx}}.$$

FIG. 5.2.
Logistic equation:
$$y = \frac{a}{(1 + ce^{-akx})}.$$

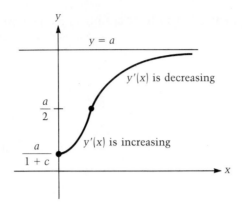

logistics equation

This important solution is called the **logistics equation.** (See Figure 5.2.)

EXERCISES

Solve each of the separable differential equations in Exercises 1 through 10.

1. $3x\dfrac{dy}{dx} - y = 0$

2. $x^2 - e^{-x} - y\dfrac{dy}{dx} = 0$

3. $xy + e^x \dfrac{dy}{dx} = 0$

4. $x\dfrac{dy}{dx} + y = 2$

5. $xy\dfrac{dy}{dx} + y\dfrac{dy}{dx} = 3$

6. $e^y \dfrac{dy}{dx} = x - x^2$

7. $x + (1 - y)\dfrac{dy}{dx} = 0$

8. $(1 + x^2)\dfrac{dy}{dx} = xy$

9. $\dfrac{dy}{dx} = y(x + e^x)$

10. $\dfrac{dy}{dx} = xy + x + y + 1$

Unlimited growth

11. (a) For the exponential growth function $y(x) = Ae^{kx}$, $A > 0$ and $k > 0$, if $y(x)$ represents the amount present at time x, what does A represent?
    (b) If $y(x)$ doubles from the time $x = 0$ to the time $x = 1$, what is k? What is $y(2)$?

5.8 SUMMARY AND REVIEW EXERCISES

1. ANTIDERIVATIVE OF A FUNCTION

Let g be a differentiable function. If $g'(x) = f(x)$ for each x in the domain of g, then g is an **antiderivative** (or **indefinite integral**) of f.

5.8 SUMMARY AND REVIEW EXERCISES

Since antiderivatives of f can differ at most by a constant,

$$\int f(x) \, dx = g(x) + C$$

is called the antiderivative of f where C is the *constant of integration*.

2. ANTIDERIVATIVE FORMULAS

(a) $\int x^p \, dx = \dfrac{x^{p+1}}{p+1} + C$ provided $p \neq -1$.

(b) $\int \dfrac{1}{x} \, dx = \ell n \, x + C$

(c) $\int [f(x)]^p f'(x) \, dx = \dfrac{[f(x)]^{p+1}}{p+1} + C$ provided $p \neq -1$.

(d) $\int \dfrac{f'(x)}{f(x)} \, dx = \ell n \, f(x) + C$

(e) $\int e^x \, dx = e^x + C$

(f) $\int e^{f(x)} f'(x) \, dx = e^{f(x)} + C$

(g) $\int k f(x) \, dx = k \int f(x) \, dx$, k a constant

(h) $\int [f(x) + g(x)] \, dx = \int f(x) \, dx + \int g(x) \, dx$

(i) $\int f(x) g'(x) \, dx = f(x) g(x) - \int f'(x) g(x) \, dx$

3. REVIEW EXERCISES

In Exercises 1 through 10, find the given antiderivatives.

1. $\int \left(5x^{3/2} - \dfrac{1}{x^2} \right) dx$

2. $\int (x^2 + 1)^2 \, dx$

3. $\int x(x^2 + 4)^{2/5} \, dx$

4. $\int x^2(x^3 + 1)^{2/3} \, dx$

5. $\int \left(e^{-x} - \dfrac{1}{2x+1} \right) dx$

6. $\int \left(e^{2x} + \dfrac{x}{x^2 + 1} \right) dx$

7. $\int x(4x + 1)^{1/2} \, dx$

8. $\int x(3x + 1)^{3/2} \, dx$

9. $\displaystyle\int \frac{\ln x}{x}\,dx$

10. $\displaystyle\int \frac{1}{x(\ln x)}\,dx$

11. The slope of the tangent line to the graph of a function f is $4x^{1/3} + 2x$ at each x in its domain. If the graph of f contains the point $(1, 5)$, what is an equation defining the function?

12. The slope of the tangent line to the graph of a function f is $6x + e^x$ at each x in its domain. Given that the graph of f contains the point $(0, 2)$, find an equation defining the function.

13. The marginal cost in dollars for producing x units of a given product each month is $C'(x) = 300 - x/2$. If the cost of producing the first ten units each month is $4500, what is the monthly cost function?

14. The marginal profit on the sale of x units of a given product is $84 - 2x$ dollars. If the profit from the sale of 10 units is $500, what is the profit (loss) from the sale of 4 units? How many units would need to be sold in order to maximize the profit and what would be the maximum profit possible?

15. The rate of change of the slope of a twice differentiable function f is -6 at each x in its domain. Given that the function has a relative maximum at the point $(2, 4)$, find an equation defining f.

16. The initial velocity of a bullet fired vertically upward is 1200 ft/sec. Find its altitude h after t seconds. [*Hint:* Since we ignore air resistance, the bullet is decelerating at 32 ft/sec² due to gravity; that is, $h''(t) = -32$.]

17. A car starting from rest accelerates at $4t$ ft/sec² for 6 seconds. What is its velocity at the end of 6 seconds? How far does it travel in 6 seconds?

18. Suppose the rate of growth of population P of a certain country t years from now is given by $e^{0.04t}$ million per year. If the population today is 40 million, what will it be in 20 years?

19. Suppose the rate of a person's heart beat t minutes after completing a series of exercises is given by $120 - 4t$ beats per minute, $0 \le t \le 15$. How many times does the person's heart beat during the 15 minutes?

20. A car starting from rest accelerates at $3t$ ft/sec² for 8 seconds. What is its velocity at the end of 8 seconds? How far does it travel in 8 seconds?

BIOGRAPHICAL SKETCH

Sonya Kovalevsky (1850–1891)

The Bettmann Archive

David Eugene Smith Collection, Rare Book and Manuscript Library, Columbia University.

Sonya Kovalevsky was born in Moscow, Russia, on January 15, 1850 and was to become one of the foremost female mathematicians of all time. Two circumstances helped her to overcome the enormous obstacles the scientific community placed in the way of any woman aspiring to become one of its members. First, she was born into a prosperous Russian family at a time when the political and intellectual climate in Russia was in favor of the emancipation of women. Second, when she went to Berlin to study she became the friend and favorite pupil of Karl Weierstrass, who is often called the father of modern analysis and is considered to have been one of the world's greatest teachers of advanced mathematics. Since she could not be admitted to the University, Weierstrass gave her copies of his lecture notes, met with her for conferences, and suggested books for her to read; this activity culminated in 1874 with her being awarded a doctorate *(in absentia)* by the University of Göttingen.

One of her research papers received the *Prix Bordin* from the Academy; the prize carried with it an award of 5000 francs, a considerable amount of money in those times. Eventually, a distinguished pupil of Weierstrass, Mittag-Leffler, secured for her a position as lecturer at the University of Stockholm, and in 1889 he succeeded in obtaining for her a life professorship.

Another famous female mathematician is Emmy Noether (1882–1935). She went to Göttingen in 1916, forty-two years after Sonya Kovalevsky had received her doctorate there. Although she could not obtain a position commensurate with her abilities, David Hilbert and Felix Klein were responsible for her staying there. Hermann Weyl, who succeeded Hilbert at Göttingen, stated that from 1930 until 1933 Emmy Noether was the "strongest center of mathematical activity" at Göttingen. Although she either adjusted to or overcame the obstacles confronting her because of her sex, she, along with many of her famous contemporaries, was unable to withstand the systematic persecution of Jews in Hitler's Germany. In 1933 she came to the United States where she was offered a position at Bryn Mawr.

CHAPTER

6

THE DEFINITE INTEGRAL AND ITS APPLICATIONS

6

6.1 INTRODUCTION

6.2 THE DEFINITE INTEGRAL

6.3 MORE ON AREA

6.4 OTHER APPLICATIONS OF THE DEFINITE INTEGRAL (OPTIONAL TOPICS)

A. AVERAGE VALUE AND THE LORENTZ CURVE

B. CONSUMER'S SURPLUS AND PRODUCER'S SURPLUS

C. PROBABILITY DENSITY FUNCTION

D. VOLUME BY THE DISK METHOD

6.5 SUMMARY AND REVIEW EXERCISES

6.1 INTRODUCTION

In this chapter, we define the definite integral (Riemann integral) of a function, discover its remarkable relationship with the antiderivative (indefinite integral) of a function, and then learn how to solve problems such as the following.

1. A manufacturer estimates that over the next 24 months 3000 units of a given product can be sold each month and that the price in x months will be $p(x) = 40 + 2(x + 1)^{1/2}$ dollars. What will be the total revenue over the next 24 months from the sale of this product? (Exercise 16, Review Exercises.)

2. What is the area of the region in the rectangular coordinate plane bounded by the graphs of $y = x^2$ and $y = 2x + 8$? (See Figure 6.1a and Example 1, Section 6.3.)

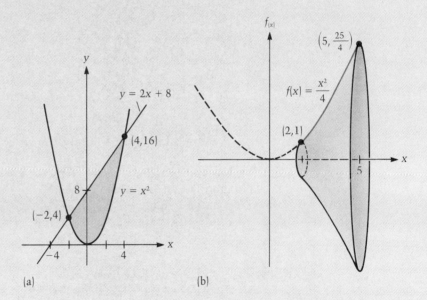

FIG. 6.1. (a) Area, (b) Volume.

3. What is the volume of the solid generated by revolving about the x-axis the region bounded by $f(x) = x^2/4$, the x-axis, and the lines $x = 2$ and $x = 5$? (See Figure 6.1b and Exercise 17, Section 6.4.)

4. Suppose a traffic engineer finds that the average velocity of traffic in miles per hour on a stretch of interstate highway between 3 o'clock in the afternoon and 8 o'clock that evening is given by $s(t) = t^3 - 18t^2 + 105t - 130$, $3 \le t \le 8$. Find the average velocity of the traffic during this time interval. (See Exercise 5, Section 6.4.) ■

6.2 THE DEFINITE INTEGRAL

In Euclidean plane geometry we learn how to assign a number called *area* to polygonal figures in the plane. For example, we learn that a triangle has area $(\frac{1}{2})bh$, where b is the length of its base and h is its altitude. A rectangle has area ℓw, where ℓ is its length and w is its width. Since any polygonal figure can be considered to be made up of triangles (see Figure 6.2), the area of any plane polygonal figure can be obtained by adding areas of triangles. The formula for finding the area of a circle is also discussed in geometry, but no attempt is made to discover a method for assigning area to an arbitrarily shaped region in the plane. Let us now begin to consider this important problem.

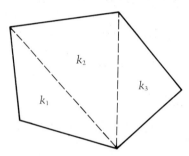

FIG. 6.2.
Area $= k_1 + k_2 + k_3$.

Suppose we wish to find the area of (assign area to) the plane region R in Figure 6.3a bounded by the graphs of the two functions f and g which are above the x-axis. If $A(R_1)$ is the area of the region bounded by the graph of f, the lines $x = a$, $x = b$, and the x-axis (see Figure 6.3b) and if $A(R_2)$ is the area of the region bounded by the graph of g, the lines $x = a$, $x = b$, and the x-axis (see Figure 6.3c), then

$$A(R) = A(R_1) - A(R_2)$$

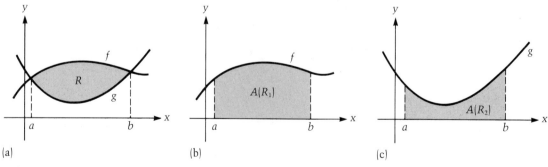

FIG. 6.3. $A(R) = A(R_1) - A(R_2)$.

6.2 THE DEFINITE INTEGRAL

where $A(R)$ is the area of the region between the graphs of f and g and the lines $x = a$ and $x = b$. Therefore, the problem of assigning area to such a region as R is reduced to assigning area to a region under the graph of a function, above the x-axis, and between two lines perpendicular to the x-axis. Let us look at a specific example.

Let f be defined by $f(x) = x^2$ and consider the region bounded by the graph of f, the x-axis, and the lines $x = 2$, $x = 6$. Suppose we divide the interval into *four equal* subintervals. The set of endpoints of the subintervals is $P = \{2, 3, 4, 5, 6\}$; any such set dividing $[2, 6]$ into *equal* subintervals is called a *regular partition* of $[2, 6]$. Since f is an increasing function on $[2, 6]$, the minimum values of f on each of the subintervals are $f(2) = 4$, $f(3) = 9$, $f(4) = 16$, and $f(5) = 25$. (See Figure 6.4.) Since the width of each subinterval is one unit, the total area of the inscribed rectangles under the graph of f is

$$f(2) + f(3) + f(4) + f(5) = 54 \text{ square units.}$$

We call 54 the *lower sum of f with respect to partition P of $[2, 6]$* and denote it by L_4. The area A of the region in question must have the property that $L_4 < A$.

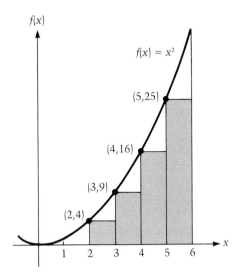

FIG. 6.4.
$L_4 = 4 + 9 + 16 + 25 = 54.$

PRACTICE PROBLEM Let $f(x) = x^2$ and let

$$P = \{2, 2.5, 3, 3.5, 4, 4.5, 5, 5.5, 6\}$$

be the regular partition of [2, 6] with eight equal subintervals. (See Figure 6.5.) Find the lower sum L_8.

Answer: $61\frac{1}{2}$. ∎

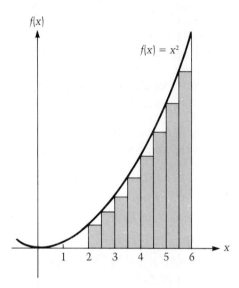

FIG. 6.5. This figure is the same as Fig. 6.4 with twice as many inscribed rectangles at the partition points 2, 2.5, 3, 3.5, 4, 4.5, 5, 5.5, and 6.

Suppose we use a regular partition of [2, 6] with n equal subintervals to find the lower sum L_n. Since this task involves some algebraic manipulations we wish to avoid, we state without proof that

$$L_n = 16 + 32\left(1 - \frac{1}{n}\right) + \frac{32}{3}\left(1 - \frac{1}{n}\right)\left(2 - \frac{1}{n}\right).$$

As a partial check we let $n = 4$ and find that

$$L_4 = 16 + 32\left(\frac{3}{4}\right) + \left(\frac{32}{3}\right)\left(\frac{3}{4}\right)\left(\frac{7}{4}\right) = 54.$$

PRACTICE PROBLEM Use the formula for L_n to verify the value for L_8 found in the preceding practice problem.

As n gets large, $(1 - 1/n)$ gets close to 1 and $(2 - 1/n)$ gets close to 2. Therefore for large n,

6.2 THE DEFINITE INTEGRAL

$$L_n = 16 + 32\left(1 - \frac{1}{n}\right) + \frac{32}{3}\left(1 - \frac{1}{n}\right)\left(2 - \frac{1}{n}\right)$$

gets close to

$$16 + 32 + \frac{32}{3}(2) = 69\frac{1}{3}.$$

The area A of the region is defined to be $69\frac{1}{3}$ square units.

In general, for any function f continuous on $[a, b]$ we define the **lower sum** of f on $[a, b]$ by

lower sum

$$L_n = \sum_{k=1}^{n} f(u_k)\,\Delta x = f(u_1)\,\Delta x + f(u_2)\,\Delta x + \cdots + f(u_n)\,\Delta x,$$

where $f(u_k)$ is the minimum value on the kth subinterval and Δx is the width of each of the n equal subintervals. (Geometrically, L_n is the total area of the n inscribed rectangles of equal width, provided $f(x) > 0$. (See Figure 6.6.) Note, however, that in general we do not assume that $f(x) > 0$ on $[a, b]$).

For a continuous function f, it can be proved that L_n approaches a real number as n gets large. This *limit* (number) is called the **Riemann integral**, or **definite integral**, of f from a to b. It is denoted by

Riemann (or definite) integral

$$\int_a^b f(x)\,dx \quad \text{or} \quad \int_a^b f.$$

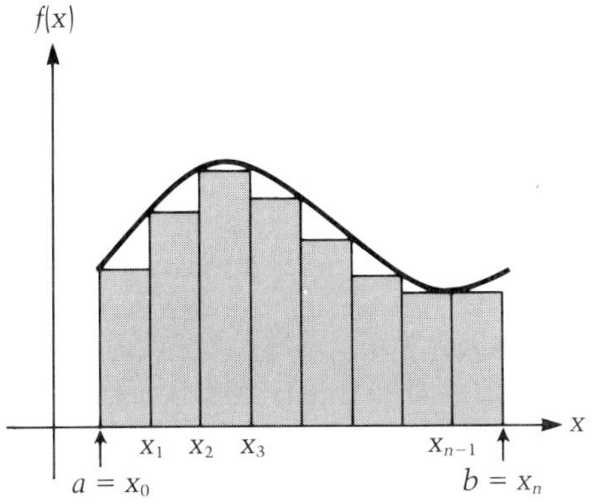

FIG. 6.6.
$L_n = [f(u_1) + f(u_2) + f(u_3) + \cdots + f(u_n)]\Delta x.$

Symbolically,

$$\lim_{n\to\infty} L_n = \int_a^b f(x)\, dx.$$

area

For a continuous function f, if $f(x) \geq 0$ for each x in $[a, b]$, then $\int_a^b f(x)\, dx$ is the **area** of the region bounded by the graph of f, the lines $x = a$, $x = b$, and the x-axis.

Using the Riemann integral to determine areas of regions in the plane is only one of its many important applications. It can be used to find the volume of a solid, length of an arc, work done by a variable force such as in stretching a spring, etc. (We shall investigate many applications in the succeeding sections.)

We already know for a continuous function f that

$$\int_a^b f(x)\, dx \quad \text{exists provided} \quad a < b.$$

We define $\int_a^b f(x)\, dx$ for $a = b$ or $a > b$ as follows:

$$\int_a^a f(x)\, dx = 0 \quad \text{and} \quad \int_a^b f(x)\, dx = -\int_b^a f(x)\, dx.$$

With this definition, the following theorem can be proved for any continuous function f. For any real numbers a, b, and c,

$$\int_a^b f(x)\, dx + \int_b^c f(x)\, dx = \int_a^c f(x)\, dx.$$

If $f(x) > 0$ and $a < b < c$, then this theorem is geometrically obvious since it states that the area of the entire region above the x-axis, below the graph of f, and between the lines $x = a$ and $x = c$ is the sum of the areas of the two regions shown in Figure 6.7 which make it up.

The student should be convinced of one thing: finding the Riemann integral of a given function from the definition is, at the very least, tedious. It should be apparent that if a function f does not have a "well-behaved" graph, then the definition would prove

6.2 THE DEFINITE INTEGRAL

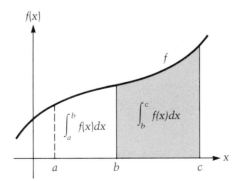

FIG. 6.7.
$\int_a^b f(x)\,dx + \int_b^c f(x)\,dx = \int_a^c f(x)\,dx$.

extremely difficult, if not impossible, to use. Let us now consider another technique for finding the Riemann integral of a specific function. The technique not only is applicable to many continuous functions but also indicates a very important application for the antiderivative of a function.

The following two steps can be used to evaluate

$$\int_2^6 x^2\,dx.$$

First find an *antiderivative* of $f(x) = x^2$; one antiderivative of f is $F(x) = x^3/3$. Second, evaluate $F(6) - F(2)$. That is, find the difference in the values of an antiderivative of the given function at the endpoints of the interval [2, 6]. This difference is the Riemann integral; that is,

$$\int_2^6 x^2\,dx = F(6) - F(2) = \frac{216}{3} - \frac{8}{3} = 69\frac{1}{3}.$$

The preceding facts are incorporated into a truly remarkable theorem, called the *fundamental theorem of calculus*, which we now state. (It indicates a most important application of the antiderivative.)

> **THEOREM 6.1 (FUNDAMENTAL THEOREM OF CALCULUS)**
> Let f be a continuous function on $[a, b]$.
>
> *Conclusion:* An antiderivative of f exists and if F is any antiderivative of f, then
>
> $$\int_a^b f(x)\,dx = F(b) - F(a).$$

Let us see why the Fundamental Theorem of Calculus is true. (Part of our proof will be justified by geometric considerations.)

Suppose f is any continuous function on $[a, b]$. Then for any u in $[a, b]$,

$$\int_a^u f(x)\, dx \quad \text{exists.}$$

Let G be the function with $[a, b]$ as its domain defined by

$$G(u) = \int_a^u f(x)\, dx$$

where $a \leq u \leq b$. We will now show that for any t in $[a, b]$

$$G'(t) = f(t).$$

From the definition of the derivative of G at t,

$$G'(t) = \lim_{\Delta x \to 0} \frac{G(t + \Delta x) - G(t)}{\Delta x},$$

provided the limit exists. Now from the definition of G,

$$G'(t) = \lim_{\Delta x \to 0} \frac{\int_a^{t+\Delta x} f(x)\, dx - \int_a^t f(x)\, dx}{\Delta x}.$$

Since

$$\int_a^t f(x)\, dx + \int_t^{t+\Delta x} f(x)\, dx = \int_a^{t+\Delta x} f(x)\, dx,$$

we have

$$\int_a^{t+\Delta x} f(x)\, dx - \int_a^t f(x)\, dx = \int_t^{t+\Delta x} f(x)\, dx.$$

Consequently,

$$G'(t) = \lim_{\Delta x \to 0} \frac{\int_t^{t+\Delta x} f(x)\, dx}{\Delta x}.$$

6.2 THE DEFINITE INTEGRAL

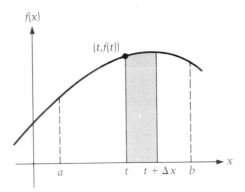

FIG. 6.8. Area $\int_t^{t+\Delta x} f(x)\, dx \doteq f(t)(\Delta x)$.

In Figure 6.8 we see that the preceding integral is the area of the shaded region and Δx is the width of the region. If Δx is close to 0, then the region is nearly rectangular. Therefore the area divided by the width is close to the "height" $f(t)$ of the "rectangle." From this, we conclude that

$$G'(t) = f(t).$$

We are now able to show that the antiderivative technique (Fundamental Theorem of Calculus) for finding

$$\int_2^6 x^2\, dx$$

is valid.

Let f be a continuous function on $[a, b]$ and let F be *any* antiderivative of f. Hence $F'(t) = f(t)$. If

$$G(u) = \int_a^u f(x)\, dx,$$

then since $G'(t) = f(t)$, the function G is also an antiderivative of f. We know that the antiderivatives F and G can differ at most by a constant C; that is,

$$G(x) = F(x) + C \quad \text{for each } x \text{ in } [a, b].$$

Since $G(a) = 0$, we conclude that $F(a) + C = 0$, which means $C = -F(a)$. Therefore

$$G(x) = F(x) - F(a)$$

for each x in $[a, b]$. In particular,
$$G(b) = F(b) - F(a).$$
That is,
$$\int_a^b f(x)\, dx = F(b) - F(a).$$

Riemann integral

EXAMPLE 1 Find $\displaystyle\int_1^3 \frac{1}{x}\, dx$.

Solution: If $f(x) = 1/x$, $1 \le x \le 3$, then one antiderivative of f is $F(x) = \ell n\, x$. Using the Fundamental Theorem of Calculus, we obtain
$$\int_1^3 \frac{1}{x}\, dx = F(3) - F(1)$$
$$= \ell n\, 3 - \ell n\, 1$$
$$= \ell n\, 3 \quad (\text{since } \ell n\, 1 = 0).$$

Finding the Riemann (definite) integral of a function by means of the Fundamental Theorem of Calculus necessitates obtaining an antiderivative. A number of important formulas are immediately available to us as a result of previously determined antiderivative formulas. Three of them are as follows.

FORMULA 6.1 $\displaystyle\int_a^b x^p\, dx = \frac{b^{p+1} - a^{p+1}}{p + 1}$ provided $p \ne -1$.

FORMULA 6.2 $\displaystyle\int_a^b kf(x)\, dx = k \int_a^b f(x)\, dx$, k a constant.

FORMULA 6.3 $\displaystyle\int_a^b [f(x) + g(x)]\, dx = \int_a^b f(x)\, dx + \int_a^b g(x)\, dx.$

Generally, when finding a Riemann integral, the notation
$$F(x)\,\Big|_a^b$$
is used to represent the difference $F(b) - F(a)$. Consider the following example.

6.2 THE DEFINITE INTEGRAL

Riemann integral

EXAMPLE 2 Find $\int_4^9 x^{1/2}\, dx$.

Solution

$$\int_4^9 3x^{1/2}\, dx = 2x^{3/2} \Big|_4^9$$
$$= 54 - 16$$
$$= 38.$$

Riemann integral

EXAMPLE 3 Evaluate the Riemann integral

$$\int_1^3 (x^2 + 2x)\, dx.$$

Solution

$$\int_1^3 (x^2 + 2x)\, dx = \left(\frac{x^3}{3} + x^2\right)\Big|_1^3 = \left(\frac{27}{3} + 9\right) - \left(\frac{1}{3} + 1\right) = 16\frac{2}{3}.$$

PRACTICE PROBLEM Evaluate $\int_1^4 (3x^2 - 2x + 1)\, dx$.

Answer: 51. ∎

Before turning to the exercises, we consider three more formulas and four additional examples.

FORMULA 6.4 $\int_a^b [f(x)]^p f'(x)\, dx = \dfrac{[f(b)]^{p+1} - [f(a)]^{p+1}}{p+1},$
provided $p \neq -1$.

FORMULA 6.5 $\int_a^b e^{f(x)} f'(x)\, dx = e^{f(b)} - e^{f(a)}.$

FORMULA 6.6 $\int_a^b \dfrac{f'(x)}{f(x)}\, dx = \ln f(b) - \ln f(a).$

Riemann integral

EXAMPLE 4 Evaluate the Riemann integral $\int_0^1 (7x+1)^{2/3}(7)\, dx$.

Solution

$$\int_0^1 (7x+1)^{2/3}(7)\,dx = \frac{3}{5}[(7x+1)^{5/3}]\Big|_0^1$$

$$= \frac{3}{5}(8^{5/3} - 1^{5/3})$$

$$= \frac{3}{5}(32 - 1)$$

$$= \frac{93}{5}.$$

Riemann integral

EXAMPLE 5 Evaluate the Riemann integral

$$\int_0^{\sqrt{15}} (x^2+1)^{1/2}x\,dx.$$

Solution

$$\int_0^{\sqrt{15}} (x^2+1)^{1/2}x\,dx = \frac{1}{2}\int_0^{\sqrt{15}} (x^2+1)^{1/2}2x\,dx$$

$$= \frac{1}{2}\cdot\frac{2}{3}(x^2+1)^{3/2}\Big|_0^{\sqrt{15}}$$

$$= \frac{1}{3}[(16)^{3/2} - 1^{3/2}]$$

$$= 21.$$

PRACTICE PROBLEM Evaluate

$$\int_0^2 x^2(x^3+1)^{3/2}\,dx.$$

Answer: $\frac{484}{15}$. ∎

Riemann integral

EXAMPLE 6 Evaluate

$$\int_2^5 \frac{x}{x^2-3}\,dx.$$

6.2 THE DEFINITE INTEGRAL

Solution

$$\int_2^5 \frac{x}{x^2 - 3} \, dx = \frac{1}{2} \int_2^5 \frac{2x}{x^2 - 3} \, dx$$

$$= \frac{1}{2} \ln(x^2 - 3) \Big|_2^5$$

$$= \frac{1}{2} [\ln 22 - \ln 1]$$

$$= \frac{1}{2} \ln 22.$$

Area

EXAMPLE 7 Sketch a graph of the function $f(x) = 4x - x^2$ and find the area of the region bounded by the graph of f, the x-axis, and the lines $x = 1$ and $x = 3$. (See Figure 6.9.)

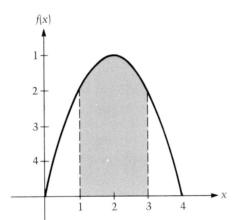

FIG. 6.9. $f(x) = 4x - x^2$.

Solution: The area of the region is

$$\int_1^3 (4x - x^2) \, dx = \left(2x^2 - \frac{x^3}{3}\right) \Big|_1^3$$

$$= (18 - 9) - \left(2 - \frac{1}{3}\right)$$

$$= \frac{22}{3} \text{ square units.}$$

To evaluate the Riemann integral

$$\int_2^6 x(4x+1)^{1/2}\,dx,$$

we can find an antiderivative by the substitution technique and then use the Fundamental Theorem of Calculus. Another method involves changing the variable and the limits of integration. The following theorem justifies this technique which is exhibited in Example 8.

THEOREM 6.2 (CHANGE OF VARIABLE) Assume g is a differentiable function on $[a, b]$ and assume g' is continuous on $[a, b]$. Let $c = g(a)$ and $d = g(b)$. If h is a continuous function on the range of g, then

$$\int_a^b h(g(x))g'(x)\,dx = \int_c^d h(x)\,dx.$$

Substitution technique

EXAMPLE 8 Evaluate

$$\int_2^6 x(4x+1)^{1/2}\,dx.$$

Solution: Let $u = 4x + 1$. Thus $x = (\tfrac{1}{4})(u-1)$ and $dx = (\tfrac{1}{4})\,du$. Since $u = 9$ when $x = 2$ and $u = 25$ when $x = 6$, we have

$$\int_2^6 x(4x+1)^{1/2}\,dx$$

$$= \int_4^{25} \frac{1}{4}(u-1)u^{1/2}\frac{1}{4}\,du$$

$$= \frac{1}{16}\int_9^{25} (u^{3/2} - u^{1/2})\,du$$

$$= \frac{1}{16}\left(\frac{2}{5}u^{5/2} - \frac{2}{3}u^{3/2}\right)\Big|_9^{25}$$

$$= \frac{1}{16}\left[\frac{2}{5}(25)^{5/2} - \frac{2}{3}(25)^{3/2} - \frac{2}{5}(9)^{5/2} + \frac{2}{3}(9)^{3/2}\right]$$

$$= \frac{1}{16}\left[2(5)^4 - \frac{2}{3}(5)^3 - \frac{2}{5}(3)^5 + 2(3)^2\right]$$

$$= \frac{2039}{30}.$$

Note: If $h(x) = (x^{3/2} - x^{1/2})/16 = x^{1/2}(x-1)/16$ and if $g(x) = 4x + 1$, then $g'(x) = 4$ and

$$h(g(x))g'(x) = (4x+1)^{1/2}(4x)(4)/16 = x(4x+1)^{1/2}.$$

Furthermore, $9 = g(2)$ and $25 = g(6)$.

EXERCISES

Set A—Riemann Integrals of Functions

In each of Exercises 1 through 26 find the value of the given Riemann integral.

1. $\int_1^4 x^4 \, dx$
2. $\int_{-1}^2 x^3 \, dx$
3. $\int_0^2 6x^2 \, dx$
4. $\int_{-1}^2 10x^4 \, dx$
5. $\int_0^{-2} (x - x^3) \, dx$
6. $\int_{-1}^0 (x^2 - x^3) \, dx$
7. $\int_1^4 (x^2 - x) \, dx$
8. $\int_1^3 (x^2 - 2x + 7) \, dx$
9. $\int_4^7 (v - 3)^{1/2} \, dv$
10. $\int_1^2 \frac{u+1}{u^3} \, du$
11. $\int_4^1 x(\sqrt{x} - 1) \, dx$
12. $\int_{-1}^{-27} u(u^{1/3} - 1) \, du$
13. $\int_0^1 (x^3 + 2x)^4 (3x^2 + 2) \, dx$
14. $\int_0^2 (4x + 1)^{3/2} \, dx$
15. $\int_0^1 (5u^2 + 1)^3 u \, du$
16. $\int_0^1 (x^2 + 4x + 4)^{1/2}(x + 2) \, dx$
17. $\int_{-3}^3 (x^3 + x) \, dx$
18. $\int_{-3}^3 (x + 3)^2 \, dx$

19. $\int_0^3 x\sqrt{x^2 + 16}\, dx$ 20. $\int_0^2 x(2x^2 + 1)^{1/2}\, dx$

21. $\int_0^2 x(3x^2 + 4)\, dx$ 22. $\int_0^1 x^2\sqrt{5x^3 + 4}\, dx$

23. $\int_0^1 e^{3x}\, dx$ 24. $\int_0^2 \frac{1}{x + 1}\, dx$

25. $\int_1^5 \frac{1}{2x + 1}\, dx$ 26. $\int_0^2 e^{-x}\, dx$

Use Theorem 6.2 to evaluate the definite integrals in Exercises 27 and 28.

27. $\int_1^5 x(3x + 1)^{3/2}\, dx$ 28. $\int_0^2 x(4x + 1)^{1/2}\, dx$

Set B—Areas of Regions in the Coordinate Plane

29. (a) Graph the function $f(x) = 2x + 1$ and then find the area of the trapezoid bounded by the graph of the function, the lines $x = 1$, $x = 5$, and the x-axis using the formula for the area of a trapezoid. Note: $A = (h/2)(b_1 + b_2)$ where b_1 and b_2 are the lengths of the bases (parallel sides) of the trapezoid and h is the altitude, the distance between the parallel lines.
    (b) Use the Riemann integral to verify the answer in part (a).

30. (a) Graph the function $f(x) = 3x + 2$ and then find the area of the trapezoid bounded by the graph of the function, the lines $x = 1$, $x = 6$, and the x-axis, using the formula for the area of a trapezoid. (See Exercise 29.)
    (b) Use the Riemann integral to verify the answer in part (a).

In each of Exercises 31 through 38 sketch a graph of the given function and find the area of the region above the x-axis and bounded by the graphs of f and the lines $x = a$ and $x = b$.

31. $f(x) = x^3$, $a = 1$ and $b = 3$
32. $f(x) = x^{1/3}$, $a = 1$ and $b = 8$
33. $f(x) = x^{3/2}$, $a = 1$ and $b = 9$
34. $f(x) = x^{2/3}$, $a = 1$ and $b = 8$
35. $f(x) = x^2 + 4x$, $a = 1$ and $b = 3$
36. $f(x) = x^{1/3} + x$, $a = 1$ and $b = 8$
37. $f(x) = \dfrac{1}{x}$, $a = 2$ and $b = 6$
38. $f(x) = e^x$, $a = 0$ and $b = 2$

6.3 MORE ON AREA

As we know, one application of the Riemann integral is in finding the area of a region in the coordinate plane which is above the x-axis, between the lines $x = a$ and $x = b$, and below the graph of a continuous function f provided $f(x) \geq 0$. The Riemann integral can also be used to obtain areas of more general regions in the plane as shown in the following examples.

Area

EXAMPLE 1 Find the area of the region bounded by the graphs of $y = x^2$ and $y = 2x + 8$. (See Figure 6.10.)

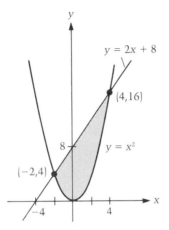

FIG. 6.10.

Solution: First we find the points of intersection of the graphs of the two functions. Substituting $y = x^2$ in the equation $y = 2x + 8$, we get

$$x^2 = 2x + 8$$
$$x^2 - 2x - 8 = 0$$
$$(x - 4)(x + 2) = 0$$
$$x = -2 \quad \text{or} \quad x = 4.$$

The area above the x-axis, between $x = -2$ and $x = 4$, and under the graph of $y = 2x + 8$ is

$$\int_{-2}^{4} (2x + 8)\, dx = (x^2 + 8x)\Big|_{-2}^{4}$$
$$= 48 - (-12)$$
$$= 60 \text{ square units}.$$

The area above the x-axis, between $x = -2$ and $x = 4$, and under the graph of $y = x^2$ is

$$\int_{-2}^{4} x^2 \, dx = \frac{x^3}{3} \Big|_{-2}^{4} = \frac{64 - (-8)}{3} = 24 \text{ square units.}$$

Hence, the area of the given region is $60 - 24 = 36$ square units.

If $f(x) \leq 0$ for x in $[a, b]$, then

$$\int_{a}^{b} [-f(x)] \, dx = -\int_{a}^{b} f(x) \, dx$$

is the *area* of the region bounded by the graph of f, the lines $x = a$, $x = b$, and the x-axis. Therefore, the total area of the regions indicated in Figure 6.11 is given by

$$\int_{a}^{c} [-f(x)] \, dx + \int_{c}^{b} f(x) \, dx.$$

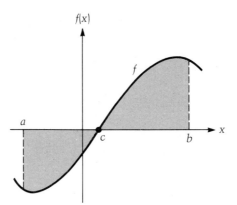

FIG. 6.11. Area = $\int_{a}^{c} [-f(x)] \, dx + \int_{c}^{b} f(x) \, dx.$

Area

EXAMPLE 2 Find the area of the regions in the coordinate plane bounded by the graph of $y = x(x - 2)(x + 3)$ and the x-axis. (See Figure 6.12.)

6.3 MORE ON AREA

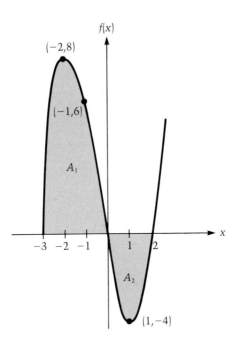

FIG. 6.12.
$f(x) = x(x + 3)(x - 2)$.

Solution: The area A_1 of the region above the x-axis is

$$A_1 = \int_{-3}^{0} x(x + 3)(x - 2)\, dx = \int_{-3}^{0} (x^3 + x^2 - 6x)\, dx$$

$$= \left(\frac{x^4}{4} + \frac{x^3}{3} - 3x^2 \right) \Big|_{-3}^{0}$$

$$= 0 - \left(\frac{81}{4} - 9 - 27 \right)$$

$$= \frac{63}{4} \text{ square units.}$$

Since the function values are negative where $0 \leq x \leq 2$, the area of the region between 0 and 2 is the negative of the Riemann integral from 0 to 2. The area A_2 of this region is

$$A_2 = -\int_{0}^{2} (x^3 + x^2 - 6x)\, dx = -\left(\frac{x^4}{4} + \frac{x^3}{3} - 3x^2 \right) \Big|_{0}^{2}$$

$$= -\left(4 + \frac{8}{3} - 12 \right)$$

$$= \frac{16}{3} \text{ square units.}$$

The total area of the indicated regions is

$$A_1 + A_2 = \frac{63}{4} + \frac{16}{3} = \frac{253}{12} \text{ square units.}$$

area

Let f and g be two continuous functions on $[a, b]$ where $f(x) \geq g(x)$ for each x in $[a, b]$. It can be shown that the **area** A of the region bounded by the lines $x = a$, $x = b$, and the graphs of the two functions is given by

$$A = \int_a^b f(x)\,dx - \int_a^b g(x)\,dx = \int_a^b [f(x) - g(x)]\,dx.$$

Note that the preceding formula is valid whether or not the graphs of f and g are above the x-axis. (See Figure 6.13.)

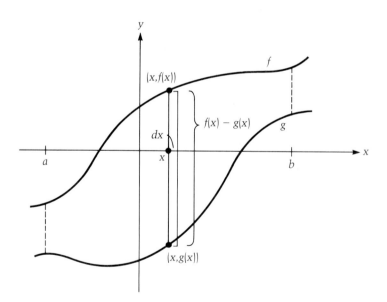

FIG. 6.13.
Area = $\int_a^b [f(x) - g(x)]\,dx$.

Area

EXAMPLE 3 Find the area of the region bounded by the graphs of $f(x) = 9 - x^2$ and $g(x) = x^2 + 1$. (See Figure 6.14.)

6.3 MORE ON AREA

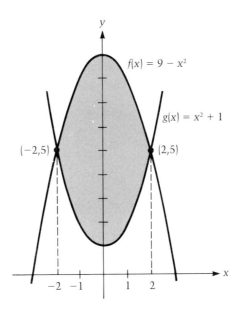

FIG. 6.14.

Solution: First we find the points of intersection of the graphs of the two functions. Since $x^2 + 1 = 9 - x^2$ where $2x^2 = 8$, $x^2 = 4$, and $x = \pm 2$, we find that the curves intersect at $(-2, 5)$ and $(2, 5)$. The area A of the enclosed region is

$$A = \int_{-2}^{2} [(9 - x^2) - (x^2 + 1)] \, dx$$

$$= \int_{-2}^{2} (8 - 2x^2) \, dx$$

$$= \left(8x - \frac{2}{3}x^3 \right) \bigg|_{-2}^{2}$$

$$= \left(16 - \frac{16}{3} \right) - \left(-16 + \frac{16}{3} \right)$$

$$= \frac{64}{3} \text{ square units.}$$

Although our work on area has been done "in the direction of the domain axis," it can also be done "in the direction of the range axis." This is particularly useful when an equation in x and y defining part of the boundary does not define a function. In particular, if $x = g(y) \geq 0$, then the area of the region bounded by

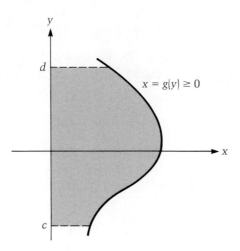

FIG. 6.15.
Area = $\int_c^d g(y)\,dy$.

the graph of this equation, the y-axis, and the lines $y = c$ and $y = d$ is given by

$$\text{Area} = \int_c^d g(y)\,dy.$$

Note: We assume that g is a function with y as independent variable and that $c < d$. (See Figure 6.15.)

Area

EXAMPLE 4 Find the area of the region bounded by the graphs of $y^2 = 4x$ and $2x - y = 4$.

First solution: (See Figure 6.16a.) We first find the simultaneous solutions of the equations $y^2 = 4x$ and $2x - y = 4$. They are the points $(1, -2)$ and $(4, 4)$. The areas of regions R_1 and R_2 are equal, and the total area of both regions is

$$2\int_0^1 2\sqrt{x}\,dx = \frac{8}{3}x^{3/2}\Big|_0^1 = \frac{8}{3} \text{ square units.}$$

The area of region R_3 is

$$\int_1^4 [2\sqrt{x} - (2x - 4)]\,dx = \left(\frac{4}{3}x^{3/2} - x^2 + 4x\right)\Big|_1^4$$

$$= \left(\frac{32}{3} - 16 + 16\right) - \left(\frac{4}{3} - 1 + 4\right)$$

$$= \frac{19}{3} \text{ square units.}$$

6.3 MORE ON AREA

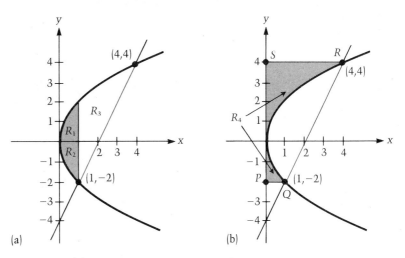

FIG. 6.16. (a) Area R_1 + area R_2 + area R_3,
$2\int_0^1 2x^{1/2}\,dx + \int_1^4 [2x^{1/2} - (2x-4)]\,dx$.
(b) Area $PQRS$ − area $R_4 = 15 - \int_{-2}^4 \frac{1}{4}y^2\,dy$.

Therefore, the area of the given region is $\frac{8}{3} + \frac{19}{3} = 9$ square units.

Second solution: (See Figure 6.16b.) "In the y-direction" we first find the area of the trapezoid $PQRS$. The area is $(\frac{1}{2})(1+4)6 = 15$ square units. The area of the region bounded by the y-axis, the graph of $y^2 = 4x$, and the lines $y = -2$ and $y = 4$ is

$$\int_{-2}^4 \frac{y^2}{4}\,dy = \frac{y^3}{12}\Big|_{-2}^4 = \frac{64+8}{12} = 6 \text{ square units.}$$

Therefore the area of the given region is $15 - 6 = 9$ square units.

Note: We could also solve $2x - y = 4$ for x, obtain $x = \frac{1}{2}(y+4)$, and conclude that

$$\text{Area} = \int_{-2}^4 \left(\frac{y+4}{2} - \frac{y^2}{4}\right)\,dy.$$

EXERCISES

Set A—Areas of Regions in the Coordinate Plane

Find the area of each region in the plane bounded by the graphs of the equations given in Exercises 1 through 16.

1. $y = x^2 - x$ and the x-axis

2. $y = (x - 2)(x - 5)$ and the x-axis
3. $y = (x + 4)(x - 2)$ and the x-axis
4. $y = x^3 + 2x$, $x = 1$, $x = 4$, and the x-axis
5. $y = x(x - 1)(x + 2)$ and the x-axis
6. $y = x(x - 2)(x + 3)$ and the x-axis
7. $y = x$ and $y = x^3$.
8. $y = x$ and $y = x^5$
9. $y = x^2$ and $y = x^3$
10. $y = x$ and $y = \dfrac{x^2}{4}$
11. $y = x^2 - 6$ and $y = 2x - 3$
12. $y = x^2 - 4$ and $y = 2x - 1$
13. $y = 2x^2 - x + 1$ and $y = 5x - 3$
14. $y = 3x^2 - 2x + 1$ and $y = 7x - 5$
15. $x = y^2$ and $x - y = 2$
16. $x = 2y^2$ and $x - 2y = 4$

Set B–Miscellaneous Exercises

Area

17. The function f is defined by $f(x) = k(x^2 - x)$, k a constant, and the area bounded by the graph of f, the x-axis, and the lines $x = 1$ and $x = 3$ is one square unit. Find k.

Area

18. The function f is defined by $f(x) = k(x^3 - x)$, k a constant, and the area bounded by the graph of f, the x-axis, and the lines $x = 1$ and $x = 2$ is one square unit. Find k.

Minimum area

19. Find the minimum area of the right triangle formed in the first quadrant by the x-axis, the y-axis, and the tangent to the curve $y = 4 - x^2$. (See Figure 6.17.)

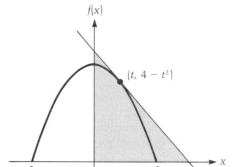

FIG. 6.17. $f(x) = 4 - x^2$.

6.4 OTHER APPLICATIONS OF THE DEFINITE INTEGRAL (OPTIONAL TOPICS)

In this section we look at several additional applications of the Riemann integral which basically are independent of one another and can be studied individually on an optional basis. Rather than give detailed discussions of some topics, such as "probability density function," which are beyond the scope of this book, we provide only the required definitions which are sufficient to indicate the breadth of the applications of the definite integral.

A. AVERAGE VALUE AND LORENTZ CURVE

For a given interval $[a, b]$ with n equal subintervals,

$$\Delta x = \frac{b - a}{n},$$

and the lower sum L_n of f on $[a, b]$ is given by

$$L_n = \sum_{k=1}^{n} f(u_k) \Delta x$$

$$= (b - a) \frac{f(u_1) + f(u_2) + f(u_3) + \cdots + f(u_n)}{n}.$$

Consequently, the arithmetic mean (average) of the n function values is $L_n \div (b - a)$. Since for large n we have

$$\frac{1}{b - a} L_n \doteq \frac{1}{b - a} \int_a^b f(x)\, dx,$$

we define the **average value** of f on $[a, b]$ as follows:

average value

> **Average value** of f on $[a, b] = \dfrac{1}{b - a} \displaystyle\int_a^b f(x)\, dx.$

Geometrically, if $f(x) > 0$ on $[a, b]$, then, since $\int_a^b f(x)\, dx$ is the area under the graph of f, above the x-axis and between $x = a$ and

$x = b$. Since

$$\int_a^b f(x)\, dx = (b - a) \times (\text{average value of } f),$$

the average value of f is the altitude of the rectangle with $b - a$ as width having an area equal to that under the graph of f. (See Figure 6.18.)

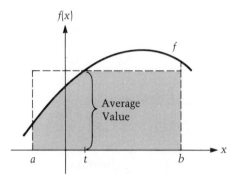

FIG. 6.18.
$\int_a^b f(x)\, dx = f(t)(b - a)$.

Average value

EXAMPLE 1 Find the average value of the function $f(x) = 3x^2 + 1$ on the interval $[-1, 5]$.

Solution: Since $b - a = 5 - (-1) = 6$, the average value is

$$\frac{1}{6} \int_{-1}^{5} (3x^2 + 1)\, dx = \frac{1}{6} (x^3 + x) \Big|_{-1}^{5}$$

$$= \frac{1}{6}[(125 + 5) - ((-1) + (-1))]$$

$$= \frac{1}{6}(130 + 2)$$

$$= 22.$$

PRACTICE PROBLEM Find the average value of $f(x) = 2x + 5$ on $[-2, 6]$.

Answer: 9. ∎

6.4 OTHER APPLICATIONS OF THE DEFINITE INTEGRAL (OPTIONAL TOPICS)

Restaurant patrons

EXAMPLE 2

(a) A restaurant operator anticipates that between 11 am and 7 pm on a given day the number of patrons in the restaurant is $f(x) = 200 - 72x^2 + 24x^3 - 2x^4$, $-1 \le x \le 7$. (Note: $x = -1$ represents 11 am, $x = 0$ represents 12 noon, etc.) Find at what time(s) the restaurant has the most patrons. What is this maximum number?

(b) At what time does the restaurant have the least number of patrons? What is this minimum number?

(c) What is the average number of patrons in the restaurant during the eight-hour period?

Solution

(a)
$$f'(x) = -144x + 72x^2 - 8x^3$$
$$= -8x(x^2 - 9x + 18)$$
$$= -8x(x-6)(x-3).$$

We see that $f'(x) = 0$ at $x = 0$, $x = 3$, and $x = 6$. To use the second derivative test for relative maxima/minima, we first determine $f''(x)$:

$$f''(x) = -144 + 144x - 24x^2.$$

We find that $f''(6) = -144$, $f''(0) = -144$, and $f''(3) = 72$. Since $f''(6) < 0$, it follows that $f(6) = 200$ is a relative maximum. Also since $f''(0) < 0$, it follows that $f(0) = 200$ is a relative maximum. Checking the endpoints of the domain, we find that $f(-1) = 102$ and $f(7) = 102$. Therefore, the restaurant has the most patrons at noon and at 6 o'clock; the maximum number is 200.

(b) Since $f''(3) > 0$, it follows that $f(3) = 38$ is a relative minimum. We conclude that at 3 o'clock, 38 patrons are the minimum number present at the restaurant during the eight-hour period.

(c) $\quad Average\ number = \dfrac{1}{8} \displaystyle\int_{-1}^{7} (200 - 72x^2 + 24x^3 - 2x^4)\, dx$

$$= \dfrac{1}{8}\left(200x - 24x^3 + 6x^2 - \dfrac{2}{5}x^5\right)\Big|_{-1}^{7}$$

$$= \dfrac{1}{8}(851.2 + 169.6)$$

$$= 127.6.$$

The average number of patrons during the eight-hour period is approximately 128.

For a fixed population of income recipients, suppose $f(x)$ is the cumulative percentage (expressed as a decimal) of income corresponding to the cumulative percentage x of income recipients, with the recipients ranked from poorest to richest. For example, let

$$f(x) = \frac{15}{16}x^2 + \frac{x}{16}, \quad 0 \leq x \leq 1,$$

be such a function. (*Note:* $0 \leq f(x) \leq 1$.) Since, for example,

$$f(0.20) = \frac{15}{16}(0.04) + \frac{1}{16}(0.20) = 0.05,$$

we find that the poorest 20% of the income recipients receive 5% of the income. Similarly, since $f(0.80) = 0.65$ and since $f(1) - f(0.80) = 1 - 0.65 = 0.35$, the richest 20% receive 35% of the income. Such a curve is called a **Lorentz curve.** Note that $g(x) = x$, $0 \leq x \leq 1$, represents equality of distribution in income since, for example, if $x = 0.30$ then $g(x) = 0.30$ and 30% of the income recipients receive 30% of the income. For any such function f, the **coefficient of inequality** (CI) is defined by

Lorentz curve

coefficient of inequality

$$\text{CI} = \frac{\text{area between the graphs of } g \text{ and } f}{\text{area under the graph of } g \text{ above the } x\text{-axis}}.$$

(See Figure 6.19.) Therefore,

$$\text{CI} = \frac{\int_0^1 [g(x) - f(x)]\, dx}{\int_0^1 g(x)\, dx}$$

$$= \frac{\int_0^1 g(x)\, dx - \int_0^1 f(x)\, dx}{\int_0^1 g(x)\, dx}.$$

Since

$$\int_0^1 g(x)\, dx = \int_0^1 x\, dx = \frac{1}{2},$$

6.4 OTHER APPLICATIONS OF THE DEFINITE INTEGRAL (OPTIONAL TOPICS)

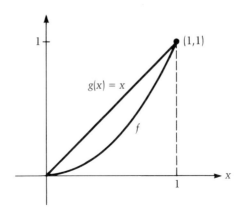

FIG. 6.19.

it follows that

$$\text{CI} = 1 - 2 \int_0^1 f(x)\,dx.$$

Income distribution

EXAMPLE 3 Find the coefficient of inequality for the income distribution given by

$$f(x) = \frac{15}{16}x^2 + \frac{x}{16}x.$$

Solution

$$\begin{aligned}
\text{CI} &= 1 - 2\int_0^1 \left(\frac{15}{16}x^2 + \frac{x}{16}\right)dx \\
&= 1 - 2\left(\frac{5}{16}x^3 + \frac{1}{32}x^2\right)\Big|_0^1 \\
&= 1 - 2\left(\frac{10}{32} + \frac{1}{32}\right) \\
&= 1 - \frac{11}{16} \\
&= \frac{5}{16}.
\end{aligned}$$

EXERCISES FOR SECTION 6.4A

Average value

1. Prove that the average value of any linear function $f(x) = mx + b$ on the interval $[c, d]$ is $f((c + d)/2)$.

Average value

2. (a) Find the average value of $f(x) = x^{1/2}$ on $[1, 9]$.
   (b) If A is the average value obtained in part (a), for what t is $f(t) = A$?

Average value

3. (a) Find the average value of $f(x) = x^2$ on $[1, 5]$.
   (b) If A is the average value obtained in part (a), for what t is $f(t) = A$?

Average value

4. (a) Find the average value of $f(x) = x^{-1/3}$ on $[-1, 8]$.
   (b) If A is the average value obtained in part (a), for what t is $f(t) = A$?

Average value

5. Suppose a traffic engineer finds that at time t the velocity of traffic in miles per hour on a stretch of interstate highway between 3 o'clock in the afternoon and 8 o'clock that evening is given by $v(t) = t^3 - 18t^2 + 105t - 130$, $3 \le t \le 8$.
   (a) What is the velocity of traffic at 4 o'clock?
   (b) At what time(s) is traffic moving the fastest?
   (c) What is the maximum velocity of traffic during the 5-hour time interval?
   (d) What is the minimum velocity of traffic during the given time interval? At what times does it occur?
   (e) Find the average velocity of the traffic during the 5-hour time interval. (That is, find the average of $v(t)$ on $[3, 8]$.)

Income distribution

6. Let
$$f(x) = \frac{19}{20}x^2 + \frac{1}{20}x$$
   be the Lorentz curve for a given income distribution.
   (a) Find $f(0.20)$ and interpret the result.
   (b) What is the percentage of income for the top 20% of the income recipients?
   (c) Find the coefficient of inequality for this income distribution.

B. CONSUMER'S SURPLUS AND PRODUCER'S SURPLUS

Suppose the demand curve for a given product is given by the function $p = D(q)$ where p is the price in dollars per unit when q is the number of units demanded. Let $\bar{p} = D(\bar{q})$ be the fixed price in dollars corresponding to a market demand of $\bar{q}$ units. At this fixed price of $\bar{p}$ dollars per unit, $\bar{p}\bar{q}$ dollars would be the amount received by the producer in selling $\bar{q}$ units. The area of the shaded rectangle in Figure 6.20a represents this amount.

6.4 OTHER APPLICATIONS OF THE DEFINITE INTEGRAL (OPTIONAL TOPICS)   315

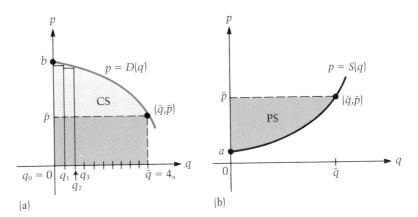

FIG. 6.20. (a) Consumer's surplus. (b) Producer's surplus.

If the interval $[0, \bar{q}]$ is divided into n subdivisions of equal width Δq as in Figure 6.20a, then, for example,

$$p_1 = D(q_1)$$

is the price consumers are willing to pay when q_1 units are available. (That is, some consumers are willing to pay more than the fixed price $\bar{p}$ to obtain the product.) For the first Δq units,

$$p_1(\Delta q) = D(q_1)\,\Delta q$$

is the total amount (approximately) that the consumer would be willing to pay for the product at this level of production. In fact, the area between 0 and q_1, under the graph of $p = D(q)$, and above the q-axis, which is given by

$$\int_0^{q_1} D(q)\,dq,$$

represents this amount. The area $\bar{p}(\Delta q)$ of the rectangle is the amount the producer receives at the fixed price $\bar{p}$. Therefore,

$$\int_0^{q_1} D(q)\,dq - \bar{p}(\Delta q)$$

represents the *consumer's surplus*, the amount saved by the consumer when the priced is fixed at $\bar{p}$. (In a perfect monopoly, the price would be set at the amount each consumer was willing to pay, and hence, there would not be a consumer's surplus.) Since the preceding argument can be extended from the first subinter-

316   CHAPTER 6   THE DEFINITE INTEGRAL AND ITS APPLICATIONS

val to each subinterval in $[0, \bar{q}]$, we see why **consumer's surplus** is defined by

consumer's surplus

$$\text{CS} = \int_0^{\bar{q}} D(q)\, dq - \bar{p}\bar{q}.$$

producer's surplus

Similarly, if $p = S(q)$ is a supply function then the **producer's surplus**, denoted by PS, is defined by

$$\text{PS} = \bar{p}\bar{q} - \int_0^{\bar{q}} S(q)\, dq.$$

As seen in Figure 6.20b, PS is represented by the area inside the rectangle above the supply curve. Generally, for a given product, $(\bar{q}, \bar{p})$ is the *equilibrium point* where the curves $p = D(q)$ and $p = S(q)$ intersect. (See Figure 6.21. We deviate here from the treatment in Section 2.7 and follow the more usual practice of letting q be the *independent variable*.)

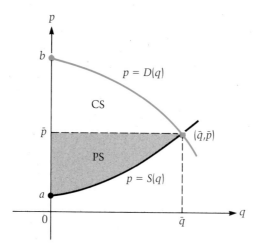

FIG. 6.21.

6.4 OTHER APPLICATIONS OF THE DEFINITE INTEGRAL (OPTIONAL TOPICS)

Consumer's surplus and producer's surplus

EXAMPLE 4 Let $p = 60 - 3q$ be the demand curve for a given product and let $p = 10 + 2q$ be the corresponding supply curve. Find the consumer's surplus when $(\bar{q}, \bar{p})$ is the equilibrium point.

Solution: (See Figure 6.22.) The equilibrium price occurs when

$$10 + 2\bar{q} = 60 - 3\bar{q}$$
$$5\bar{q} = 50$$
$$\bar{q} = 10.$$

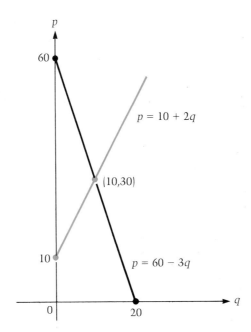

FIG. 6.22.

Substituting in either the demand or supply equation, we find $\bar{p} = 30$. Consequently,

$$\text{CS} = \int_0^{10} (60 - 3q)\, dq - 300$$
$$= \left(60q - \frac{3}{2}q^2\right)\Big|_0^{10} - 300$$
$$= 600 - 150 - 300$$
$$= 150.$$

Similarly,

$$PS = 300 - \int_0^{10} (10 + 2q)\, dq$$

$$= 300 - (10q + q^2)\Big|_0^{10}$$

$$= 300 - 200$$

$$= 100.$$

If $p = D(q)$ is a demand function for which each p determines a unique q (as in Figure 6.20a), then $q = \overline{D}(p)$ where $\overline{D}$ is the inverse function of D. [This equation can often be found by solving $p = D(q)$ for q.] In this case,

$$CS = \int_{\bar{p}}^{b} \overline{D}(p)\, dp \qquad \text{where} \quad b = D(0).$$

Similarly, as can be seen in Figure 6.20b, if $\overline{S}$ is the inverse of S, then

$$PS = \int_a^{\bar{p}} \overline{S}(p)\, dp \qquad \text{where} \quad a = S(0).$$

Inverse function

EXAMPLE 5 Check the solutions in Example 4 by the preceding formulas.

Solution: Since $p = D(q)$ is given by $p = 60 - 3q$,

$$D(0) = 60 = b.$$

Solving $p = 60 - 3q$ for q, we get $q = \overline{D}(p)$; it is given by

$$q = 20 - \frac{p}{3}.$$

Therefore,

$$CS = \int_{30}^{60} \left(20 - \frac{p}{3}\right) dp$$

$$= \left(20p - \frac{1}{6}p^2\right)\Big|_{30}^{60}$$

$$= (1200 - 600) - (600 - 150)$$

$$= 600 - 450$$

$$= 150.$$

6.4 OTHER APPLICATIONS OF THE DEFINITE INTEGRAL (OPTIONAL TOPICS)

Since $p = S(q)$ is given by $p = 10 + 2q$, $S(0) = 10 = a$. Solving $p = 10 + 2q$ for q, we get $q = \overline{S}(p)$; it is given by

$$q = \frac{p}{2} - 5.$$

Therefore,

$$\begin{aligned} PS &= \int_{10}^{30} \left(\frac{p}{2} - 5\right) dp \\ &= \left(\frac{p^2}{4} - 5p\right)\Big|_{10}^{30} \\ &= (225 - 150) - (25 - 50) \\ &= 75 + 25 \\ &= 100. \end{aligned}$$

EXERCISES FOR SECTION 6.4B

Consumer's and producer's surplus

7. Let $D(q) = 40 - 2q$ be the demand curve for a given product and let $S(q) = 10 + q/2$ be the corresponding supply curve. Find the consumer's surplus and producer's surplus when $(\bar{p}, \bar{q})$ is the equilibrium point.

Consumer's and producer's surplus

8. Let $D(q) = 100 - 4q$ be the demand curve for a given product and let $S(q) = 10 + 2q$ be the corresponding supply curve. Find the consumer's surplus and producer's surplus when $(\bar{p}, \bar{q})$ is the equilibrium point.

9. Use the techniques discussed in Example 5 to check your answers to Exercise 7.

10. Use the techniques discussed in Example 5 to check your answers to Exercise 8.

C. PROBABILITY DENSITY FUNCTION

Suppose we consider the height of all individuals in the world who are over the age of twenty-one and suppose we find that the shortest person is a inches in height and the tallest is b inches. Furthermore, suppose we can determine a continuous function f on $[a, b]$ such that $f(x)$ represents the probability that a person selected at random is x inches tall. Since the "sum" of all such probabilities must be 1, this is equivalent to having

$$\int_a^b f(x)\, dx = 1.$$

Furthermore, if a person is selected at random, the probability that the individual is between c and d inches tall, $a \le c \le d \le b$, is given by

$$\int_c^d f(x)\, dx.$$

Although we do not pursue in depth this application of calculus, let us give two important definitions which are important to the development of probability and consider problems pertinent to them.

> **DEFINITION** A continuous function f is called a **probability density function** on $[a, b]$ provided $f(x) \ge 0$ on $[a, b]$ and
>
> $$\int_a^b f(x)\, dx = 1.$$
>
> For a probability density function f on $[a, b]$, the **cumulative distribution function** F on $[a, b]$ is defined by
>
> $$F(x) = \int_a^x f(u)\, du \quad \text{where} \quad a \le x \le b.$$

EXERCISES FOR SECTION 6.4C

Probability density function

11. Prove that $f(x) = (\frac{3}{38})(x^2 - x)$ is a probability density function on $[2, 4]$.

Cumulative distribution function

12. Find the cumulative distribution function for the probability density function given in Exercise 11.

Probability density function

13. Prove that $f(x) = x^2/39$ is a probability density function on $[2, 5]$.

Probability density function

14. If $f(x) = k(x^3 - x^2)$ is a probability density function on $[1, 2]$, what is the value for the constant k?

Cumulative distribution function

15. Find the cumulative distribution function for the function f given in Exercise 13.

Cumulative distribution function

16. Find the cumulative distribution function for the function f given in Exercise 14.

6.4 OTHER APPLICATIONS OF THE DEFINITE INTEGRAL (OPTIONAL TOPICS)

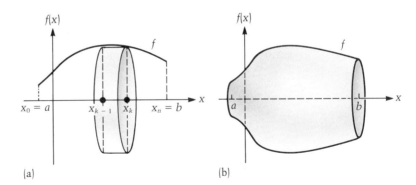

FIG. 6.23. (a) $\pi[f(u_k)]^2 \Delta x$.
(b) $\sum_{k=1}^{n} \pi[f(u_k)]^2 \Delta x$.

D. VOLUME BY DISK METHOD

Let f be a continuous function on $[a, b]$ where $f(x) \geq 0$ for each x in the interval. The Riemann integral can be used to assign volume to the solid generated by revolving about the x-axis the region bounded by the lines $x = a$, $x = b$, the x-axis, and the graph of f. (See Figure 6.23a.) For a regular partition of $[a, b]$, if $f(u_k)$ is the minimum value of f in the kth subinterval, then the volume of the cylinder (cylindrical disk) obtained by rotating about the x-axis the rectangle with $f(u_k)$ as length and Δx as width is $\pi[f(u_k)]^2 \Delta x$. Therefore,

$$L_n = \sum_{k=1}^{n} \pi[f(u_k)]^2 \Delta x$$

is the volume of the solid generated by revolving all the inscribed rectangles about the x-axis. (See Figure 6.23b.) The function F defined on $[a, b]$ by $F(x) = \pi[f(x)]^2$ is continuous and L_n is its *lower sum* on $[a, b]$. Hence, $\lim_{n \to \infty} L_n$ exists and is equal to

$$\int_a^b F(x)\, dx = \pi \int_a^b [f(x)]^2\, dx.$$

volume

Therefore, the **volume** of the solid generated by revolving about the x-axis the region bounded by the lines $x = a$, $x = b$, the x-axis and the graph of f is given by

$$V = \pi \int_a^b [f(x)]^2\, dx.$$

disk method

Using this integral to find the volume of a solid of revolution is called the **disk method.**

Volume of a sphere

EXAMPLE 6 Use the disk method to find the volume of a sphere of radius r.

Solution: If we revolve about the x-axis the semicircle given by

$$f(x) = (r^2 - x^2)^{1/2}, \quad -r \leq x \leq r,$$

it generates a sphere of radius r. (See Figure 6.24.) Using the disk method to find the volume, we get

$$V = \pi \int_{-r}^{r} [(r^2 - x^2)^{1/2}]^2 \, dx = \pi \int_{-r}^{r} (r^2 - x^2) \, dx$$

$$= \pi \left(r^2 x - \frac{x^3}{3} \right) \Big|_{-r}^{r}$$

$$= \pi \left[\left(r^3 - \frac{r^3}{3} \right) - \left(-r^3 + \frac{r^3}{3} \right) \right]$$

$$= \frac{4}{3} \pi r^3 \text{ cubic units.}$$

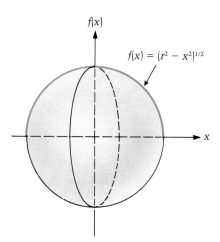

FIG. 6.24.

PRACTICE PROBLEM Use the disk method to find the volume of the solid generated by revolving about the x-axis the region bounded by $y = 4 - x^2$ and the x-axis.

Answer: $512\pi/15$ cubic units. ∎

6.4 OTHER APPLICATIONS OF THE DEFINITE INTEGRAL (OPTIONAL TOPICS)

EXAMPLE 7 Derive the formula for the volume of a right circular cone with radius r and height h.

Solution: (See Figure 6.25.) The equation of the line containing the origin and the point (h, r) is

$$f(x) = \frac{r}{h}x.$$

Revolving about the x-axis the region bounded by $f(x) = (r/h)x$, the x-axis, and $x = h$ generates a cone with radius r and height h. Using the disk method to obtain the volume V, we find

$$V = \pi \int_0^h \left(\frac{r}{h}x\right)^2 dx$$

$$= \frac{\pi r^2}{h^2} \int_0^h x^2 \, dx$$

$$= \frac{\pi r^2}{h^2} \left(\frac{x^3}{3}\right)\bigg|_0^h$$

$$= \frac{1}{3}\pi r^2 h.$$

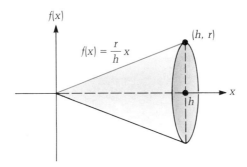

FIG. 6.25.
$V = \int_0^h \left(\frac{r}{h}x\right)^2 dx.$

Volume by disk method

EXERCISES FOR SECTION 6.4D

In Exercises 17 through 20, use the disk method to find the volume of the solid obtained by revolving about the x-axis the region bounded by

17. $f(x) = x^2/4$, $x = 2$, $x = 5$, and the x-axis.

18. $f(x) = x^2 + 2$, the x-axis, and the lines $x = 1$ and $x = 4$.
19. $f(x) = x^{3/2}$, the x-axis, and the lines $x = 1$ and $x = 9$.
20. $f(x) = x - x^2$ and the x-axis.

6.5 SUMMARY AND REVIEW EXERCISES

1. THE RIEMANN INTEGRAL

For a continuous function f on $[a, b]$ with n subintervals of equal length Δx, let $f(u_k)$ be the minimum value of f on the kth subinterval and let

$$L_n = \sum_{k=1}^{n} f(u_k) \Delta x = \Delta x [f(u_1) + f(u_2) + \cdots + f(u_n)].$$

The limit of the sequence L_n exists and

$$\lim_{n \to \infty} L_n = \int_a^b f(x) \, dx$$

is the **Riemann integral** (or **definite integral**) of f from a to b.

(a) *Fundamental Theorem of Calculus* Let f be a continuous function on $[a, b]$. Then an antiderivative of f exists, and if F is any antiderivative of f, then

$$\int_a^b f(x) \, dx = F(b) - F(a).$$

(b) If a, b, and c are numbers in an interval on which f is continuous, then

$$\int_a^b f(x) \, dx + \int_b^c f(x) \, dx = \int_a^c f(x) \, dx.$$

2. AREA

(a) For a continuous function f, if $f(x) \geq 0$ for each x in $[a, b]$, then $\int_a^b f(x) \, dx$ is the **area** of the region bounded by the graph of f, the lines $x = a$, $x = b$, and the x-axis.

(b) If f and g are two continuous functions on $[a, b]$ for which $f(x) \leq g(x)$ for each x in $[a, b]$, then the **area** A of the region bounded by the lines $x = a$, $x = b$, and the graphs of the two functions is given by

$$A = \int_a^b [g(x) - f(x)] \, dx.$$

3. **VOLUME**

(a) Let f be a continuous function on $[a, b]$ for which $f(x) \geq 0$ for each x in $[a, b]$. The **volume** of the solid generated by revolving about the x-axis the region bounded by the lines $x = a$, $x = b$, the x-axis, and the graph of f is given by

$$V = \pi \int_a^b [f(x)]^2 \, dx.$$

This is called the **disk method** for finding the volume of the solid of revolution.

4. **OTHER APPLICATIONS OF THE RIEMANN INTEGRAL**

(a) If f is a continuous function on $[a, b]$, then

$$\textbf{Average value of } f \text{ on } [a, b] = \frac{1}{b - a} \int_a^b f(x) \, dx.$$

(b) For a demand function $p = D(q)$, if $\bar{p}$ is a fixed market price corresponding to a fixed market demand $\bar{q}$, then

$$\textbf{Consumer's Surplus} = \int_0^{\bar{q}} [D(q) - \bar{p}] \, dq.$$

(c) For a supply function $p = S(q)$, if $\bar{p}$ is a fixed market price corresponding to a fixed market demand $\bar{q}$, then

$$\textbf{Producer's Surplus} = \int_0^{\bar{q}} [\bar{p} - S(q)] \, dq.$$

(d) A continuous function f is called a **probability density function** on $[a, b]$ provided $f(x) \geq 0$ on $[a, b]$ and $\int_a^b f(x) \, dx = 1$. The **cumulative distribution** function F on $[a, b]$ for a probability density function f on $[a, b]$ is defined by $F(x) = \int_a^x f(u) \, du$.

5. REVIEW EXERCISES

In Exercises 1 through 10, evaluate the given Riemann integrals.

1. $\int_1^4 (5x^4 - x^2) \, dx$

2. $\int_{-2}^2 (2x - x^3) \, dx$

3. $\int_0^4 (5x^{3/2} - x^{1/2}) \, dx$

4. $\int_0^8 (6x^{1/3} - 1) \, dx$

5. $\int_0^1 (e^{3x} - x^2) \, dx$

6. $\int_1^e \left(\frac{1}{x} + x\right) dx$

7. $\int_0^2 x(4x + 1)^{1/2} \, dx$

8. $\int_1^5 x(3x + 1)^{3/2} \, dx$

9. $\int_1^3 \left(e^{-x} + \frac{1}{x}\right) dx$

10. $\int_e^4 \frac{\ln x}{x} \, dx$

11. Find the area of the region above the x-axis, between the lines $x = 0$ and $x = 4$, and below the graph of $f(x) = (2x + 1)^{1/2}$.

12. Find the area of the region above the x-axis, between the lines $x = 0$ and $x = 2$, and below the graph of $f(x) = x(x^2 + 4)^{1/3}$.

13. Find the area of the region bounded by the graphs of $x = 1$, $x = e$, the x-axis, and $f(x) = x(\ln x)$.

14. Find the area of the region bounded by the graphs of $y = 2x$ and $y = 8 - x^2$.

15. Find the area of the region bounded by the graphs of $y = 2x - 1$ and $y = -x^2 + 2x + 3$.

16. (a) A manufacturer estimates that over the next 8 months 200 units of a given product can be sold each month and that the price in x months will be $p(x) = 40 + 2(x + 1)^{1/2}$ dollars. Find the sum

    $$200\left[p\left(\frac{1}{2}\right) + p\left(\frac{3}{2}\right) + p\left(\frac{5}{2}\right) + \cdots + p\left(\frac{15}{2}\right)\right]$$

    and interpret the answer.
    (b) Explain why

    $$\int_0^8 200[40 + 2(x + 1)^{1/2}] \, dx$$

    would approximate total revenue over the next 24 months.
    (c) Find the Riemann integral in part (b) and compare the answer with that in part (a).

BIOGRAPHICAL SKETCH

Georg F. B. Riemann

The Bettmann Archive

Riemann was born on September 17, 1826 in Hanover, Germany. His health, as with many of his five brothers and sisters, was never very good and an untimely death at age 40 probably prevented Riemann from becoming the "Einstein" of the nineteenth century. Although he did not publish a great number of papers, his very inventive mind produced an unusual amount of original mathematics which inaugurated entirely new fields and opened new areas for productive research.

He entered the University of Göttingen in 1846. Before finishing at Göttingen, he went to Berlin and studied with Jacobi, Dirichlet, Eisenstein, and Steiner. After two years in Berlin, he returned to Göttingen where he obtained a doctor's degree in 1851. In 1859 he obtained a professorship at Göttingen, a position previously held by Dirichlet.

Riemann produced a function that is continuous at all irrational numbers but discontinuous at all rational numbers. Examples like this had a significant effect on the development of analysis. His definition of the (Riemann) integral replaced the one used by Cauchy and Dirichlet. His research in functions of a complex variable resulted in the outstanding contribution of Riemann surfaces. Riemann's last paper (1859), in which the distribution of primes was investigated, contained what is now called the Riemann hypothesis. The Riemann hypothesis is a conjecture about convergence of the (Riemann) zeta function, a conjecture that has never been proved or disproved to this day. The zeta function is defined by the infinite series

$$\zeta(s) = 1 + \frac{1}{2^s} + \frac{1}{3^s} + \frac{1}{4^s} + \frac{1}{5^s} + \cdots$$

where s is a complex number so chosen that the series converges. Riemann conjectured that all the imaginary zeros of the zeta function have their real parts equal to one half. This conjecture is to classical analysis what Fermat's last theorem is to number theory.

CHAPTER

MULTIVARIABLE CALCULUS AND ITS APPLICATIONS

7

7.1 INTRODUCTION
7.2 RECTANGULAR COORDINATE SYSTEM IN THREE DIMENSIONS
7.3 FUNCTIONS OF TWO VARIABLES
7.4 PARTIAL DERIVATIVES
7.5 TANGENT PLANES AND THE TOTAL DIFFERENTIAL
7.6 APPLIED MAXIMA AND MINIMA PROBLEMS
7.7 MAXIMA AND MINIMA WITH CONSTRAINTS
7.8 METHOD OF LEAST SQUARES (OPTIONAL)
7.9 SUMMARY AND REVIEW EXERCISES

7.1 INTRODUCTION

In this chapter we generalize and expand to three dimensions many of the ideas in differential calculus and analytic geometry associated with two dimensions. We first discuss one way to introduce a coordinate system into three dimensions and then look at such concepts as distance between two points in three dimensions, midpoint of a line segment, graphs of equations, and so on. Next we discuss functions of two variables and consider important ideas and applications similar to those associated with the derivative of a function of one variable. Some of the problems we shall learn how to solve are:

1. A company's profit in thousands of dollars from the sale of dress shirts and sport shirts at x and y dollars each, respectively, is given by

$$P(x, y) = 8x - 14y + 6xy - 4x^2 - 2y^2 + 210,$$

provided $10 \leq x \leq 16$ and $12 \leq y \leq 19$. Find the price of each type of shirt which will maximize profits and find the maximum profit. Also determine the price of each type of shirt which will minimize profits and find the minimum profit possible. (See Example 3, Section 7.6.)

2. A club holds a bingo game each week to raise money for charity. With a $2.00 charge per card and a policy of returning $\frac{1}{5}$ (i.e., 20%) of the total amount taken in as prize money, the club sells 1700 cards each week. It is estimated that if x is the number of 25-cent increases in the price of each card and if y, $0.20 \leq y \leq 0.80$, is the payout, then the number of cards sold each week will be given by

$$N(x, y) = 1200 - 100x + 2500y.$$

What should the price be for each card and what should the percentage payout be in order to maximize the profit for the club? (See Example 4, Section 7.6.)

3. Suppose a rectangular packing box containing 9 ft^3 is to be manufactured from three different materials. The lid costs 6 cents/ft^2, the sides cost 3 cents/ft^2, and the bottom costs 10 cents/ft^2. Find the most economical dimensions for the box. (See Example 5, Section 7.6.) ■

7.2 RECTANGULAR COORDINATE SYSTEM IN THREE DIMENSIONS

rectangular coordinate system

Let us now consider how the **cartesian (rectangular) coordinate system** for three dimensions is constructed. Take any plane and construct the familiar two-dimensional rectangular xy-coordinate system in this plane, which is called the *xy-plane*. Next construct a number line, generally called the z-axis, perpendicular to the xy-plane at its origin $(0, 0)$, and let the origin of the z-axis coincide with the point $(0, 0)$. (See Figure 7.1.) Now, for example, the ordered triple of real numbers $(3, 4, 0)$ is paired with the point $(3, 4)$ in the xy-plane. In general, $(a, b, 0)$ are the three-dimensional coordinates of the point (a, b) in the xy-plane. If the positive z-axis is "up," then $(3, 4, 5)$ is the ordered triple assigned to the point five units directly above the point $(3, 4, 0)$. Similarly, $(3, 4, -6)$ are the coordinates of the point six units below the point $(3, 4, 0)$. For the point $(3, 4, 5)$, 3 is called the *x-coordinate*, 4 is the *y-coordinate*, and 5 is the *z-coordinate*.

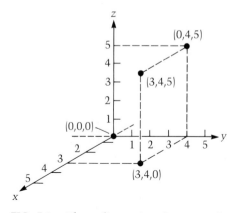

FIG. 7.1. Three-dimensional rectangular coordinate system.

There are essentially two different ways to construct a rectangular coordinate system in three dimensions since the positive z-axis can be oriented in two different directions from the xy-plane. When we look "down" on the xy-plane from the direction of the positive z-axis, if the positive y-axis is 90° clockwise from the positive x-axis, then the system is called a *left-hand* coordinate system. When we look "down" on the xy-plane from the positive z-axis, if the positive y-axis is 90° counterclockwise from the positive x-axis, then the system is called a *right-hand* coordinate system. Generally, we shall use only a right-hand coordinate system. (See Figure 7.2.)

7.2 RECTANGULAR COORDINATE SYSTEM IN THREE DIMENSIONS

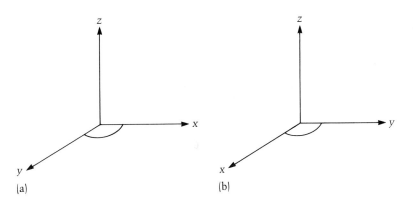

FIG. 7.2. (a) Left-hand system. (b) Right-hand system.

If $P(x_1, y_1, z_1)$ and $Q(x_2, y_2, z_2)$ are two points in three dimensions, then the perpendicular projections of the two points onto the xy-plane are $R(x_1, y_1, 0)$ and $S(x_2, y_2, 0)$, respectively. (See Figure 7.3.) From the formula for the distance between two points in two dimensions, we obtain

$$|RS| = \sqrt{(x_2 - x_1)^2 + (y_2 - y_1)^2}.$$

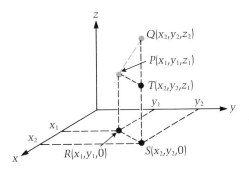

FIG. 7.3. Distance, $|PQ| = \sqrt{(x_2 - x_1)^2 + (y_2 - y_1)^2 + (z_2 - z_1)^2}$.

In the rectangle $RPTS$, $|PT| = |RS|$ and T has coordinates (x_2, y_2, z_1). Furthermore, $|TQ| = |z_2 - z_1|$ and by the Pythagorean theorem we obtain

$$|PQ| = \sqrt{|PT|^2 + |QT|^2}.$$

Consequently, the **distance** between P and Q is given by

distance formula

$$|PQ| = \sqrt{(x_2 - x_1)^2 + (y_2 - y_1)^2 + (z_2 - z_1)^2}.$$

334 CHAPTER 7 MULTIVARIABLE CALCULUS AND ITS APPLICATIONS

The student should recognize the similarity in the formulas for distance between two points in two dimensions and between two points in three dimensions.

Distance between two points

EXAMPLE 1 Find the length of the line segment with $A(-3, 1, 5)$ and $B(1, 5, 7)$ as endpoints.

Solution

$$|AB| = \sqrt{(-3 - 1)^2 + (1 - 5)^2 + (5 - 7)^2}$$
$$= \sqrt{16 + 16 + 4}$$
$$= \sqrt{36} = 6.$$

If $P(x_1, y_1, z_1)$ and $Q(x_2, y_2, z_2)$ are two points in three dimensions, then the coordinates of the midpoint of the line segment joining P and Q are found by a formula similar to the one for two dimensions. (See Figure 7.4.) The **midpoint** is given by

midpoint formula

$$\left(\frac{x_1 + x_2}{2}, \frac{y_1 + y_2}{2}, \frac{z_1 + z_2}{2}\right).$$

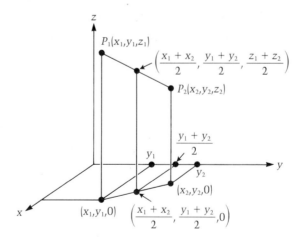

FIG. 7.4. Midpoint: $\left(\dfrac{x_1 + x_2}{2}, \dfrac{y_1 + y_2}{2}, \dfrac{z_1 + z_2}{2}\right)$.

Midpoint

EXAMPLE 2 Find the midpoint of the line segment with $A(-3, 1, 5)$ and $B(1, 5, 7)$ as endpoints.

Solution: The midpoint is

$$\left(\frac{-3+1}{2}, \frac{1+5}{2}, \frac{5+7}{2}\right) = (-1, 3, 6).$$

PRACTICE PROBLEM A line segment has $P(-3, -1, 6)$ and $Q(5, 3, 14)$ as endpoints. Find its length and its midpoint.

Answer: 12; (1, 1, 10). ∎

equation of a plane

Any linear equation $ax + by + cz = d$, where a, b, and c not all zero, can be graphed in three dimensions. For example, the graph of the linear equation $3x + 4y + z = 12$ is the set of points whose ordered triples (x, y, z) satisfy the equation. It can be shown that the *graph of any linear equation in x, y, and z is a* **plane.** The points on the graph of the plane $3x + 4y + z = 12$, where $z = 0$ are those points on the line of intersection of the given plane and the xy-plane. The equation of this line in the xy-plane is $3x + 4y = 12$. Any *curve or line* that is the intersection of a three-dimensional surface and a plane is called a **trace** of the surface in the plane. For example, $3x + 4y = 12$ is an equation of the trace of the surface $3x + 4y + z = 12$ in the xy-plane. An equation of the trace of a surface in the xz-plane is found by letting $y = 0$. Thus $3x + z = 12$ is an equation of the trace of $3x + 4y + z = 12$ in the xz-plane. Similarly, an equation of the trace of a surface in the yz-plane is found by letting $x = 0$. Thus $4y + z = 12$ is an equation of the trace of the given plane in the yz-plane. The traces of the plane $3x + 4y + z = 12$ in each of the coordinate planes where $x \geq 0$, $y \geq 0$, $z \geq 0$ are graphed in Figure 7.5. Since

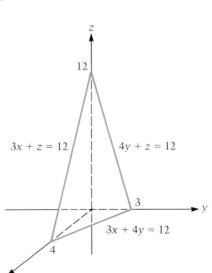

FIG. 7.5. $3x + 4y + z = 12$ for $x \geq 0$, $y \geq 0$, $z \geq 0$.

the traces are straight lines, they are graphed by connecting the intercepts of each line. (Graphing traces is the best way to discover what a three-dimensional surface looks like. This should become clearer as we consider other examples in this and the next section.)

Graph of a plane

EXAMPLE 3 Sketch the graph of $6x + 4y + 2z = 11$ by sketching the traces of this plane in the coordinate planes where $x \geq 0$, $y \geq 0$, $z \geq 0$.

Solution: (See Figure 7.6.) The trace of the plane in the xy-plane is the line $6x + 4y = 11$, the trace of the plane in the xz-plane is the line $6x + 2z = 11$, and the trace of the plane in the yz-plane is $4y + 2z = 11$.

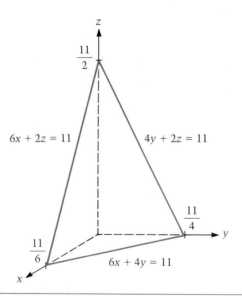

FIG. 7.6.
$6x + 4y + 2z = 11.$

Let us now investigate what the graph of the equation $z = x^2 + y^2$ looks like. Since $x^2 + y^2 > 0$ for any real numbers x and y except for $x = 0$ and $y = 0$, the graph of the surface is above the xy-plane except for the point $(0, 0, 0)$. Observe that the points on the graph of $z = x^2 + y^2$ where $z = 9$, for example, are those points on the intersection of the surface and the plane $z = 9$. The trace of the surface in this plane parallel to the xy-plane is the curve whose equation is $x^2 + y^2 = 9$. It is the circle in the plane $z = 9$ with center on the z-axis and radius 3. The trace of the surface in the xz-plane, obtained by letting $y = 0$, is the parabola $z = x^2$, and the trace of the surface in the yz-plane, obtained by letting $x = 0$, is the parabola $z = y^2$. This surface can be obtained by rotating either of the parabolas $z = x^2$ or $z = y^2$ about the z-axis. The surface is called a *paraboloid of revolution*. (See Figure 7.7.)

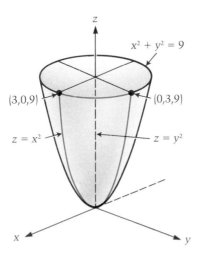

FIG. 7.7. $z = x^2 + y^2$.

Graph of a cylinder

EXAMPLE 4 Sketch the graph of $x^2 + y^2 = 25$ in three dimensions.

Solution: Since the graph of $x^2 + y^2 = 25$ in the xy-plane is a circle with center at the origin and radius 5, this circle is the trace of the surface in the xy-plane. Since we are given that the equation represents a surface in three dimensions, it could be considered in the form $x^2 + y^2 + (0)(z) = 25$. Thus the trace in any plane $z = c$ parallel to the xy-plane is a circle with center on the z-axis and radius 5. The surface is a right circular cylinder. Its graph for $z \geq 0$ is shown in Figure 7.8.

Paraboloid of revolution

EXAMPLE 5 Sketch the graph of the equation $z = x^2 + y^2 - 4$.

Solution: This surface is the same as the paraboloid of revolution $z = x^2 + y^2$ given before Example 4 except that the "low point" is at $(0, 0, -4)$ instead of $(0, 0, 0)$. The trace in the xy-plane is the circle $x^2 + y^2 = 4$; the trace in the yz-plane is the parabola $z = y^2 - 4$; and the trace in the xz-plane is the parabola $z = x^2 - 4$. (See Figure 7.9.)

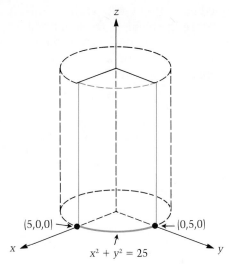

FIG. 7.8. $x^2 + y^2 = 25$.

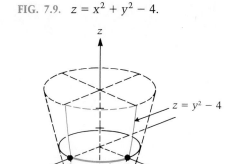

FIG. 7.9. $z = x^2 + y^2 - 4$.

Right circular cone

EXAMPLE 6 Sketch the graph of $x^2 + y^2 = z^2$.

Solution: The trace of the surface in the plane $z = 4$ is a circle with center on the z-axis and radius 2. In fact, the trace in any plane $z = a$, $a \neq 0$, is a circle. The traces of the surface in the yz-plane are the intersecting lines $y = \pm z$ and the traces of the surface in the xz-plane are the intersecting lines $x = \pm z$. (See Figure 7.10. The graph is a right circular cone.)

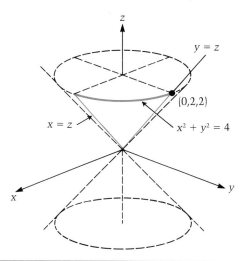

FIG. 7.10. $x^2 + y^2 = z^2$.

EXERCISES

Set A–Rectangular Coordinate System in Three Dimensions

In Exercises 1 through 4, graph each of the points given in a right-hand rectangular coordinate system:

Graphing points
1. $A(1, 2, 1)$, $B(3, 6, 5)$, and $C(4, 2, 9)$.
2. $A(3, 5, 8)$, $B(1, 1, 1)$, and $C(-2, 8, 5)$.
3. $A(2, 4, 1)$, $B(3, 6, 8)$, and $C(6, 8, 2)$.
4. $A(3, 3, 3)$, $B(-1, 6, 3)$, and $C(2, 5, -3)$.

In Exercises 5 through 8, find the length of each of the three line segments formed by connecting the points in:

Lengths of segments
5. Exercise 1.
6. Exercise 2.
7. Exercise 3.
8. Exercise 4.

In Exercises 9 through 12, find the midpoint of each of the three line segments formed by connecting the points in:

Midpoints
9. Exercise 1.
10. Exercise 2.
11. Exercise 3.
12. Exercise 4.

Pythagorean theorem
13. Use the distance formula and the Pythagorean theorem to prove that $A(4, 9, 8)$, $B(6, 8, 14)$, and $C(5, 5, 7)$ are vertices of a right triangle.

Pythagorean theorem
14. Determine whether $A(2, 3, 5)$, $B(4, 1, 6)$, and $C(8, 4, -5)$ are vertices of a right triangle.

Set B–Graphs of Equations in Three Dimensions

In Exercises 15 through 18, sketch the graph of the given plane.

Graphs
15. $3x + 5y + 6z = 30$.
16. $6x + 4y + 3z = 24$.
17. $z = 3x - 4y + 12$.
18. $z = -2x + 5y - 10$.
19. Sketch the graph of the surface $z = 9 - x^2 - y^2$.
20. Sketch the graph of the surface $z = 16 - x^2 - 4y^2$.

7.3 FUNCTIONS OF TWO VARIABLES

In the preceding chapters the functions we considered were generally functions of one variable. We now turn our attention to functions of two variables. Functions of two variables arise naturally in real-world problems. For example, suppose a certain steel company manufactures I-beams and large cylinders, and suppose an I-beam requires 1.3 tons of steel and a cylinder requires 2.4 tons of steel. If x denotes the number of I-beams to be produced and y denotes the number of cylinders to be produced, then the total amount A of steel needed is given by

$$A = 1.3x + 2.4y.$$

function of two variables

A **function of two variables** is a set of ordered pairs where no two first elements are the same, but the first element in each ordered pair is itself an ordered pair of real numbers. For example, if

$$f = \{((x, y), z) \mid z = x^2 + 4y^2, x \text{ and } y \text{ real numbers}\},$$

then f is a function of two variables. If $(x, y) = (2, 3)$, then

$$z = 2^2 + 4(3)^2 = 40,$$

the ordered pair $((2, 3), 40)$ is an element in f, and 40 is the number in the range paired with $(2, 3)$ in the domain. To be consistent with our previous use of the functional notation, the value of f at $(2, 3)$ should be denoted by $f((2, 3))$; however, it is standard practice to omit one set of parentheses and write $f(2, 3) = 40$. In general, either

$$f(x, y) = x^2 + 4y^2 \quad \text{or} \quad z = x^2 + 4y^2$$

is used to define this function. Unless stated otherwise, the domain of such a function f is the set of all ordered pairs of real numbers (x, y) such that $f(x, y)$ is a real number.

graph of a function

For a real-valued function f of two variables, if $z = f(x, y)$, then the **graph** of f is the graph of all ordered triples (a, b, c) in a three-dimensional coordinate system for which $c = f(a, b)$. For example, the graph of the function defined by

$$f(x, y) = 12 - 3x - 4y \quad \text{or} \quad z = 12 - 3x - 4y$$

is the plane shown in Figure 7.5.

Recall that the equation $x^2 + y^2 = 25$ represents a circle in *two dimensions* but does not define a function. Similarly, not all equations in x, y, and z define functions of two variables. For example, the collection of all points in the three-dimensional cartesian coordinate system at a distance of 5 units from the origin is a sphere consisting of all points whose coordinates satisfy the

7.3 FUNCTIONS OF TWO VARIABLES

equation $x^2 + y^2 + z^2 = 25$. (See Figure 7.11a.) The graph of this equation is *not* a function since there exist at least two different points (x, y, z) satisfying the equation with the same (x, y); for example, $(0, 3, 4)$ and $(0, 3, -4)$ satisfy the equation. However, if we solve the equation for z by taking the positive square root, we get

$$z = \sqrt{25 - x^2 - y^2}.$$

Now, since each pair of numbers x and y satisfying $x^2 + y^2 \leq 25$ determines exactly one z when $z = (25 - x^2 - y^2)^{1/2}$, this equation defines a function. Its graph is a hemisphere. (See Figure 7.11b.) The domain of the function is the set of points in the xy-plane on and interior to the circle $x^2 + y^2 = 25$; the range contains all real numbers z such that $0 \leq z \leq 5$.

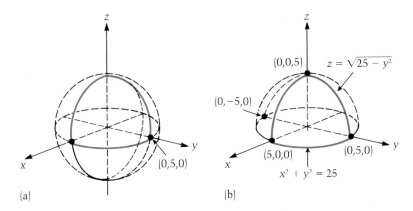

FIG. 7.11.
(a) $x^2 + y^2 + z^2 = 25$.
(b) $z = \sqrt{25 - x^2 - y^2}$

In practical problems, the functional models constructed to solve them are often functions of more than one variable. For example, the cost C of producing and selling a certain product depends upon labor costs, plant costs, costs of materials, transportation costs, etc. Fortunately, much of what we shall learn about functions of two variables is directly applicable to functions of more than two variables. (More about this later.)

Function of two variables

EXAMPLE 1 Let f be the function defined by

$$f(x, y) = 2xy - 3x^{1/2} + y^2.$$

(a) Find $f(4, 5)$.

(b) Find $f(5, 5)$.

(c) Approximate $f(5, 5) - f(4, 5)$.

Solution

(a) $f(4, 5) = 2(4)(5) - 3(4)^{1/2} + 5^2 = 40 - 6 + 25 = 59$.

(b) $f(5, 5) = 2(5)(5) - 3(5)^{1/2} + 5^2 = 50 - 3\sqrt{5} + 25 = 75 - 3\sqrt{5}$.

(c) $f(5, 5) - f(4, 5) = 16 - 3\sqrt{5} \doteq 9.2918$.

Volume of a cone

EXAMPLE 2 The volume V of a right circular cone is a function of its radius r and altitude h and is given by

$$V(r, h) = \frac{\pi}{3} r^2 h$$

where $r > 0$ and $h > 0$.

(a) Find the volume of a right circular cone with radius 6 cm and altitude 10 cm.

(b) If the altitude remains 10 cm but the radius is increased from 6 to 6.03 cm, what is the change ΔV in the volume V?

Solution

(a) $V(6, 10) = (\pi/3)(6)^2(10) = 120\pi$ cm^3.

(b) Since $V(6.03, 10) = (\pi/3)(6.03)^2(10) = 121.203\pi$ cm^3, the change ΔV in volume is $\Delta V = 121.203\pi - 120\pi = 1.203\pi$ cm^3.

Cost function

EXAMPLE 3 Suppose a rectangular box containing 600 cm^3 is to be manufactured from three different materials. The lid costs 6 cents/cm^2, the sides cost 4 cents/cm^2, and the bottom costs 8 cents/cm^2. Express the total cost of each box as a function of the length x and width y of the base. (See Figure 7.12.)

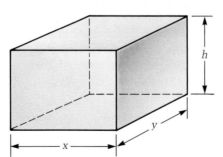

FIG. 7.12.

Solution: Since $V = xyh$ where h is the height of the box, $600 = xyh$ and

$$h = \frac{600}{xy}.$$

Let us now find the area and cost in cents for each part of the box.

	Area	Cost
Base	xy	$8xy$
Two sides	$2xh$	$4(2x)\left(\dfrac{600}{xy}\right) = \dfrac{4800}{y}$
Two sides	$2yh$	$4(2y)\left(\dfrac{600}{xy}\right) = \dfrac{4800}{x}$
Top	xy	$6xy$

Therefore the cost in cents for the material needed to make the box is

$$C(x, y) = 14xy + \frac{4800}{y} + \frac{4800}{x}.$$

PRACTICE PROBLEM Find the cost function for the box in Example 3 provided the lid costs 8 cents/cm^2, the sides cost 5 cents/cm^2, and the bottom costs 10 cents/cm^2

Answer: $C(x, y) = 18xy + 6000/x + 6000/y$ cents. ■

Sales function

EXAMPLE 4 A car dealer's cost for a Slowback model is $5000 and for a Fastback model, her cost is $7000. If x, $200 \leq x \leq 1000$, is the profit (amount over cost) on the Fastback model and y, $200 \leq y \leq 1000$, is the profit on the Slowback model, the dealer finds that the number of Slowback models sold per month is given by

$$S(x, y) = 500 + 0.1x - 0.2y,$$

and the number of Fastback models sold per month is given by

$$F(x, y) = 300 - 0.3x + 0.1y.$$

Note, for example, that as the profit x on the Fastback increases, the number of Fastbacks sold decreases provided the profit y on the Slowbacks remains the same. Furthermore, an increase in y (the profit on the Slowbacks) with x fixed increases the sales of the Fastbacks. (That is, some buyers switch to the more expensive car as the price of the cheaper car increases.)

With a $500 profit on each car, the monthly sales are

$$S(500, 500) = 500 + 50 - 100 = 450 \text{ Slowback models}$$

and

$$F(500, 500) = 300 - 150 + 50 = 200 \text{ Fastback models.}$$

Profit function

EXAMPLE 5

(a) Find the dealer's profit function for the two models of automobiles discussed in Example 4.

(b) Find the total profit if each model can be sold at $1000 above cost.

Solution

(a) Since the dealer makes x dollars profit on each Fastback model, $xF(x, y)$ is the profit per month on these models. Similarly, $yS(x, y)$ is the profit per month on the Slowback models. Consequently, the monthly profit function is

$$P(x, y) = xF(x, y) + yS(x, y)$$
$$= 300x - 0.3x^2 + 0.1xy + 500y + 0.1xy - 0.2y^2$$
$$= 300x - 0.3x^2 + 0.2xy + 500y - 0.2y^2.$$

(b) $P(1000, 1000)$

$$= 300,000 - 300,000 + 200,000 + 500,000 - 200,000$$
$$= \$500,000.$$

Note: This is the dealer's markup profit. Because of the cost of doing business, the actual profit would be less.

PRACTICE PROBLEM

(a) In Examples 4 and 5, determine the percentage change in total sales for both cars, assuming that the profit on each model were increased from $500 to $1000.

(b) What would be the percentage change in the total profit resulting from this 100% increase in profits on each car?

Answer: (a) From 650 to 500, a *decrease* of 23%. (b) From $350,000 to $500,000, an *increase* of 42.9%. ∎

Let f be a function of two variables. As might be expected from our experience with functions of one variable, the *limit* of a

7.3 FUNCTIONS OF TWO VARIABLES

limit of a function

function of two variables is an important and useful concept. Roughly speaking, the **limit** of f at (a, b) is a real number L provided we can find a circle in the xy-plane (domain) with (a, b) as its center such that $f(x, y)$ is arbitrarily close to L for each point (x, y) other than (a, b) in the domain of f and within the circle. As with functions of one variable, it is usually much easier to guess the limit of a function than it is to prove its existence. For example, it should be obvious that

$$\lim_{(x,y)\to(2,3)} \frac{x^2 + y^2}{3x^2 - y^2} = \frac{13}{3},$$

but verifying this fact using a precise limit definition is not trivial. (Although a precise definition of limit and proofs of limit theorems are essential to a rigorous development of calculus for functions of two variables, we can, fortunately, proceed without such an investigation by accepting a few basic theorems.)

continuity of a function

Continuity for a function of two variables is defined in much the same way as it is defined for a function of one variable. A function f of two variables is said to be **continuous** at a point (a, b) if (a, b) is in the domain of f and if

$$\lim_{(x,y)\to(a,b)} f(x, y) = f(a, b).$$

A function such as $f(x, y) = x^2 + y^2$ is continuous for each ordered pair of real numbers. Its graph has no "breaks" or "jumps." (See Figure 7.7) A function such as

$$g(x, y) = \frac{x^2 - 2xy}{y}$$

is continuous at all ordered pairs (x, y) for which $y \ne 0$. In fact, the quotient of any two polynomial functions in x and y is continuous at any point (a, b), provided the value of the denominator at (a, b) is not 0.

EXERCISES

Set A—Function Values

For each function in Exercises 1 through 10 find $f(3, 4)$.

1. $f(x, y) = 2x - 3y + 4$
2. $f(x, y) = 2x^3 - y^2$
3. $f(x, y) = x^2 + y^2 + 5$
4. $f(x, y) = x^2 + 2xy + y^2 + 4$

5. $f(x, y) = x^3 - 2x + 3xy^2$
6. $f(x, y) = xy + x^2 - 9$
7. $f(x, y) = \dfrac{3x + xy}{x + 2y}$
8. $f(x, y) = \dfrac{2x - y}{xy + y}$
9. $f(x, y) = 2^x + \ln xy$
10. $f(x, y) = y^x + 4x - 3y$

Set B—Functional Models for Practical Problems

Volume of a box
11. (a) Express the volume of a box with a square base as a function of the length x of each side of the base and the altitude y.
    (b) Use part (a) to find $V(4, 7)$.

Volume of a box
12. (a) If the length of the rectangular base of a box is twice its width x and if y is its altitude, express the volume V as a function of x and y.
    (b) Use part (a) to find $V(4, 5)$.

Change in volume
13. (a) Refer to Exercise 11 and find $V(4.02, 7)$.
    (b) Find the difference $V(4.02, 7) - V(4, 7)$.

Change in volume
14. (a) Refer to Exercise 12 and find $V(4, 4.98)$.
    (b) Find the difference $V(4, 4.98) - V(4, 5)$ and interpret the significance of the sign in the answer.

Total cost of a box
15. Suppose a rectangular box containing 300 cm³ is to be manufactured from three different materials. The lid costs 4 cents/cm², the sides cost 5 cents/cm², and the bottom costs 8 cents/cm². Express the total cost of each box as a function of the length x and width y of the base.

Total cost of a box
16. Suppose a rectangular box containing 400 cm³ is to be manufactured from two different materials. The lid and sides cost 4 cents/cm² and the bottom costs 10 cents/cm². Express the total cost of each box as a function of the length x and width y of the base.

Surface area of a box
17. A box open at the top contains 400 in³. Express its surface area as a function of the length x and width y of its base.

In Exercises 18 through 22 use the following information. A car dealer's cost for a Slowback model is $6000 and for a Fastback model is $8000. If x, $200 \le x \le 2000$, is the profit (amount over cost) on the Fastback model and y, $200 \le y \le 2000$, is the profit on the Slowback model, the dealer finds that the number of Slowback models sold per year is given by

$$S(x, y) = 800 + 0.1x - 0.2y,$$

and the number of Fastback models sold per year is given by

$$F(x, y) = 600 - 0.3x + 0.1y.$$

Sales level

18. (a) How many Slowback models would be sold per year if the profit on the Slowback model were $500 and the profit on the Fastback were $1000?
    (b) How many Slowback models would be sold per year if the profit on the Slowback model were $1000 and the profit on the Fastback model were $1000?

Sales level

19. (a) How many Fastback models would be sold per year if the profit on the Slowback model were $500 and the profit on the Fastback model were $500?
    (b) How many Fastback models would be sold per year if the profit on the Slowback model were $500 and the profit on the Fastback model were $1000?

Profit function

20. (a) Find the dealer's profit function for the two models.
    (b) Find the profit per year if each model can be sold at $1500 above cost.

Change in sales

21. (a) What would be the change in total sales for both cars if the profit on each model were increased from $500 to $1000?
    (b) What would be the percentage change in total sales for both cars if the profit on each model were increased from $500 to $1000?

Change in profits

22. (a) What would be the change in total profit if the profit on each model were increased from $500 to $1000?
    (b) What would be the percentage change in total profits for both cars if the profit on each model were increased from $500 to $1000?

7.4 PARTIAL DERIVATIVES

We have seen that the derivative is a powerful tool in applications for functions of one variable. The same is true for functions of two variables, but the situation is somewhat more complicated. Suppose (a, b) is a point in the domain of a function f of two variables. Assume that the plane $x = a$ (parallel to the yz-plane) and the plane $y = b$ (parallel to the xz-plane) intersect the graph of f. (See Figure 7.13.) Then each plane determines a curve *(trace)* on the surface of f. This leads to two "derivatives," one giving the slope of the line tangent to the curve in the plane $x = a$ and the other giving the slope of the line tangent to the curve in the plane $y = b$. Let us consider a specific example.

If $f(x, y) = \sqrt{25 - x^2 - y^2}$, $x^2 + y^2 \leq 25$, then the graph of the function f is a hemisphere. If we let $x = 4$, then

$$f(4, y) = \sqrt{9 - y^2}$$

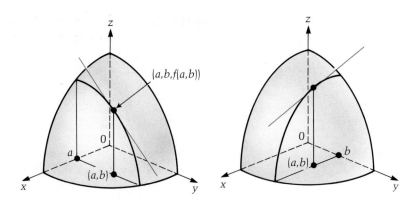

FIG. 7.13.

is the equation of the intersection of the plane $x = 4$ and the hemisphere. The intersection, represented by a function of the one variable y, is a semicircle with radius 3. (See Figure 7.14.) If we let $H(y) = f(4, y)$, then the derivative of $H(y) = (9 - y^2)^{1/2}$ is

$$H'(y) = \frac{-y}{(9 - y^2)^{1/2}}.$$

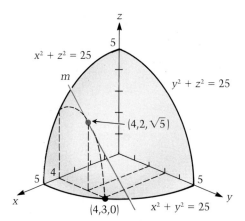

FIG. 7.14.

The derivative obtained by differentiating a function of two variables such as

$$f(x, y) = (25 - x^2 - y^2)^{1/2}$$

with respect to one of the variables while treating the other as a constant is called a **partial derivative** of f. If we *differentiate with*

7.4 PARTIAL DERIVATIVES

respect to y and treat x as a constant, then the partial derivative of f is said to be *with respect to y*; it is denoted by

partial derivative with respect to y

$$f_y(x, y), \quad f_y, \quad \text{or} \quad \frac{\partial f}{\partial y}.$$

If $f(x, y) = (25 - x^2 - y^2)^{1/2}$, then

$$f_y(x, y) = \frac{1}{2}(25 - x^2 - y^2)^{-1/2}(-2y) = \frac{-y}{(25 - x^2 - y^2)^{1/2}}.$$

Note that

$$f_y(4, y) = \frac{-y}{(9 - y^2)^{1/2}}$$

which, of course, is $H'(y)$ obtained earlier. The partial derivative of f with respect to y evaluated at $(4, 2)$ is

$$f_4(4, 2) = \frac{-2}{\sqrt{5}} = \frac{-2\sqrt{5}}{5}.$$

This is the slope of the tangent line, in the plane $x = 4$, to the hemisphere at the point $(4, 2, \sqrt{5})$. (See Figure 7.14.)

The **partial derivative** of f *with respect to x* is obtained by *differentiating with respect to x while treating y as a constant*. It is denoted by

partial derivative with respect to x

$$f_x(x, y), \quad f_x, \quad \text{or} \quad \frac{\partial f}{\partial x}.$$

For the function $f(x, y) = (25 - x^2 - y^2)^{1/2}$,

$$f_x(x, y) = \frac{-x}{(25 - x^2 - y^2)^{1/2}}.$$

The following are definitions of the **first partial derivatives** of a function f of two variables.

first partial derivatives

$$f_x(x, y) = \lim_{h \to 0} \frac{f(x + h, y) - f(x, y)}{h},$$

$$f_y(x, y) = \lim_{k \to 0} \frac{f(x, y + k) - f(x, y)}{k}.$$

Fortunately, we rarely need to resort to the definitions of the partial derivatives. To find $f_x(x, y)$ for the function

$$f(x, y) = x^3 + 2xy^2,$$

we treat y as a constant and proceed with the techniques for differentiating a function of one variable. Therefore,

$$f_x(x, y) = 3x^2 + 2y^2.$$

Similarly, by treating x as a constant and differentiating, we obtain

$$f_y(x, y) = 4xy.$$

Partial derivatives

EXAMPLE 1 $f(x, y) = x^2 + 3xy + 6$; find $f_x(x, y)$ and $f_y(x, y)$.

Solution: $f_x(x, y) = 2x + 3y$ and $f_y(x, y) = 3x$.

Partial derivatives

EXAMPLE 2 $f(x, y) = e^{xy^2}$; find $f_x(x, y)$ and $f_y(x, y)$.

Solution

$$f_x(x, y) = e^{xy^2}(y^2) \quad \text{and} \quad f_y(x, y) = e^{xy^2}(2xy).$$

Partial derivatives

EXAMPLE 3 $f(x, y) = (x/y) + (y^2/x)$; find

$$\frac{\partial f}{\partial x} \quad \text{and} \quad \frac{\partial f}{\partial y}.$$

Solution

$$\frac{\partial f}{\partial x} = \frac{1}{y} - \frac{y^2}{x^2} \quad \text{and} \quad \frac{\partial f}{\partial y} = -\frac{x}{y^2} + \frac{2y}{x}.$$

7.4 PARTIAL DERIVATIVES

PRACTICE PROBLEM Find $f_x(x, y)$ and $f_y(x, y)$ for $f(x, y) = x^3 + xy^{-2}$.

Answer: $f_x(x, y) = 3x^2 + y^{-2}$ and $f_y(x, y) = -2xy^{-3}$. ∎

Since a function f of two variables has two first partial derivatives, we can obtain four partial derivatives of the first partial derivatives. These are called **second partial derivatives** of f and are defined as follows.

second partial derivatives

$$\frac{\partial}{\partial x}\left(\frac{\partial f}{\partial x}\right) = \frac{\partial^2 f}{\partial x^2} = f_{xx} \text{ is the partial of } \frac{\partial f}{\partial x} \text{ with respect to } x.$$

$$\frac{\partial}{\partial y}\left(\frac{\partial f}{\partial x}\right) = \frac{\partial^2 f}{\partial y\, \partial x} = f_{xy} \text{ is the partial of } \frac{\partial f}{\partial x} \text{ with respect to } y.$$

$$\frac{\partial}{\partial x}\left(\frac{\partial f}{\partial y}\right) = \frac{\partial^2 f}{\partial x\, \partial y} = f_{yx} \text{ is the partial of } \frac{\partial f}{\partial y} \text{ with respect to } x.$$

$$\frac{\partial}{\partial y}\left(\frac{\partial f}{\partial y}\right) = \frac{\partial^2 f}{\partial y^2} = f_{yy} \text{ is the partial of } \frac{\partial f}{\partial y} \text{ with respect to } y.$$

Important Note: Observe that in the notation f_{xy} the order for performing the partial differentiation is obtained by reading from left to right, while in the notation

$$\frac{\partial^2 f}{\partial y\, \partial x},$$

it is obtained by reading from right to left. (In a sense the second notation is an operator notation where the order of operation is determined by the variable closest to the function.)

The second partial f_{xx} is the derivative of f_x where y is treated as a constant. This is essentially the second derivative for the function of one variable representing the trace of the surface in a fixed plane parallel to the xz-plane. Therefore f_{xx} gives information about the concavity of this curve. If $f_{xx} > 0$, then the curve bends up, and if $f_{xx} < 0$, then the curve bends down. Similarly, f_{yy} gives information about the concavity of a trace of the surface in a plane parallel to the yz-plane.

Second partials

EXAMPLE 4 If $f(x, y) = x^2 + 3y^3 + 4xy + 13$, then the first partial derivatives of f are

$$f_x(x, y) = 2x + 4y \quad \text{and} \quad f_y(x, y) = 9y^2 + 4x.$$

The second partial derivatives of f are

$$f_{xx}(x, y) = 2, \quad f_{xy}(x, y) = 4, \quad f_{yx}(x, y) = 4, \quad f_{yy}(x, y) = 18y.$$

For the function in Example 4 we found that $f_{xy} = f_{yx}$. The equality of the "mixed" partial derivatives is more than coincidental. It can be proved that for any function $f(x, y)$ if f_x and f_y exist and if f_{xy} exists and is continuous, then f_{yx} exists and $f_{yx} = f_{xy}$.

PRACTICE PROBLEM Find the first and second partial derivatives of

$$f(x, y) = x^4y^2 - 3xy^3 - 7x + 5y + 11.$$

Answer

$$f_x(x, y) = 4x^3y^2 - 3y^3 - 7, \quad f_y(x, y) = 2x^4y - 9xy^2 + 5,$$
$$f_{xx}(x, y) = 12x^2y^2, \quad f_{yy}(x, y) = 2x^4 - 18xy,$$
$$f_{xy}(x, y) = f_{yx}(x, y) = 8x^3y - 9y^2. \quad \blacksquare$$

EXERCISES

Set A—Partial Derivatives

Find the two first partial derivatives and the four second partial derivatives of each of the functions in Exercises 1 through 20.

1. $f(x, y) = 3xy + x$
2. $f(x, y) = xy^2 + 2xy$
3. $f(x, y) = x^2 + 3y^2 + 7x$
4. $f(x, y) = x^2y + 7xy + 6y + 17$
5. $f(x, y) = x^2y^3 + 6x + 5y$
6. $f(x, y) = x^4y^2 + 6x^3y - 3xy$

7.4 PARTIAL DERIVATIVES

7. $f(x, y) = ye^{2x}$
8. $f(x, y) = x(\ln xy)$
9. $f(x, y) = (x + 2y)^{5/2}$
10. $f(x, y) = (3x + y)^{7/3}$
11. $f(x, y) = (x + y)^4 + (y - 2)^3$
12. $f(x, y) = (3x + y)^3 - (1 - x)^4$
13. $f(x, y) = \dfrac{x}{x^2 + y^2}$
14. $f(x, y) = \dfrac{-y}{x^2 - y^2}$
15. $f(x, y) = (x^2 + y^2)^{5/2}$
16. $f(x, y) = (x^2 + y)^{1/2}$
17. $f(x, y) = \dfrac{x^2 + y}{xy}$
18. $f(x, y) = \dfrac{x - y^3}{xy}$
19. $f(x, y) = xe^{xy}$
20. $f(x, y) = e^{xy} + x^2 + 3xy$
21. For what point (x, y) in the domain of

$$C(x, y) = 16xy + \frac{2000}{x} + \frac{2000}{y}$$

are both $C_x(x, y)$ and $C_y(x, y)$ equal to 0?

22. For what point (x, y) in the domain of

$$P(x, y) = 300x + 500y - 0.3x^2 - 0.2y^2 + 0.2xy$$

are both $P_x(x, y)$ and $P_y(x, y)$ equal to 0?

23. Refer to the function $C(x, y)$ given in Exercise 21.
    (a) What is the sign of C_{xx} for each point in the domain?
    (b) What is the sign of C_{yy} for each point in the domain?
    (c) What are the signs of C_{xy} and C_{yx} for each point in the domain?

24. Refer to the function $P(x, y)$ given in Exercise 22.
    (a) What is the sign of P_{xx} for each point in the domain?
    (b) What is the sign of P_{yy} for each point in the domain?
    (c) What are the signs of P_{xy} and P_{yx} for each point in the domain?

7.5 TANGENT PLANES AND THE TOTAL DIFFERENTIAL

For a function of one variable given by $y = f(x)$, recall that the *differential dy*, defined by

$$dy = f'(x)\, dx.$$

approximates the change in y, denoted by Δy, corresponding to a change dx in x. (See Section 2.6.) In exactly the same way, if $z = f(x, y)$, then

$$f_x(x, y)\, dx$$

approximates the change Δz in z corresponding to a change dx in x, *provided y* is unchanged (held constant). For example, if

$$z = f(x, y) = 2x^{1/2}y - x^2,$$

then

$$f(4, 2) = -8.$$

To approximate $f(4.01, 2)$, we note that y remains constant and that x increases from 4 to 4.01. Therefore,

$$\Delta z = f(4.01, 2) - f(4, 2) \doteq f_x(4, 2)(0.01),$$

and

$$f(4.01, 2) \doteq f(4, 2) + f_x(4, 2)(0.01).$$

Since $f_x(x, y) = x^{-1/2}y - 2x$, we have $f_x(4, 2) = 1 - 8 = -7$. Hence,

$$f(4.01, 2) \doteq -8 + (-7)(0.01) = -8.07.$$

As a check, we find that $f(4.01, 2) = -8.0701$, correct to four decimal places.

PRACTICE PROBLEM For $f(x, y) = 2x^{3/2}y - 2x$, use the partial derivative f_x at $(4, 2)$ to approximate $f(4.002, 2)$.

Answer: 24.02. ∎

It should be clear that for a function f of two variables, if x remains constant and y changes by a small amount dy, then the corresponding change in $z = f(x, y)$ can be approximated by

$$f_y(x, y)\, dy.$$

7.5 TANGENT PLANES AND THE TOTAL DIFFERENTIAL

total differential

In fact, for small changes in x and y, the corresponding change Δz in z is approximated by what is called the **total differential** of $z = f(x, y)$. It is denoted by dz and is defined by

$$dz = f_x(x, y)\, dx + f_y(x, y)\, dy$$

or

$$dz = \frac{\partial f}{\partial x} dx + \frac{\partial f}{\partial y} dy.$$

Therefore,

$$dz \doteq \Delta z = f(x + dx, y + dy) - f(x, y)$$

for small changes dx in x and dy in y.

Total differential

EXAMPLE 1 Let $f(x, y) = x^2 y$. Use the total differential of f at $(4, 7)$ to approximate $f(4.02, 7.03)$.

Solution: Since $f_x = 2xy$, $f_y = x^2$ and $dz = f_x\, dx + f_y\, dy$, at the point $(4, 7)$ the change in x is $dx = 0.02$, the change in y is $dy = 0.03$, and the total differential is

$$dz = 2(4)(7)(0.02) + (4)^2(0.03) = 1.60.$$

Since $dz \doteq \Delta z$, $dz \doteq f(4.02, 7.03) - f(4, 7)$ and

$$f(4.02, 7.03) \doteq f(4, 7) + dz = 112 + 1.60 = 113.60.$$

Note: The actual change in z is $\Delta z = 1.607612$.

PRACTICE PROBLEM Assuming that $f(x, y) = x^2 y + 3xy$, find $f(4.03, 2.98) - f(4, 3)$. Then use the total differential to approximate $f(4.03, 2.98) - f(4, 3)$.

Answer: 0.426082; 0.43. ∎

Change in profit

EXAMPLE 2 Suppose that the total profit function in thousands of dollars from selling sport shirts at x dollars each, $10 \le x \le 16$, and dress shirts at y dollars each, $12 \le y \le 19$, is

$$P(x, y) = 8x + 14y + 6xy - 4x^2 - 2y^2.$$

Use the differential to determine whether an increase in the selling price of the sport shirts from $11 to $11.25 and a decrease in the selling price of the dress shirts from $14 to $13.75 increases or decreases the profit and by how much.

Solution: Since $P(x, y) = 8x + 14y + 6xy - 4x^2 - 2y^2$,

$$P_x(x, y) = 8 + 6y - 8x \quad \text{and} \quad P_y(x, y) = 14 + 6x - 4y.$$

Furthermore,

$$\Delta P \doteq dP = P_x(11, 14)\, dx + P_y(11, 14)\, dy$$

where $dx = 0.25$ and $dy = -0.25$. Therefore,

$$dP = (4)(0.25) + (24)(-0.25) = -5.$$

The changes in the selling prices of the two types of shirts will *decrease* profit by approximately $5000.

Note

$$\Delta P = P(11.25, 13.75) - P(11, 14) = 326.25 - 332 = -5.75.$$

If the output Q of a factory is a function of capital K and labor L, then

$$\frac{\partial Q}{\partial K} \quad \text{and} \quad \frac{\partial Q}{\partial L}$$

marginal productivity

are the *marginal productivity of capital* and *marginal productivity of labor*, respectively. Marginal productivity of capital approximates the change in output Q when L remains fixed and the change in K is $dK = 1$ unit. Similarly, marginal productivity of labor approximates the change dQ in output Q when K remains fixed and the change in L is $dL = 1$ unit.

If the total production cost C of a product is given in terms of x dollars for labor and y dollars for materials, then

$$\frac{\partial C}{\partial x} \quad \text{and} \quad \frac{\partial C}{\partial y}$$

marginal costs

are the *marginal cost of labor* and the *marginal cost of materials*, respectively. Each approximates the change in C for a one-unit change in each of the independent variables when the other remains unchanged. Similar statements apply for a revenue function $R(x, y)$ and a profit function $P(x, y)$.

7.5 TANGENT PLANES AND THE TOTAL DIFFERENTIAL

Let f be a function whose partial derivatives exist at (a, b). Consider the linear function (plane) $z = g(x, y)$ defined by

tangent plane

$$g(x, y) = f(a, b) + f_x(a, b)(x - a) + f_y(a, b)(y - b).$$

This is the equation of the **tangent plane** to f at (a, b). To make this intuitively clear, observe that $g(a, b) = f(a, b)$. Then note that since $g_x(x, y) = f_x(a, b)$ and $g_y(x, y) = f_y(a, b)$,

$$g_x(a, b) = f_x(a, b) \quad \text{and} \quad g_y(a, b) = f_y(a, b).$$

Therefore, the plane not only contains the point $(a, b, f(a, b))$ on f but also has the "correct slopes" in the x- and y-directions.

Equation of tangent plane

EXAMPLE 3 Find an equation of the tangent plane to the surface

$$f(x, y) = x^3 - 3xy^2$$

at the point $(1, -3)$.

Solution: $f_x(x, y) = 3x^2 - 3y^2$ and $f_y(x, y) = -6xy$. Thus, if $z = g(x, y)$ is the tangent plane at $(1, -3)$, then

$$z = f(1, -3) + f_x(1, -3)(x - 1) + f_y(1, -3)(y + 3)$$
$$= -26 - 24(x - 1) + 18(y + 3)$$

or

$$24x - 18y + z = 52.$$

If we assume that the tangent plane

$$g(x, y) = f(a, b) + f_x(a, b)(x - a) + f_y(a, b)(y - b)$$

approximates the graph of f at a point (x, y) near the point (a, b), then $f(x, y) \doteq g(x, y)$ and

$$f(x, y) - f(a, b) \doteq f_x(a, b)\, dx + f_y(a, b)\, dy = dz$$

where $dx = x - a$ and $dy = y - b$. Consequently, as with a function of one variable, using the (total) differential to approximate a change in the function value of f is a linear approximation.

EXERCISES

Set A—Approximating Function Values Using the Total Differential

1. (a) Let $f(x, y) = x^2y + 3xy^2$. Use the total differential at $(3, 3)$ to approximate $f(3.1, 2.98)$.
   (b) Find by direct computation the error in this approximation.

2. (a) Let $f(x, y) = x^2y - 2y^2 + x$. Use the total differential at $(2, 2)$ to approximate $f(2.01, 1.97)$.
   (b) Find by direct computation the error in this approximation.

3. (a) Find the product 2.037×7.98.
   (b) Use the total differential at $(2, 8)$ to approximate this product.

4. (a) Find the product 4.013×8.074.
   (b) Use the total differential at $(4, 8)$ to approximate this product.

5. (a) Let $f(x, y) = x^{1/2}y + 3x + 2y$. Use the total differential at $(4, 3)$ to approximate $f(4.01, 3.02)$.
   (b) Use tables or a calculator to approximate $f(4.01, 3.02)$. Find the difference between this answer and that in part (a).

6. (a) Assuming that $f(x, y) = x^{1/2}y + 3xy^{2/3}$, use the total differential at $(4, 8)$ to approximate $f(4.03, 7.98)$.
   (b) Use tables or a calculator to approximate $f(4.03, 7.98)$. Find the difference between this answer and that in part (a).

Set B—Equations of Tangent Planes

In Exercises 7 through 12 find an equation of the tangent plane to the given surface at the point where $x = 2$ and $y = -1$.

7. $f(x, y) = x^2 - xy$
8. $f(x, y) = x^3y$
9. $f(x, y) = x^2y - xy^2$
10. $f(x, y) = x^4y + x^2y^2$
11. $f(x, y) = \dfrac{x + y^2}{x^2 - y}$
12. $f(x, y) = \dfrac{x}{x^2 + y^2}$

Set C—Applications

Change in volume

13. (a) Each side of the base of a box with a square bottom is to be increased from 12.2 in to 12.23 in, and its height is to be increased from 17.3 in to 17.302 in. Find the actual change in the volume of the box.

7.5 TANGENT PLANES AND THE TOTAL DIFFERENTIAL

Change in volume

(b) Use the total differential at (12.2, 17.3) to approximate the change in the volume.

Change in volume

14. (a) Each side of the base of a box with a square bottom is to be decreased from 11.4 cm to 11.36 cm and its height is to be increased from 8.6 cm to 8.62 cm. Find the actual change in the volume of the box.
    (b) Use the total differential at (11.4, 8.6) to approximate the change in the volume.

Change in volume

15. (a) The height of a cone is to be increased from 6 cm to 6.03 cm and the radius is to be decreased from 4 cm to 3.99 cm. Find the actual change in the volume of the cone. *Note:* $V = (\pi/3)r^2 h$.
    (b) Use the total differential at (4, 6) to approximate the change in the volume.

Material in a can

16. Suppose a metal can without a lid is to have an inner radius of 3 cm and an inner height of 7 cm. If the can is to be 0.1 cm thick, approximate the amount of metal, in cubic centimeters, necessary to make the can.

Change in profit

In Exercises 17 through 19, refer to the automobile dealer's profit function in Example 5, Section 7.3, given by

$$P(x, y) = 300x - 0.3x^2 + 0.2xy + 500y - 0.2y^2.$$

17. (a) Find $P(800, 525) - P(800, 500)$ and explain its significance.
    (b) Use a partial derivative to approximate the answer to part (a).

18. (a) Find $P(825, 500) - P(800, 500)$ and explain its significance.
    (b) Use a partial derivative to approximate the answer to part (a).

19. (a) Use the total differential to approximate

    $$P(750, 525) - P(700, 500).$$

    (b) Find the actual change in profit.

Marginal productivity of labor

20. (a) If the units of output Q of a factory depend upon K units of capital and L units of labor, what would you expect the sign of $\dfrac{\partial Q}{\partial L}$ to be? Explain.
    (b) If $\dfrac{\partial^2 Q}{\partial L^2} < 0$, explain the significance. Should this result be expected?

7.6 APPLIED MAXIMA AND MINIMA PROBLEMS

Before discussing how to solve problems related to maxima and minima of functions of two variables, we must define *relative extrema* for such functions. (Recall for functions of one variable that $f(t)$ is a relative maximum of f if there exists an open interval containing t and in the domain of f such that $f(x) \leq f(t)$ for each x in the open interval.) For a point (a, b) in the domain of a function f of two variables, all points interior to a circle with (a,b) as center is called a *neighborhood* of (a, b). If there exists a neighborhood of (a, b) contained in the domain of f such that

$$f(x, y) \leq f(a, b)$$

relative maximum

for each (x, y) in the neighborhood, then $f(a, b)$ is called a **relative maximum** of f. Similarly, if there exists a neighborhood of (a, b) contained in the domain of f such that

$$f(x, y) \geq f(a, b)$$

relative minimum

for each (x, y) in the neighborhood, then $f(a, b)$ is called a **relative minimum** of f.

Essentially, a relative maximum is an interior high point on the graph of f and a relative minimum is an interior low point on the graph of f. If $f(a, b) \geq f(x, y)$ for *every* (x, y) in the domain of f then $f(a, b)$ is the **absolute maximum** of f. Similarly, if $f(a, b) \leq f(x, y)$ for *every* (x, y) in the domain of f, then $f(a, b)$ is the **absolute minimum** of f. As with functions of one variable, absolute extrema may not be relative extrema since they can occur on the boundary of the domain of a function. As with functions of one variable, we may have to consider separately points on the boundary of the domain of a function when determining its absolute extrema.

absolute maximum

absolute minimum

In Chapter 3 we found for a differentiable function f of one variable that a necessary condition for the existence of a relative maximum or minimum at t is that $f'(t) = 0$. We also discovered that if $f'(t) = 0$ and $f''(t) \neq 0$, then the function has either a relative maximum or a relative minimum value at t. We now turn our attention to finding both necessary and sufficient conditions for a function of two variables to have a *relative maximum or minimum* at some (interior) point (a, b) in its domain.

It should be geometrically obvious from Figure 7.15 that a *necessary condition for a function f of two variables to have a relative maximum (or minimum) at (a, b) is that*

7.6 APPLIED MAXIMA AND MINIMA PROBLEMS

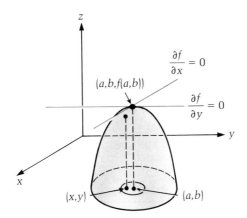

FIG. 7.15.

$$f_x(a, b) = f_y(a, b) = 0$$

provided the partial derivatives exist. We see that if $f(a, b)$ is a relative maximum for the function, then the trace of the surface in the plane $y = b$ is the graph of a function of x which must have a relative maximum at $x = a$. Consequently, $f_x(a, b) = 0$. By a similar argument for the trace in the plane $x = a$, we conclude that $f_y(a, b) = 0$.

From our experience with functions of one variable, it might be expected that a point (a, b) where both first partial derivatives exist and are equal to 0 is a "candidate" for a relative maximum or minimum point. To see why $f_x(a, b) = f_y(a, b) = 0$ is not a sufficient condition for a relative maximum or minimum at (a, b), consider the point $(c, d, f(c, d))$ in Figure 7.16, which is called a *saddle point* for the surface. At the saddle point,

saddle point

$$f_x(c, d) = f_y(c, d) = 0,$$

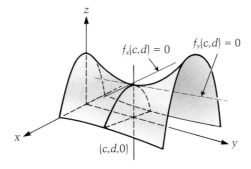

FIG. 7.16.

but f has neither a relative maximum nor a relative minimum at (c, d). We see that function values are greater than $f(c, d)$ "in the y-direction" and less than $f(c, d)$ "in the x-direction."

Sufficient conditions for a function of two variables to have a relative maximum or minimum value at (a, b) are given in the following theorem.

> **THEOREM 7.1** Let f be a function of two variables defined in some neighborhood of (a, b) having continuous second partial derivatives in the neighborhood. Let
>
> $$A = f_{xx}(a, b) \qquad B = f_{xy}(a, b), \qquad \text{and} \qquad C = f_{yy}(a, b).$$
>
> Assume that $f_x(a, b) = f_y(a, b) = 0$ and that $AC - B^2 > 0$.
>
> *relative maxima and relative minima*
>
> If $A < 0$ (or $C < 0$), then f has a **relative (local) maximum** at (a, b).
>
> If $A > 0$ (or $C > 0$), then f has a **relative (local) minimum** at (a, b).

Notes

1. Although a proof of Theorem 7.1 is not given, some observations should make it seem reasonable and easier to remember. If $AC - B^2 > 0$, then $AC > B^2$ and AC is positive. Therefore A and C have the same sign, which means the traces in the planes $x = a$ and $y = b$ have the same concavity. If A is negative, then C is negative, the concavity is downward, and $f(a, b)$ is a local maximum. If A is positive, then C is positive, the concavity is upward, and $f(a, b)$ is a local minimum.

2. If $AC - B^2 < 0$, then the function has a saddle point at (a, b).

3. If $AC - B^2 = 0$, then f may or may not have a local maximum or minimum at (a, b). Two examples are given in the exercises to verify this fact. (The situation is similar for a function of one variable where $f'(t) = 0$ and $f''(t) = 0$.)

Relative extrema

EXAMPLE 1 Let $f(x, y) = x^4 - 2x^2 + 3y^2$. Find any relative maxima and minima for this function.

7.6 APPLIED MAXIMA AND MINIMA PROBLEMS

Solution

$$f_x = 4x^3 - 4x \quad \text{and} \quad f_y = 6y.$$

Furthermore,

$$f_{xx} = 12x^2 - 4, \; f_{xy} = 0, \text{ and } f_{yy} = 6.$$

Since $f_x = 4x(x^2 - 1) = 4x(x - 1)(x + 1)$ and $f_y = 6y$,

$$f_x = f_y = 0$$

at the points $(0, 0)$, $(1, 0)$, and $(-1, 0)$. Now let us consider each of these three points separately.

1. At $(0, 0)$, $f_{xx} = -4$, $f_{xy} = 0$, and $f_{yy} = 6$. Therefore,

$$AC - B^2 = -24 < 0,$$

and there is a saddle point at $(0, 0)$.

2. At $(1, 0)$, $f_{xx} = 8$, $f_{xy} = 0$, and $f_{yy} = 6$. Therefore,

$$AC - B^2 = 48 > 0.$$

Since $f_{xx} > 0$, it follows that $f(1, 0) = -1$ is a relative minimum.

3. At $(-1, 0)$, $f_{xx} = 8$, $f_{xy} = 0$, and $f_{yy} = 6$. Therefore,

$$AC - B^2 = 48 > 0.$$

Since $f_{xx} > 0$, it follows that $f(-1, 0) = -1$ is a relative minimum.

Note: f has the same relative minimum value at two different points.

Relative extrema

EXAMPLE 2 Let $f(x, y) = 3x^2/2 + xy^3 - y$. Find all relative maxima and minima for the function.

Solution

$$f_x = 3x + y^3 \quad \text{and} \quad f_y = 3xy^2 - 1;$$
$$f_{xx} = 3, \quad f_{yy} = 6xy, \quad \text{and} \quad f_{xy} = 3y^2.$$

Setting the first partials equal to zero, we obtain

$$3x + y^3 = 0, \qquad (1)$$
$$3xy^2 - 1 = 0. \qquad (2)$$

From Eq. (1), $x = -y^3/3$. Substituting in Eq. (2), we obtain

$$-\frac{3y^5}{3} - 1 = 0, \qquad y^5 = -1, \qquad y = -1.$$

Substituting $y = -1$ in $x = -y^3/3$ yields $x = \tfrac{1}{3}$. Therefore $f(\tfrac{1}{3}, -1)$ is the only "candidate" for a relative maximum or minimum. Now computing $AC - B^2$ as given in Theorem 7.1, we obtain

$$AC - B^2 = 3(-2) - (3)^2 = -15 < 0.$$

Therefore, f has a saddle point at $(\tfrac{1}{3}, -1)$, and f has no relative extrema.

Maximum profit from shirt sales

EXAMPLE 3 A company's profit in thousands of dollars from the sale of dress shirts and sport shirts at x and y dollars each, respectively, is given by

$$P(x, y) = 8x - 14y + 6xy - 4x^2 - 2y^2 + 210,$$

provided $10 \leq x \leq 16$ and $12 \leq y \leq 19$. Find the price of each type of shirt which will maximize profits and find the maximum profit. Also determine the price of each type of shirt which will minimize profits and find the minimum profit possible.

Solution

$$P_x = 8 + 6y - 8x \quad \text{and} \quad P_y = -14 + 6x - 4y.$$

Setting $P_x = P_y = 0$ yields

$$8x - 6y = 8, \qquad (1)$$
$$6x - 4y = 14. \qquad (2)$$

Multiplying each side of Eq. (1) by 2 and each side of Eq. (2) by 3 we get

$$16x - 12y = 16, \qquad (3)$$
$$18x - 12y = 42. \qquad (4)$$

Subtracting Eq. (3) from Eq. (4) we get

$$2x = 26,$$

$$x = 13.$$

Substituting $x = 13$ in Eq. (1) yields $y = 16$. Therefore, (13, 16) is the only interior point in the domain for which f can have a relative extremum.

Since $P_{xx} = -8$, $P_{yy} = -4$, and $P_{xy} = 6$, we find that $AC - B^2 = -4 < 0$. This means there is a saddle point at (13, 16): furthermore, since there are no relative extrema, this means that the maximum value and the minimum value of the profit function must occur on the boundary of the domain. Let us consider now what happens to the function values on the boundary of the domain. (See Figure 7.17.)

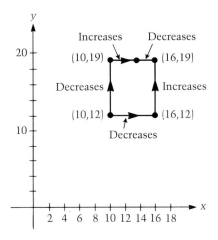

FIG. 7.17.

Along the boundary where $y = 12$, since $P_x = 8 + 6y - 8x$,

$$P_x(x, 12) = 80 - 8x.$$

Since $80 - 8x < 0$ for $10 < x < 16$, the function values decrease along the segment from (10, 12) to (16, 12). Similarly, along the boundary where $y = 19$, $P_x(x, 19) = 122 - 8x$. Since $122 - 8x > 0$ for $10 < x < 15.25$, the function values increase along the segment from (10, 19) to (15.25, 19). Since $122 - 8x < 0$ for $15.25 < x < 16$, the function values decrease along the segment from (15.25, 19) to (16, 19). From the fact that

$$P_y(10, y) = 46 - 4y < 0 \quad \text{for} \quad 12 < y < 19$$

and

$$P_y(16, y) = 82 - 4y > 0 \quad \text{for} \quad 12 < y < 19,$$

we conclude that p decreases along the boundary from (10, 12) to (10, 19) and increases along the boundary from (16, 12) to (16, 19).

The function values at the corners of the domain of $P(x, y)$ and at (15.25, 19) are:

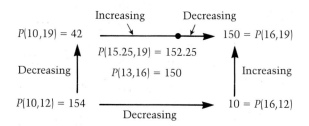

We see that the maximum profit is $154,000; it is attained when the sport shirt is priced at $10, and the dress shirt is priced at $12. The minimum profit is $10,000; it is attained when the sport shirt is priced at $16, and the dress shirt is priced at $12.

Maximizing profits on a bingo game

EXAMPLE 4 A club holds a bingo game each week to raise money for charity. With a $2.00 charge per card and a policy of returning $\frac{1}{5}$ (i.e., 20%) of the total amount taken in as prize money, the club sells 1700 cards each week. It is estimated that if x is the number of 25-cent increases in the price of each card and if y, $0.20 \leq y \leq 0.80$, is the payout, then the number of cards sold each week will be given by

$$N(x, y) = 1200 - 100x + 2500y.$$

This means that for each 25-cent increase in the price of a card, the number of cards sold will decrease by 100; furthermore, for each 0.10 increase (10% increase) in the payout, the number of cards sold will increase by 250. For example,

$$N(0, 0.2) = 1200 + 500 = 1700$$

means that 1700 cards will be sold at $2.00 with a payout of 20%, and

$$N(2, 0.8) = 1200 - 200 + 2000 = 3000$$

means that 3000 cards will be sold at $2.50 with a payout of 80%.

7.6 APPLIED MAXIMA AND MINIMA PROBLEMS

What should be the price for each card and what should be the percentage payout in order to maximize the profit for the club?

Solution: A card will sell for $2 + x/4$ dollars and the club will take in

$$(1200 - 100x + 2500y)\left(2 + \frac{x}{4}\right)$$

dollars. Since the payout is y, the amount retained by the club is $1 - y$. Hence the profit function is

$$P(x, y) = (1 - y)(1200 - 100x + 2500y)\left(2 + \frac{x}{4}\right).$$

The first partial derivatives of the profit function are

$$P_x = (1 - y)\left[(1200 - 100x + 2500y)\left(\frac{1}{4}\right) + (-100)\left(2 + \frac{x}{4}\right)\right],$$

$$P_y = \left(2 + \frac{x}{4}\right)[(-1)(1200 - 100x + 2500y) + (1 - y)(2500)].$$

Setting $P_y = 0$, the first factor yields $x = -8$, which we ignore. Setting the second factor equal to 0 yields

$$1200 - 100x + 2500y = 2500 - 2500y.$$

$$-100x + 5000y = 1300,$$

$$x - 50y = -13. \tag{1}$$

Setting $P_x = 0$, the first factor yields $y = 1$; we ignore this solution since it makes $x = 65$ and $N(x, y)$ negative. Setting the second factor in P_x equal to 0 yields

$$1200 - 100x + 2500y = 800 + 100x,$$

$$-200x + 2500y = -400,$$

$$2x - 25y = 4. \tag{2}$$

Multiplying each side of Eq. (1) by 2 and subtracting from Eq. (2) yields

$$75y = 30,$$

$$y = \frac{2}{5} = 0.40.$$

Substituting $y = \frac{2}{5}$ into Eq. (1) yields

$$x = 7.$$

We leave for the exercises a proof that $P(7, 0.4)$ is not only a relative maximum of the function but also the absolute maximum in its domain. Therefore, the club should charge

$$2 + \frac{7}{4} = \$3.75$$

for each card and set the payout at 40% to obtain the maximum amount for its charity. The maximum profit is

$$P(7, 0.40) = \$3375.$$

Minimum cost of a box

EXAMPLE 5 Suppose a rectangular packing box containing 9 ft^3 is to be manufactured from three different materials. The lid costs 6 cents/ft^2, the sides cost 3 cents/ft^2, and the bottom costs 10 cents/ft^2. Find the most economical dimensions for the box.

Solution: The volume $V = xyz$ where x is the width of the rectangular bottom and top, y is its length, and z is the height of the box. The cost of the bottom is $10xy$ cents, two sides cost $3xz$ cents each, the other two sides cost $3yz$ cents each, and the cost of the top is $6xy$ cents. Therefore the total cost T is given by

$$T = 10xy + 6xz + 6yz + 6xy.$$

Since volume $= 9 = xyz$, $z = 9/xy$ and the total cost as a function of x and y is

$$T(x, y) = 10xy + 6x\left(\frac{9}{xy}\right) + 6y\left(\frac{9}{xy}\right) + 6xy$$

$$= 16xy + \frac{54}{y} + \frac{54}{x}.$$

The first partial derivatives of $T(x, y)$ are

$$T_x(x, y) = 16y - \frac{54}{x^2} \quad \text{and} \quad T_y(x, y) = 16x - \frac{54}{y^2}.$$

7.6 APPLIED MAXIMA AND MINIMA PROBLEMS

To solve for the simultaneous solution of the system

$$16y - \frac{54}{x^2} = 0, \tag{1}$$

$$16x - \frac{54}{y^2} = 0, \tag{2}$$

we solve Eq. (1) for y and substitute in Eq. (2). From Eq. (1) we get

$$y = \frac{27}{8x^2},$$

and substituting in Eq. (2) yields

$$16x - (54)\frac{64x^4}{27^2} = 0,$$

$$27x - 8x^4 = 0,$$

$$x(27 - 8x^3) = 0,$$

$$x = 0 \quad \text{or} \quad 27 - 8x^3 = 0.$$

For this practical problem, $x = 0$ can be ignored. If $27 - 8x^3 = 0$, then $8x^3 = 27$, $x = \sqrt[3]{\frac{27}{8}}$, and $x = \frac{3}{2}$ is the only real solution for x. Substituting $x = \frac{3}{2}$ in $y = 27/8x^2$, we get $y = \frac{3}{2}$.

The second partial derivatives of T are

$$T_{xx}(x, y) = \frac{108}{x^3}, \quad T_{yy}(x, y) = \frac{108}{y^3}, \quad \text{and} \quad T_{xy}(x, y) = 16.$$

At $x = \frac{3}{2}$ and $y = \frac{3}{2}$,

$$AC - B^2 = (32)(32) - 256 = 768 > 0.$$

Since $T_{xx}(\frac{3}{2}, \frac{3}{2}) > 0$, $T(x, y)$ has a relative minimum at that point. Since $z = 9/xy$, $z = 4$. The domain of T consists of all reals $x > 0$ and $y > 0$. For x near 0 or y near 0, $T(x, y)$ becomes arbitrarily large. Also as x gets large or y gets large, $T(x, y)$ becomes arbitrarily large; therefore the relative minimum is the absolute minimum. To minimize the cost, the dimensions of the base should be 1.5 feet on each side and the height should be 4 feet.

EXERCISES

Set A–Relative Extrema for Functions of Two Variables

Find the relative maxima and minima (if they exist) for each of the functions in Exercises 1 through 10.

1. $f(x, y) = x^2 + 2xy + y^2 + 4$
2. $f(x, y) = x^2 - 2xy + y^2 - 6$
3. $f(x, y) = 12xy - 6x^2y - 3xy^2$
4. $f(x, y) = 2x^2 - 4x + xy^2 - 1$
5. $f(x, y) = 25x + 36y + \dfrac{49}{x} + \dfrac{64}{y}$
6. $f(x, y) = xy + \dfrac{2}{x} + \dfrac{4}{y}$
7. $f(x, y) = 6x - xy - x^2y^2$
8. $f(x, y) = x^2 + 3xy + 3y^2 - 6x - 3y - 6$
9. $f(x, y) = x^3 - 3x^2 + 2y^2$
10. $f(x, y) = x^3 + xy$
11. Use the functions $f(x, y) = x^4 + y^4$ and $g(x, y) = x^3 + y^3$ to verify that a function may or may not have a relative extremum at a point where the first partials are equal to 0 and $AC - B^2 = 0$. (This verifies remark 3 after Theorem 7.1.)

Set B–Applications for Functions of Two Variables

Minimizing cost

12. The cost function for the rectangular box discussed in Example 3, Section 7.3, was given by

$$C(x, y) = 14xy + \dfrac{4800}{y} + \dfrac{4800}{x}.$$

Find the dimensions of the box that minimize its cost. (Recall that $z = 600/xy$.)

Monthly profits

13. The automobile dealer's monthly profit function discussed in Example 5, Section 7.3, was

$$P(x, y) = 300x - 0.3x^2 + 0.2xy + 500y - 0.2y^2,$$
$$0 \le x \le 2000 \quad \text{and} \quad 0 \le y \le 2000.$$

What profit on each model would maximize total profit?

Maximum box dimensions

14. The largest rectangular package that may be sent by parcel post must meet the specification that the sum of the length and girth (perimeter of a cross section) is 84 inches. Find the

dimensions of the rectangular package of greatest volume that may be sent parcel post.

Minimizing cost of box

15. A paneled box is to be open at the top and have a volume of 500 cm³. What are the most economical dimensions requiring the minimum amount (in square centimeters) of paneling?

Minimizing cost of box

16. Suppose a rectangular box containing 150 cm³ is to be manufactured from three different materials. The lid costs 4 cents/cm², the sides cost 5 cents/cm², and the bottom costs 8 cents/cm². Find the most economical dimensions for the box.

Maximum box dimensions

17. If the largest rectangular package that may be sent by parcel post must meet the requirement that the sum of the length and girth (perimeter of a cross section) is 108 inches, what are the dimensions of the rectangular package of greatest volume that may be sent parcel post?

Minimum distance from origin to a plane

18. Use partial derivatives to find the shortest distance from the origin $(0, 0, 0)$ to the plane $z = 2x - 2y + 8$. [*Hint:* The distance from the origin to any point (x, y, z) in the plane is $d(x, y) = \sqrt{x^2 + y^2 + (2x - 2y + 8)^2}$.]

Profit from bingo game

19. (a) For $P(x, y)$ in Example 4, find P_{xx}, P_{xy}, and P_{yy}.
    (b) Show that $P_{xx}P_{yy} - (P_{xy})^2 > 0$ at $(7, 0.4)$.
    (c) Show that $P_{xx}(7, 0.4) < 0$ and conclude that $P(7, 0.4)$ is a relative maximum of P.

Profit from bingo game

20. (a) For $P(x, y)$ in Example 4, graph its domain in the xy-coordinate plane.
    (b) Let $F(x) = P(x, 0.2)$ and find the maximum value of F where $x \geq 0$.
    (c) Let $G(x) = P(x, 0.8)$ and find the maximum value of G where $x \geq 0$.
    (d) Let $H(y) = P(0, y)$ and find the maximum value of H, $0.2 \leq y \leq 0.8$.
    (e) Use Exercise 19 and the preceding parts of this exercise to conclude that $P(7, 0.4)$ is the absolute maximum of the profit function P given in Example 4 of this section.

7.7 MAXIMA AND MINIMA WITH CONSTRAINTS

In the preceding section, solving Example 5 essentially required finding the minimum value of the function

$$f(x, y, z) = 16xy + 6xz + 6yz$$

under the condition that $xyz = 9$. The condition $xyz = 9$ is called a **constraint** on x, y, and z. A powerful technique for obtaining the solutions to such maxima and minima problems is due to Joseph Lagrange (1736–1813) and is called the method of the **Lagrange multiplier.** We exhibit this technique by first considering a function of two variables with a given constraint on the variables.

Suppose we want to find the minimum value of the function $f(x, y) = x^2 + y^2$ subject to the constraint (condition) that

$$3x + 2y - 6 = 0.$$

To use the method of Lagrange multipliers, we let g be the constraint function defined by

$$g(x, y) = 3x + 2y - 6.$$

Then we find the simultaneous solutions of the three equations

$$f_x(x, y) = \lambda g_x(x, y),$$
$$f_y(x, y) = \lambda g_y(x, y),$$
$$g(x, y) = 0.$$

Each solution is a "candidate" for minimizing or maximizing f with the given constraint. For our example, the three equations are

$$2x = 3\lambda, \qquad (1)$$
$$2y = 2\lambda, \qquad (2)$$
$$3x + 2y - 6 = 0. \qquad (3)$$

From Eqs. (1) and (2) we get $x = 3y/2$. Substituting in Eq. (3) we find

$$\frac{9}{2}y + 2y - 6 = 0,$$
$$13y = 12,$$
$$y = \frac{12}{13}.$$

Substituting this value of y in $x = 3y/2$, we obtain $x = \frac{18}{13}$. Therefore $(\frac{18}{13}, \frac{12}{13})$ is a candidate in the domain of f for a maximum or minimum value for the function with the given constraint. Often

7.7 MAXIMA AND MINIMA WITH CONSTRAINTS

the practical aspects of a problem will make it easy to identify the nature of such a point. For this problem we use geometrical considerations. It should be clear from Figure 7.18 that with the given constraint, f has a *minimum* at the point since it is the low point on the trace where the plane $3x + 2y - 6 = 0$ intersects the surface $f(x, y) = x^2 + y^2$. The minimum value is $f(\frac{18}{13}, \frac{12}{13}) = \frac{36}{13}$.

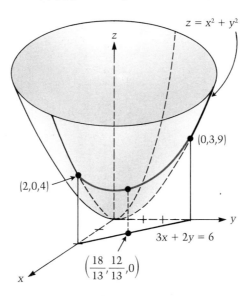

FIG. 7.18. Minimum value.

Let us verify the solution to the preceding problem by first solving the constraint equation $3x + 2y - 6 = 0$ for x then substituting in $x^2 + y^2$. (A maximum/minimum problem for a function of several variables with one constraint on the variables is equivalent to a maximum/minimum problem for a function with one less variable.) We obtain the function H of one variable defined by

$$H(y) = \left(\frac{6 - 2y}{3}\right)^2 + y^2.$$

Differentiating H yields

$$H'(y) = 2\left(\frac{6 - 2y}{3}\right)\left(-\frac{2}{3}\right) + 2y = -\frac{4}{9}(6 - 2y) + 2y.$$

Setting $H'(y)$ equal to 0 and solving for y, we obtain

$$-24 + 8y + 18y = 0,$$

$$y = \frac{12}{13}.$$

Substituting this value for y in $x = (6 - 2y)/3$, we find that

$$x = \frac{18}{13}.$$

and the values for x and y agree with those obtained using the method of Lagrange multipliers. *Note:* Since

$$H''(y) = \frac{8}{9} + 2 > 0,$$

we can conclude that $H(y)$ is a minimum at $y = \frac{12}{13}$.

The method of Lagrange multipliers is also applicable to finding the maximum or minimum value of a function $f(x, y, z)$ of three variables with a constraint $g(x, y, z) = 0$. For example, to find the "candidates" in the domain for a relative extremum of a function f of three variables subject to the constraint $g(x, y, z) = 0$, we solve the following **Lagrange system** of four equations in x, y, and z.

Lagrange system

$$f_x = \lambda g_x, \tag{1}$$
$$f_y = \lambda g_y, \tag{2}$$
$$f_z = \lambda g_z, \tag{3}$$
$$g(x, y, z) = 0. \tag{4}$$

In the following maximum/minimum problem, we make one more comparison between not using and using the method of Lagrange multipliers to find the extreme value of a function.

Maximum value with constraints

EXAMPLE 1 Find the maximum value of $f(x, y, z) = xyz$ if x, y, and z are positive and are subject to the additional constraint $2xy + xz + 3yz = 72$.

Solutions

(1) Solve $2xy + xz + 3yz = 72$, for, say, z.

$$(x + 3y)z = 72 - 2xy$$

$$z = \frac{72 - 2xy}{x + 3y}$$

7.7 MAXIMA AND MINIMA WITH CONSTRAINTS

Substituting in $f(x, y, z) = xyz$, we obtain a function F of two variables defined by

$$F(x, y) = xy\left(\frac{72 - 2xy}{x + 3y}\right)$$

$$= \frac{72xy - 2x^2y^2}{x + 3y}$$

The first partial of F with respect to x is

$$F_x(x, y) = \frac{(x + 3y)(72y - 4xy^2) - (72xy - 2x^2y^2)}{(x + 3y)^2}$$

$$= \frac{72xy - 4x^2y^2 + 216y^2 - 12xy^3 - 72xy + 2x^2y^2}{(x + 3y)^2}$$

$$= \frac{-2x^2y^2 - 12xy^3 + 216y^2}{(x + 3y)^2}$$

$$= \frac{-2y^2(x^2 + 6xy - 108)}{(x + 3y)^2}.$$

The first partial of F with respect to y is

$$F_y(x, y) = \frac{(x + 3y)(72x - 4x^2y) - (72xy - 2x^2y^2)(3)}{(x + 3y)^2}$$

$$= \frac{72x^2 - 4x^3y + 216xy - 12x^2y^2 - 216xy + 6x^2y^2}{(x + 3y)^2}$$

$$= \frac{-2x^2(3y^2 + 2xy - 36)}{(x + 3y)^2}.$$

Since x, y, and z are positive, in order for $F_x(x, y)$ and $F_y(x, y)$ to be 0 we must have

$$x^2 + 6xy - 108 = 0 \tag{1}$$

and

$$3y^2 + 2xy - 36 = 0. \tag{2}$$

Multiplying each side of Eq. (2) by 3 and subtracting the result from Eq. (1), we get

$$x^2 - 9y^2 - 108 + 108 = 0,$$
$$x^2 = 9y^2,$$
$$x = \pm 3y.$$

Since x and y are positive, $x \neq -3y$. Substituting $x = 3y$ in Eq. (2) yields
$$3y^2 + 6y^2 - 36 = 0,$$
$$9y^2 = 36,$$
$$y = 2.$$

Since $x = 3y$ and $z = (72 - 2xy)/(x + 3y)$, we conclude that $x = 6$ and $z = 4$.

By using Theorem 7.1 or by carefully considering the given conditions, we conclude that the maximum of f with the given constraint is obtained when $x = 6$, $y = 2$, and $z = 4$. The maximum value is $f(6, 2, 4) = 48$.

(2) *(Lagrange Multiplier)* We let
$$g(x, y, z) = 2xy + xz + 3yz - 72.$$

Then the Lagrange system is

$$f_x = \lambda g_x \Rightarrow yz = \lambda(2y + z), \tag{3}$$
$$f_y = \lambda g_y \Rightarrow xz = \lambda(2x + 3z), \tag{4}$$
$$f_z = \lambda g_z \Rightarrow xy = \lambda(x + 3y), \tag{5}$$
$$g(x, y, z) = 0 \Rightarrow 2xy + xz + 3yz - 72 = 0. \tag{6}$$

Multiplying Eq. (3) by x and Eq. (4) by y and then subtracting Eq. (4) from Eq. (3) yields $xz = 3yz$. Since $z \neq 0$, we obtain
$$x = 3y. \tag{7}$$

Multiplying Eq. (4) by y and Eq. (5) by z and then subtracting Eq. (5) from Eq. (4), we obtain
$$z = 2y. \tag{8}$$

Substituting Eqs. (7) and (8) in Eq. (6) we get
$$6y^2 + 6y^2 + 6y^2 = 72,$$
$$y^2 = 4,$$
$$y = 2.$$

Therefore, $x = 3y = 6$ and $z = 2y = 4$, the same solution previously obtained.

7.7 MAXIMA AND MINIMA WITH CONSTRAINTS

Maximum box dimensions

EXAMPLE 2 If the largest rectangular package that may be sent by parcel post must meet the requirement that the sum of the length and girth (perimeter of a cross section) is 84 inches, find the dimensions of the rectangular package of greatest volume that may be sent parcel post. (See Exercise 14, Section 7.6.)

Solution: If x is the length, y is the width, and z is the height of the box, we want to find the dimensions which will maximize

$$V(x, y, z) = xyz$$

with the constraint $x + 2y + 2z = 84$. Let

$$g(x, y, z) = x + 2y + 2z - 84$$

and solve the system

$$V_x = \lambda g_x \Rightarrow yz = \lambda, \tag{1}$$
$$V_y = \lambda g_y \Rightarrow xz = 2\lambda, \tag{2}$$
$$V_z = \lambda g_z \Rightarrow xy = 2\lambda, \tag{3}$$
$$g(x, y, z) = 0 \Rightarrow x + 2y + 2z = 84. \tag{4}$$

From Eqs. (2) and (3) we find that $y = z$, and from Eqs. (1) and (2) we find that $x = 2y$. Substituting $z = y$ and $x = 2y$ in Eq. (4), we get

$$2y + 2y + 2y = 84,$$
$$6y = 84,$$
$$y = 14.$$

Therefore, $x = 2y = 28$ and $z = y = 14$. The physical conditions of the problem indicate that a maximum exists. Since there is only one solution to the Lagrange system, we conclude that the maximum volume is

$$28 \times 14 \times 14 = 5488 \text{ in}^3.$$

Minimizing cost of box

EXAMPLE 3 Suppose a rectangular box containing 150 cm³ is to be manufactured from three different materials. The lid costs 4 cents/cm², the sides cost 5 cents/cm², and the bottom costs 8 cents/cm². Find the most economical dimensions for the box. (See Exercise 16, Section 7.6.)

Solution: Let x be the length, y be the width, and z be the height of the box. The problem is to minimize the cost

$$C(x, y, z) = 12xy + 10xz + 10yz$$

with the constraint $xyz = 150$. Letting

$$g(x, y, z) = xyz - 150,$$

we solve the following Lagrange system of equations.

$$C_x = \lambda g_x \Rightarrow 12y + 10z = \lambda yz \quad (1)$$
$$C_y = \lambda g_y \Rightarrow 12x + 10z = \lambda xz \quad (2)$$
$$C_z = \lambda g_z \Rightarrow 10x + 10y = \lambda xy \quad (3)$$
$$g(x, y, z) = 0 \Rightarrow xyz - 150 = 0 \quad (4)$$

Multiplying Eq. (1) by x and Eq. 2 by (y) yields

$$12xy + 10xz = \lambda xyz,$$
$$12xy + 10yz = \lambda xyz.$$

Subtracting the last equation from the preceding one yields

$$x = y. \quad (5)$$

Multiplying Eq. (2) by y and Eq. (3) by z yields

$$12xy + 10yz = \lambda xyz,$$
$$10xz + 10yz = \lambda xyz.$$

Subtracting the last equation from the preceding one yields

$$y = \frac{5}{6}z \quad (6)$$

Therefore, $x = (\frac{5}{6})z$ and substituting in Eq. (4), we get

$$\left(\frac{5}{6}z\right)\left(\frac{5}{6}z\right)z = 150,$$
$$z^3 = 216,$$
$$z = 6.$$

7.7 MAXIMA AND MINIMA WITH CONSTRAINTS

Since $x = 5$, $y = 5$, $z = 6$ is the only solution to the Lagrange system and since a minimum obviously exists, the minimum cost for such a box is

$$C(5, 5, 6) = 300 + 300 + 300 = 900 \text{ cents} = \$9.00.$$

EXERCISES

Set A—Maxima and Minima Using the Lagrange Multiplier

Maximum value

1. (a) Let $f(x, y) = 4xy$, $x > 0$, $y > 0$, and $x^2 + y^2 = 32$; find the maximum value of f using the Lagrange multiplier.
   (b) Find the maximum value of f by solving the second equation for y, substituting in the first equation, and using the techniques for determining the maximum of a function of one variable.

Minimum value

2. (a) For $f(x, y) = x^2 + y^2$ where $x + 2y = 1$, find the minimum value of f using the Lagrange multiplier.
   (b) Find the minimum value of f with the given constraint by solving the second equation for x, substituting in the first, and using methods for determining the minimum of a function of one variable.

Minimum value

3. (a) For $f(x, y) = 3x^2 + y^2$ where $x + 4y = 6$, find the minimum value of f using the Lagrange multiplier.
   (b) Find the minimum value of f with the given constraint by solving the second equation for x, substituting in the first, and using methods for determining the minimum of a function of one variable.

Maximum value

In Exercises 4 and 5, use the Lagrange multiplier to find the maximum value of:

4. $f(x, y) = 15x - 10y - 2x^2 - y^2$ where $3x + 2y = 150$.
5. $f(x, y, z) = xyz$ where x, y, z are positive and $x + y + z = 108$.

Minimum value

In Exercises 6 and 7, use the Lagrange multiplier to find the minimum value of:

6. $f(x, y, z) = x^2 y^2 z^2$ where $x + y + z = 1$.
7. $f(x, y, z) = x^2 + y^2 + z^2$ where $x + y + 2z = 4$.

Set B—Applications Using the Lagrange Multiplier

Use the Lagrange multiplier to solve each of the following problems.

Minimum cost for a box

8. A box is to have a volume of 1536 in³. The material for the top and bottom costs 2 cents/in², the material for the front

Minimum cost of a box

and back costs 3 cents/in^2, and the material for the sides costs 4 cents/in^2. What dimensions should the box have if the manufacturer wants to minimize its cost?

9. Use the Lagrange multiplier to solve the problem in Example 5, Section 7.6.

Maximizing profits

10. The automobile dealer whose monthly profit function was

$$P(x, y) = 300x - 0.3x^2 + 0.2xy + 500y - 0.2y^2$$

decides that the combined profit on the sale of one each of the two models should be $3047. What *price* should she put on each car in order to maximize profits? (See Example 5, Section 7.3.)

7.8 METHOD OF LEAST SQUARES (OPTIONAL)

Let us consider another important idea with many applications: it leads to a problem which can be solved using multivariable calculus. The goal is to "fit" a polynomial in some prescribed way to a given set of points in the coordinate plane. As we shall see, such an approximating polynomial may contain all the points, some of the points, or none of the points. In practical applications, the nature of the problem often indicates whether a linear function (line), a quadratic function (parabola), some higher-degree polynomial function, an exponential function, a trigonometric function, or a combination of such elementary functions should be used to fit a curve to a set of points. One general use of such a function is to interpolate or extrapolate additional data from data obtained in a practical situation. Let us consider the simplest case first.

Suppose we are given the set of points (2, 1), (3, 3), (4, 5), and (5, 6). (See Figure 7.19.) The graph of this set of points (or the

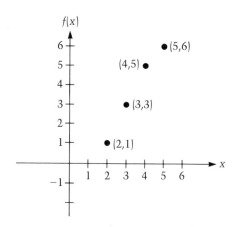

FIG. 7.19.

7.8 METHOD OF LEAST SQUARES (OPTIONAL)

nature of the problem from which the points were obtained) might prompt us to seek a linear function as an approximating polynomial.

First we must decide what will be considered to be the "best" linear approximation for the set of points. One of the most popular definitions of "best fit" involves the squares of the vertical distances between the given points and the corresponding points on the line we seek. For a linear function $f(x) = a + bx$, the vertical distance between each point (x_i, y_i), $i = 1, 2, 3, \ldots, n$, in the set of n points, and the corresponding point $(x_i, f(x_i))$ on the line is

$$f(x_i) - y_i.$$

The difference $f(x_i) - y_i$ will be positive or negative according to whether the point is below or above the line. If two different points (x_i, y_i) and (x_k, y_k) were the same vertical distance (regardless how great) from the line but on opposite sides of the line, then the sum

$$[f(x_i) - y_i] + [f(x_k) - y_k]$$

would be 0. In this situation the sum of the numbers would not indicate how close the line was to the points. But the sum of the squares

$$[f(x_i) - y_i]^2 + [f(x_k) - y_k]^2,$$

which are both nonnegative, would be a "measure" of how close the line was to the two points. The sum would be greater when the points are farther from the line, and it would be 0 if and only if the line contained both points.

The linear function $f(x) = a + bx$ that "best fits" a set of n points $(x_1, y_1), (x_2, y_2), \ldots, (x_n, y_n)$ is defined to be the one that makes the sum

$$\sum_{k=1}^{n} [f(x_k) - y_k]^2 = [f(x_1) - y_1]^2 + [f(x_2) - y_2]^2 + \cdots$$
$$\cdots + [f(x_n) - y_n]^2$$

a minimum. (*Note:* The slope of the line is b and the y-intercept is a.) For any given finite set of points, such a linear function exists and is unique; it is called the **linear least squares approximation** for the set of points.

linear least squares approximation

Let $(x_1, y_1), (x_2, y_2), (x_3, y_3),$ and (x_4, y_4) be the four points $(2, 1), (3, 3), (4, 5),$ and $(5, 6)$, respectively. Let $f(x) = a + bx$ and let

$$F(a, b) = \sum_{i=1}^{4} [f(x_k) - y_k]^2.$$

Then

$$F(a, b) = [f(2) - 1]^2 + [f(3) - 3]^2 + [f(4) - 5]^2 + [f(5) - 6]^2$$

and

$$F(a, b) = [a + 2b - 1]^2 + [a + 3b - 3]^2 + [a + 4b - 5]^2 + [a + 5b - 6]^2.$$

The function F is a function of the two variables a and b which we wish to minimize. Recall that a necessary condition for F to have a minimum (or maximum) is that the first partial derivatives F_a and F_b be 0. Since it is usually easier to find F_a and F_b directly from the preceding equation without simplifying, we determine the partial derivatives as follows.

$$F_a(a, b) = 2(a + 2b - 1) + 2(a + 3b - 3) + 2(a + 4b - 5) + 2(a + 5b - 6).$$

Consequently,

$$F_a(a, b) = 8a + 28b - 30.$$

Similarly,

$$F_b(a, b) = 4(a + 2b - 1) + 6(a + 3b - 3) + 8(a + 4b - 5) + 10(a + 5b - 6).$$

Consequently,

$$F_b(a, b) = 28a + 108b - 122.$$

To find the values for a and b that will minimize F, we set each partial derivative equal to 0 and solve for a and b.

$$F_a(a, b) = 0 \quad \text{implies} \quad 4a + 14b = 15,$$

and

$$F_b(a, b) = 0 \quad \text{implies} \quad 14a + 54b = 61.$$

Solving the preceding system, we find that

$$a = -\frac{11}{5} \quad \text{and} \quad b = \frac{17}{10}.$$

7.8 METHOD OF LEAST SQUARES (OPTIONAL)

Since the partial derivatives are 0 only for $a = -\frac{11}{5}$ and $b = \frac{17}{10}$, we know that if F has a minimum or maximum, then it is at $(-\frac{11}{5}, \frac{17}{10})$. Obviously F has no maximum. Either from the geometric aspects of the problem or Theorem 7.1, we conclude that F has a minimum value at the point. Therefore, the linear least squares approximation is

$$f(x) = \frac{17}{10}x - \frac{11}{5}.$$

Furthermore, $F(-\frac{11}{5}, \frac{17}{10}) = \frac{3}{10} = 0.3$; thus, the sum of the squares of the vertical distances from the given points to the line is 0.3. (See Figure 7.20.)

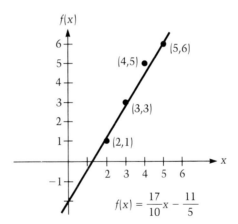

FIG. 7.20. $f(x) = \frac{17}{10}x - \frac{11}{5}$

Let us now attack the problem of obtaining the linear least squares approximation $f(x) = a + bx$ for any set of n points $(x_1, y_1), (x_2, y_2), \ldots, (x_n, y_n)$. The approach is the same as finding the linear least squares approximation for the four points. Let

$$F(a, b) = \sum_{k=1}^{n} [f(x_k) - y_k]^2.$$

Consequently,

$$F(a, b) = \sum_{k=1}^{n} [a + bx_k - y_k]^2.$$

Now we obtain the partial derivatives of F to establish the values for a and b that minimize F:

$$\frac{\partial F}{\partial a} = 2 \sum_{k=1}^{n} (a + bx_k - y_k),$$

$$\frac{\partial F}{\partial b} = 2 \sum_{k=1}^{n} (a + bx_k - y_k)x_k.$$

Setting each partial derivative equal to 0, we obtain

$$na + \left(\sum_{k=1}^{n} x_k\right)b = \sum_{k=1}^{n} y_k,$$

$$\left(\sum_{k=1}^{n} x_k\right)a + \left(\sum_{k=1}^{n} x_k^2\right)b = \sum_{k=1}^{n} x_k y_k.$$

normal equations

The preceding system of equations is called the set of **normal equations** for the linear least squares approximation. Solving the system for a and b and using Σx_k to represent $\Sigma_{k=1}^{n} x_k$, for example, we obtain

$$b = \frac{n(\Sigma x_k y_k) - (\Sigma x_k)(\Sigma y_k)}{n(\Sigma x_k^2) - (\Sigma x_k)^2} \quad \text{and} \quad a = \frac{\Sigma y_k - b(\Sigma x_k)}{n}.$$

The preceding solution for b and a is used in the following example.

Linear least squares

EXAMPLE 1 On a standardized test with a top score of 720 points it was found that the cumulative averages y_k for 13 students scoring x_k were as given in the accompanying table. Find a and b for the linear least squares approximation $f(x) = a + bx$ for these data and use it to predict the cumulative averages for students scoring 620 and 720 on the test.

7.8 METHOD OF LEAST SQUARES (OPTIONAL)

Test Score	Cumulative Average	x_k^2	$x_k y_k$
$x_1 = 510$	$y_1 = 2.1$	260,100	1071.0
$x_2 = 510$	$y_2 = 2.2$	260,100	1122.0
$x_3 = 516$	$y_3 = 2.4$	266,256	1238.4
$x_4 = 524$	$y_4 = 2.2$	274,576	1152.8
$x_5 = 528$	$y_5 = 2.3$	278,784	1214.4
$x_6 = 534$	$y_6 = 2.3$	285,156	1228.2
$x_7 = 536$	$y_7 = 2.2$	287,296	1179.2
$x_8 = 542$	$y_8 = 2.4$	293,764	1300.8
$x_9 = 546$	$y_9 = 2.4$	298,116	1310.4
$x_{10} = 546$	$y_{10} = 2.5$	298,116	1365.0
$x_{11} = 560$	$y_{11} = 2.6$	313,600	1456.0
$x_{12} = 572$	$y_{12} = 2.7$	327,184	1544.4
$x_{13} = 576$	$y_{13} = 2.8$	331,776	1612.8
7000	31.1	3,774,824	16795.4

$$b = \frac{(13)(16{,}795.4) - (7000)(31.1)}{(13)(3{,}774{,}824) - (7000)^2} = \frac{640.2}{72{,}712} \doteq 0.0088,$$

$$a = \frac{31.1 - (0.0088)(7000)}{13} = -2.3486.$$

Since $f(x) = a + bx = -2.3486 + 0.0088x$,

$$f(620) = 3.1 \quad \text{and} \quad f(720) = 3.99$$

would be the predicted cumulative averages for students scoring 620 and 720, respectively.

---

quadratic least squares approximation

The **quadratic least squares approximation** for the set of points $(x_1, y_1), (x_2, y_2), \ldots, (x_n, y_n)$ is the function

$$f(x) = a + bx + cx^2 \quad \text{such that} \quad \sum_{k=1}^{n} [f(x_k) - y_k]^2$$

is a minimum. Consider the following example.

Quadratic least squares

EXAMPLE 2 Find the quadratic least squares approximation for the set of points (2, 1), (3, 3), (4, 5), and (5, 6).

Solution: Let $f(x) = a + bx + cx^2$ and let

$$F(a, b, c) = \sum_{k=1}^{4} [a + bx_k + cx_k^2 - y_k]^2.$$

Substituting the four separate points in the sum and differentiating, we get

$$F_a = 2(4a + 14b + 54c - 15),$$
$$F_b = 2(14a + 54b + 224c - 61),$$
$$F_c = 2(54a + 224b + 978c - 261).$$

Setting the partial derivatives equal to 0 and solving the resulting system of equations, we obtain $a = -\frac{99}{20}$, $b = \frac{69}{20}$, and $c = -\frac{1}{4}$. Therefore,

$$f(x) = -\frac{x^2}{4} + \frac{69}{20}x - \frac{99}{20}$$

is the quadratic least squares approximation. Note that

$$F\left(-\frac{99}{20}, \frac{69}{20}, -\frac{1}{4}\right) = 0.25.$$

Since the sum of the squares of the vertical distances from the given points to the quadratic polynomial is 0.25, which is less than 0.3 obtained in the linear least squares approximation for this set, the quadratic least squares *in this case* is a better fit than the linear least squares approximation.

EXERCISES

Set A–Least Squares Approximations

Graphing

1. Graph the points $(2, 1)$, $(3, 3)$, $(4, 5)$, and $(5, 6)$, the linear least squares approximation $f(x) = \frac{17x}{10} - \frac{11}{5}$, and the quadratic least squares approximation $h(x) = -\frac{x^2}{4} + \frac{69x}{20} - \frac{99}{20}$ in the same coordinate system. (See figure 7.20)

Linear least squares

2. (a) Determine the linear least squares approximation for the points $(0, 0)$, $(1, 1)$, $(2, 2)$, $(3, 4)$, and $(4, 8)$.
   (b) Graph the points and the linear least squares approximation.
   (c) Find the sum of the squares of the vertical distances between the points and the linear approximation.

Linear least squares

3. (a) Determine the linear least squares approximation for the points $(-1, -1)$, $(0, 0)$, $(1, 1)$, and $(2, 6)$.
   (b) Graph the points and the linear least squares approximation.
   (c) Find the sum of the squares of the vertical distances between the points and the linear approximation.

Quadratic least squares

4. (a) Determine the quadratic least squares approximation for the points $(0, 0)$, $(1, 1)$, $(2, 2)$, $(3, 4)$, and $(4, 8)$.
   (b) Graph the points and the quadratic least squares approximation.
   (c) Find the sum of the squares of the vertical distances between the points and the quadratic approximation. (See Exercise 2.)

Quadratic least squares

5. (a) Determine the quadratic least squares approximation for the points $(-1, -1)$, $(0, 0)$, $(1, 1)$, and $(2, 6)$.
   (b) Graph the points and the quadratic least squares approximation.
   (c) Find the sum of the squares of the vertical distances between the points and the quadratic approximation. (See Exercise 3.)

7.9 SUMMARY AND REVIEW EXERCISES

1. **DISTANCE FORMULA**

   The **distance** between two points $P(x_1, y_1, z_1)$ and $Q(x_2, y_2, z_2)$ in three dimensions is given by

   $$|PQ| = \sqrt{(x_2 - x_1)^2 + (y_2 - y_1)^2 + (z_2 - z_1)^2}.$$

2. **MIDPOINT FORMULA**

   The **midpoint** between two points $P(x_1, y_1, z_1)$ and $Q(x_2, y_2, z_2)$ in three dimensions is given by

   $$\left(\frac{x_1 + x_2}{2}, \frac{y_1 + y_2}{2}, \frac{z_1 + z_2}{2} \right).$$

3. **GRAPH OF A FUNCTION OF TWO VARIABLES**

   For a real-valued function f of two variables, if $z = f(x, y)$, then the **graph** of f is the graph of all ordered triples (a, b, c) in three dimensions for which $c = f(a, b)$.

| 4. CONTINUITY OF A FUNCTION OF TWO VARIABLES | A function f of two variables is **continuous** at a point (a, b) if (a, b) in the domain of f and if

$$\lim_{(x,y)\to(a,b)} f(x, y) = f(a, b).$$ |

| 5. PARTIAL DERIVATIVES | The **first partial derivatives** of a function f of two variables are defined by

$$\frac{\partial f}{\partial x} = f_x(x, y) = \lim_{h\to 0} \frac{f(x + h, y) - f(x, y)}{h},$$

$$\frac{\partial f}{\partial y} = f_y(x, y) = \lim_{k\to 0} \frac{f(x, y + k) - f(x, y)}{k}.$$

The **second partial derivatives** of f are defined as follows:

$\dfrac{\partial}{\partial x}\left(\dfrac{\partial f}{\partial x}\right) = \dfrac{\partial^2 f}{\partial x^2} = f_{xx}$ is the partial of $\dfrac{\partial f}{\partial x}$ with respect to x,

$\dfrac{\partial}{\partial y}\left(\dfrac{\partial f}{\partial x}\right) = \dfrac{\partial^2 f}{\partial y\, \partial x} = f_{xy}$ is the partial of $\dfrac{\partial f}{\partial x}$ with respect to y,

$\dfrac{\partial}{\partial x}\left(\dfrac{\partial f}{\partial y}\right) = \dfrac{\partial^2 f}{\partial x\, \partial y} = f_{yx}$ is the partial of $\dfrac{\partial f}{\partial y}$ with respect to x,

$\dfrac{\partial}{\partial y}\left(\dfrac{\partial f}{\partial y}\right) = \dfrac{\partial^2 f}{\partial y^2} = f_{yy}$ is the partial of $\dfrac{\partial f}{\partial y}$ with respect to y. |

| 6. THE TOTAL DIFFERENTIAL | Let $z = f(x, y)$ be a function of two variables. For small changes dx and dy in x and y, respectively, the corresponding change Δz in z is approximated by the **total differential** dz defined by

$$dz = f_x(x, y)\, dx + f_y(x, y)\, dy = \frac{\partial f}{\partial x}\, dx + \frac{\partial f}{\partial y}\, dy.$$ |

| 7. TANGENT PLANE | Let f be a function whose first partial derivatives exist at (a, b). The equation of the **tangent plane** to f at (a, b) is given by

$$z = f(a, b) + f_x(a, b)(x - a) + f_y(a, b)(y - b).$$ |

8. MAXIMA AND MINIMA

(a) For a point (a, b) in the domain of a function f of two variables, all points interior to a circle with (a, b) as its center is called a **neighborhood** of (a, b).

(b) If there exists a neighborhood of (a, b) contained in the domain of a function f of two variables such that

$$f(x, y) \leq f(a, b)$$

for each (x, y) in the neighborhood, then $f(a, b)$ is a **relative maximum** value of f.

(c) If there exists a neighborhood of (a, b) contained in the domain of a function f of two variables such that

$$f(x, y) \geq f(a, b)$$

for each (x, y) in the neighborhood, then $f(a, b)$ is a **relative minimum** value of f.

(d) If there is a point (a, b) in the domain of f such that $f(a, b) \geq f(x, y)$ for every (x, y) in the domain, then $f(a, b)$ is the **absolute maximum** of f. Similarly, if $f(a, b) \leq f(x, y)$ for every (x, y) in the domain of f, then $f(a, b)$ is the **absolute minimum** of f.

(e) Let f be a function of two variables defined in some neighborhood of (a, b) and having continuous second partial derivatives in the neighborhood, and let

$$A = f_{xx}(a, b), \quad B = f_{xy}(a, b), \quad \text{and} \quad C = f_{yy}(a, b).$$

Assume that $f_x(a, b) = f_y(a, b) = 0$ and that $AC - B^2 > 0$.

(1) If $A < 0$, then f has a **relative (local) maximum** at (a, b).
(2) If $A > 0$, then f has a **relative (local) minimum** at (a, b).

(f) Let $z = f(x, y)$ be a function of two variables with $z = g(x, y)$ being a *constraint* on the variables. The **Lagrange system** is defined by

$$f_x(x, y) = \lambda g_x(x, y),$$
$$f_y(x, y) = \lambda g_y(x, y),$$
$$g(x, y) = 0.$$

Each solution is a "candidate" for minimizing or maximizing f with the given constraint.

8. REVIEW EXERCISES

In Exercises 1 and 2, (a) find the midpoint of the segment PQ and (b) find the length of the segment.

1. $P(-2, 4, 1)$ and $Q(0, 6, 2)$.
2. $P(4, -1, 6)$ and $Q(6, 3, 8)$.

Sketch the graph of the function in each of Exercises 3 and 4.

3. $f(x, y) = 15 - 3x - 5y$
4. $f(x, y) = 32 - 2x^2 - 2y^2$

In Exercises 5 and 6, find all first and second partial derivatives of the given functions.

5. $f(x, y) = x^2 y^3 + \ln xy + y$
6. $f(x, y) = xy^4 + e^{xy} + 3x$

In Exercises 7 and 8, find an equation of the plane tangent to the given surface at the given point on the surface.

7. $f(x, y) = 2xy^2 - xy + x^2$ at $(1, 2, 7)$
8. $f(x, y) = 3x^2 y + 5xy - 2y^3$ at $(2, 1, 20)$

Use the total differential in Exercises 9 and 10 to approximate the given function value.

9. For $f(x, y) = x^{1/2} y^{2/3} + 3xy$, approximate $f(4.01, 7.98)$.
10. For $f(x, y) = 3x^{2/7} y - 5xy^{1/2}$, approximate $f(1.01, 3.98)$.

In Exercises 11 through 14, find any relative maxima and relative minima and identify.

11. $f(x, y) = x^2 + 4xy + 2y^2 + 1$
12. $f(x, y) = 4x^2 - 8xy + 2y^2 + 4$
13. $f(x, y) = x + 4y + \dfrac{25}{x} + \dfrac{16}{y}$
14. $f(x, y) = x^3 + 3x^2 - y^2$

15. The largest rectangular package that may be sent by parcel post must satisfy the requirement that the sum of the length and girth (perimeter of a cross section) is 100 inches. Find the dimensions of the rectangular package of greatest volume that may be sent by parcel post.

16. A manufacturer finds that $x = \left(10 - \dfrac{p}{5}\right)1000$ units of Product A can be sold per month at price p dollars apiece where $10 \leq p \leq 45$ and that $y = \left(5 - \dfrac{q}{10}\right)1000$ units of Product B can be sold per month at q dollars apiece where $10 \leq q \leq 40$. The cost per month to manufacture these units is given

by $C(x, y) = 10x + 20y + 1000$. Find the selling price for each product that will maximize the total profits. [Hint: $R(x, y) = xp(x) + yq(y)$.]

17. A rectangular box of one cubic foot with an open top is to be made from material for which the bottom costs twice as much per square foot as the sides. Find the dimensions of the least expensive box.

18. Prove that the rectangular box of fixed volume with minimum surface area must be a cube.

19. Use the Lagrange multiplier to find the minimum distance from the point (4, 2, 6) to the plane $2x + 2y + z = 6$. [Hint: Minimize $D(x, y, z) = [d(x, y, z)]^2$ where

$$d(x, y, z) = \sqrt{(x - 4)^2 + (y - 2)^2 + (z - 6)^2}$$

with the constraint $g(x, y, z) = 2x + 2y + z - 6 = 0$.]

20. For any output production function $P(x, y)$ where x items of Product A are produced at a dollars apiece and y items of Product B are produced at b dollars apiece, show that if $ax + by = K$ dollars is the budgetary constraint, then $P(x, y)$ is maximized when

$$\frac{P_x}{P_y} = \frac{a}{b}.$$

(This is called the *Law of Equimarginal Productivity*.) [Hint: Use the Lagrange multiplier technique.]

Joseph Louis Lagrange

The Bettmann Archive

Lagrange was born at Turin in 1736. In 1755, at the age of 19, he became a professor of mathematics at Turin. He organized a research society at Turin which was to become the Turin Academy of Sciences. The fine mathematical quality of the early publications of the academy was probably due to Lagrange's influence and uncredited contributions. Such generosity with his excessive talent was typical of this shy and modest man. His first publication resulted from a communication he sent to the great Euler who immediately recognized the merit of his results and urged Lagrange to publish his findings.

In 1766 when Euler left Berlin for St. Petersburg, he recommended that Lagrange be invited to the Berlin Academy. Lagrange accepted the invitation and remained there until the death of Frederick the Great in 1786. He went to Paris in 1787 with an invitation from Louis XIV who made him a "veteran pensioner" of the French Academy of Sciences. Lagrange was a member of the committee that perfected the metric system of weights and measures, a reform proposed by Talleyrand in 1790. Except for his work on this committee, Lagrange, along with his contemporaries Legendre and Laplace, took little part in the political life of France after the Revolution. However, Lagrange did make the "intemperate" statement when the great chemist Lavoisier was beheaded that "it took the mob only a moment to remove his head; a century will not suffice to reproduce it."

The investigations and accomplishments of Lagrange include important original discoveries in the calculus of variations, a proof of the famous "four square" problem (every positive integer can be expressed as the sum of four or fewer squares), and a study of rational functions of the roots of polynomial equations. He was also one of the early mathematicians to apply calculus to the theory of probability. He developed a technique, known today as the method of Lagrange multipliers, for maximizing functions of several variables subject to certain constraints.

CHAPTER

8

SEQUENCES AND MATHEMATICS OF FINANCE

8

8.1 INTRODUCTION
8.2 ARITHMETIC SEQUENCES
8.3 GEOMETRIC SEQUENCES
8.4 INTEREST: SIMPLE AND COMPOUND
8.5 CONTINUOUS INTEREST
8.6 INTEREST ON PERIODIC SAVINGS
8.7 INSTALLMENT PAYMENTS AND RETIREMENT INCOME
8.8 SUMMARY AND REVIEW EXERCISES

8.1 INTRODUCTION

In this chapter we consider two special types of sequences of numbers and investigate their uses in practical problems. Some of the problems we shall learn how to solve are:

1. Suppose $4000 is deposited for ten years at 6% interest compounded quarterly. What would be the total amount on deposit after 10 years?

2. Suppose a person decides to pay off a loan of $6000 over a 3-year period by making equal monthly payments. If the annual interest rate on the loan is 12% compounded monthly, what would be the monthly payments?

3. Suppose $150 is deposited in a savings account at the beginning of each month, and the annual interest rate is 8% compounded monthly. What would be the total amount on deposit at the end of 10 years? At the end of 10 years, for how many months could $300 a month be withdrawn from the account provided the amount left on deposit continued to draw the same rate of interest?

4. Suppose a person borrows $6000 to buy a car and plans to pay off the loan in 48 equal monthly installments. The lender states that there is an *add-on interest rate* of 8.5% per year. What would be the monthly payments? What is the *annual percentage rate* for this loan? (The answer to the second question is 15.17%.)

5. How does one compute the monthly mortgage payments in the accompanying table which appeared in the business sec-

How Much a $50,000 30-Year Mortgage Costs You

Interest rate, %	Monthly payment, $	Total payout, $
8	366.89	132,080.40
9	402.32	144,835.20
10	438.79	157,964.40
11	476.16	171,417.60
12	514.31	185,151.60
13	553.10	199,116.00
14	592.44	213,278.40
15	632.22	227,559.20
16	672.38	242,056.80
17	712.84	256,622.40
18	753.54	271,274.40

tion of a newspaper? How would the monthly payments change if the 30-year mortgage were replaced by (a) a 20-year mortgage, (b) a 25-year mortgage?

6. Suppose a 30-year-old individual can invest money at an interest rate of 12% compounded monthly for an indefinite period of time. How much must he or she invest each month to be a millionaire by the age of 60? (Make a guess.) ■

8.2 ARITHMETIC SEQUENCES

Consider the sequence of odd positive integers: 1, 3, 5, 7, 9, 11, 13, ... What is the 50th term of this sequence? To answer this question, note that the second term can be obtained by adding two to the first term; the third by adding 2 twos to the first term; the fourth by adding 3 twos to the first term; and the 50th term can be obtained by adding 49 twos to the first term. Therefore, the 50th odd positive integer is

$$1 + (49)(2) = 99.$$

Furthermore, the nth term of this sequence is

$$1 + (n - 1)(2) = 2n - 1.$$

The sum S of the first 50 odd positive integers can be found in several ways. One is to list them all from 1 to 99 in a column and then add. Aside from the risk of making a mistake in addition, just writing down the integers would be a long and boring task.

A second and more efficient way to find the sum of the first 50 odd positive integers uses the fact that the order in which numbers are added does not affect the sum. If

$$S = 1 + 3 + 5 + \cdots + 95 + 97 + 99, \qquad (1)$$

then

$$S = 99 + 97 + 95 + \cdots + 5 + 3 + 1. \qquad (2)$$

Adding Eqs. (1) and (2) yields

$$2S = 100 + 100 + 100 + \cdots + 100 + 100 + 100.$$

8.2 ARITHMETIC SEQUENCES

Since the preceding sum contains fifty terms each of which is 100, it follows that

$$2S = 50(100)$$

and

$$S = 25(100) = 2500.$$

Suppose we want to find the sums of given sequences. For what kinds of sequences could we use the above technique? A careful investigation shows that if for a given sequence the difference between each pair of consecutive terms is the same, then the above technique will always produce a sum with identical terms and the sum can be found as in the example. Let us now consider such sequences.

A sequence of numbers $a_1, a_2, a_3, \ldots$ in which the difference between each pair of consecutive terms is the same is called an **arithmetic sequence** (or *arithmetic progression*). In an arithmetic sequence with a_1 as its **first term,** the second term a_2 is $a_1 + d$ where d is the **common difference** between the terms. The third term a_3 is $a_1 + 2d$, the fourth term a_4 is $a_1 + 3d$, the fifth term a_5 is $a_1 + 4d$, and the nth term a_n of an arithmetic sequence is given by

arithmetic sequence

common difference

nth term of arithmetic sequence

$$a_n = a_1 + (n - 1)d.$$

For example, 4, 7, 10, 13, 16, ... is an arithmetic sequence with first term $a_1 = 4$ and common difference $d = 3$; the 14th term, for example, is

$$a_{14} = 4 + (13)(3) = 4 + 39 = 43.$$

Arithmetic sequence

EXAMPLE 1 If the first term of an arithmetic sequence is 11 and its common difference is -3, find its 31st term.

Solution: $a_{31} = a_1 + (30)d = 11 + (30)(-3) = -79.$

Arithmetic sequence

EXAMPLE 2 If the first term of an arithmetic sequence is 4 and the 23rd term is 50, what is the common difference?

Solution: Since $a_{23} = a_1 + (22d)$,

$$50 = 4 + (22)d,$$
$$22d = 46,$$
$$d = \frac{23}{11}.$$

Arithmetic sequence

EXAMPLE 3 For an arithmetic sequence with first term 2, common difference 3, and nth term 137, find the number of terms in the sequence.

Solution: Since $a_n = a_1 + (n-1)d$, it follows that

$$137 = 2 + (n-1)3,$$
$$3n - 3 = 135,$$
$$3n = 138,$$
$$n = 46.$$

PRACTICE PROBLEM

(a) In an arithmetic sequence, $a_1 = 3$ and $d = \frac{3}{2}$; find a_{31}.

(b) In an arithmetic sequence, $a_1 = 21$ and $a_{20} = 2$; find d.

(c) In an arithmetic sequence, $a_1 = 6$, $d = \frac{5}{8}$, and $a_n = 36$; find n.

Answer: (a) 48. (b) -1. (c) 49. ■

Savings plan

EXAMPLE 4 Suppose a person decides to save $3.00 on Jan. 1 and to increase the amount saved each day thereafter by 75 cents each day. How much will he or she save on Jan. 21 of that year?

Solution: The savings pattern is an arithmetic sequence with $a_1 = 3$, $d = 0.75$, and $n = 21$. Consequently,

$$a_{21} = 3 + (20)(0.75) = 3 + 15 = 18.$$

Hence, $18.00 would be saved on Jan. 21.

PRACTICE PROBLEM Following the savings pattern described in Example 4, how much would the person save on Jan. 30?

Answer: $24.75. ■

8.2 ARITHMETIC SEQUENCES

The technique used to find the sum of the first 50 odd positive integers can be used to derive the formula for the sum of any arithmetic sequence of n terms. The sum s_n of n terms of an arithmetic sequence with first term a_1 and common difference d is given by

$$s_n = a_1 + [a_1 + d] + \cdots + [a_1 + (n-2)d] + [a_1 + (n-1)d].$$

Writing in terms of the preceding sum in reverse order, we get

$$s_n = [a_1 + (n-1)d] + [a_1 + (n-2)d] + \cdots + [a_1 + d] + a_1.$$

Adding the two above equalities yields

$$2s_n = [2a_1 + (n-1)d] + [2a_1 + (n-1)d] + \cdots + [2a_1 + (n-1)d].$$

Since this sum contains n terms of the form $[2a_1 + (n-1)d]$, we have

$$2s_n = n[2a_1 + (n-1)d]$$

and conclude that

sum of an arithmetic sequence

$$s_n = \frac{n}{2}[2a_1 + (n-1)d]$$

is the formula for the **sum** of n *terms of any arithmetic sequence with* a_1 *as* first term *and* d *as* common difference.

Total savings

EXAMPLE 5 Find the total amount saved in 21 days by the person whose savings plan is described in Example 4.

Solution: In Example 4, $n = 21$, $a_1 = 3$, and $d = 0.75$. Therefore,

$$s_{21} = \frac{21}{2}[2(3) + (20)(0.75)]$$

$$= 10.5[6 + 15] = 220.5.$$

Hence the person would save $220.50.

PRACTICE PROBLEM Find the total amount saved in 30 days by the person whose savings plan is described in Example 4.

Answer: $416.25. ■

Gross income

EXAMPLE 6 Suppose an editor secures a job with a starting salary of $16,000 a year. She receives an $800 raise each year thereafter for 20 years. Calculate her total gross income from the job for the 20-year period.

Solution: The salary schedule is an arithmetic sequence with $a_1 = 16{,}000$, $d = 800$, and $n = 20$. Therefore,

$$s_{20} = \frac{20}{2}[32{,}000 + (19)(800)]$$
$$= 10[32{,}000 + 15{,}200]$$
$$= 472{,}000.$$

The total gross income for the 20-year period would be $472,000.

Since the nth term (last term) of an arithmetic sequence is given by

$$a_n = a_1 + (n-1)d,$$

another important formula for the sum of an arithmetic sequence is obtained from the fact that

$$2a_1 + (n-1)d = a_1 + [a_1 + (n-1)d] = a_1 + a_n.$$

It is

sum of an arithmetic sequence

$$s_n = \frac{n}{2}(a_1 + a_n).$$

From the preceding formula, we see that *the sum of an arithmetic sequence with n terms is the number of terms times the arithmetic mean (average) of the first and last terms.*

PRACTICE PROBLEM Find a formula for $1 + 2 + 3 + \cdots + n$, the sum of the first n positive integers.

Answer: $n(n + 1)/2$. ∎

Savings plan

EXAMPLE 7 In a given year a child saves 1 cent the first day, 2 cents the second, 3 cents the third, 4 cents the fourth, etc. How many days will it take the child to save a total of $40.95?

Solution: Since $40.95 is 4095 cents, we need to find the integer n, if it exists, such that

$$1 + 2 + 3 + \cdots + n = 4095.$$

Using the formula from the preceding practice problem yields

$$\frac{n(n + 1)}{2} = 4095$$

$$n^2 + n = 8190$$

$$n^2 + n - 8190 = 0.$$

The quadratic equation can be solved either by the quadratic formula or by factoring. By factoring we obtain

$$(n + 91)(n - 90) = 0.$$

Although the quadratic equation has $n = -91$ and $n = 90$ as its two solutions, the positive solution is the only tenable one for this problem. Hence, 90 days are required to save $40.95. *Note:* If the goal were to save $41.00, an extra day would be needed. In this case, the resulting quadratic equation would not have an integral solution; the answer would be the smallest integer greater than its positive solution.

PRACTICE PROBLEM For the savings plan described in Example 7, how long would it take the child to save $100?

Answer: 141 days. ∎

Finance charges

EXAMPLE 8 A person plans to pay off a $1200 debt by paying $100 at the end of each month plus a finance charge of 1.5% of the balance due before the payment is deducted. What will be the total finance charges?

Solution

nth month	Amount owed at the end of nth month	Finance charges paid at the end of the nth month
1	1200	0.015(1200) = 18.00
2	1100	0.015(1100) = 16.50
3	1000	0.015(1000) = 15.00
⋮	⋮	⋮
11	200	0.015(200) = 3.00
12	100	0.015(100) = 1.50

The sequence of finance charges is an arithmetic sequence containing 12 terms with first term 18.00 and last term 1.50. (*Note:* $d = -1.50$.) Hence the sum S is

$$S = \frac{12}{2}(18.00 + 1.50)$$

$$= 6(19.50)$$

$$= \$117.$$

(If only the finance charges were paid and the debt was not reduced, the total finance charges for the year would be

$$0.015(1200)(12) = \$216.$$

Since $\frac{216}{1200} = 0.18$, the person pays an 18% annual interest rate.)

PRACTICE PROBLEM In Example 8, suppose the person had decided to pay $200 plus a finance charge at the end of each month. What would have been the finance charges?

Answer: $63. ∎

EXERCISES

Set A–Arithmetic Sequences

For each of the arithmetic sequences in Exercises 1 through 10, (a) find the common difference, (b) list the next 3 terms, and (c) find the sum of the first 21 terms of the sequence.

8.2 ARITHMETIC SEQUENCES

1. 8, 11, 14, ...
2. 2, 4, 6, ...
3. 1, 5, 9, ...
4. 1.5, 3.0, 4.5, ...
5. 2.3, 3.4, 4.5, ...
6. −4, 1, 6, ...
7. 3, −2, −7, ...
8. 1, −3, −7, ...
9. $\frac{2}{3}, \frac{5}{6}, 1, \ldots$
10. $\frac{2}{9}, \frac{1}{2}, \frac{7}{9}, \ldots$

11. An arithmetic sequence has 4 as its first term and 5 as its common difference.
    (a) What is the 50th term?
    (b) What is the sum of the first 50 terms of this sequence.

12. An arithmetic sequence has 3 as its first term and $\frac{4}{3}$ as the common difference.
    (a) What is the 50th term?
    (b) What is the sum of the first 50 terms?

13. In an arithmetic sequence having 5 as its first term and 98 as its 31st term, what is the common difference?

14. An arithmetic sequence has first term 6 and common difference $-\frac{2}{3}$.
    (a) What is the 20th term?
    (b) What is the sum of the first 20 terms?

15. Given: 5, 8, 11, ..., 65 is an arithmetic sequence.
    (a) What is the common difference?
    (b) How many terms are there in the sequence if 65 is the last term?
    (c) What is the sum of the terms of the sequence?

16. Given: 5, $\frac{11}{3}$, ..., −11 is an arithmetic sequence.
    (a) What is the common difference?
    (b) How many terms are there in the sequence if −11 is the last term?
    (c) Find the sum of the terms of the sequence.

17. What is the sum of the first 30 multiples of 3?

18. For the arithmetic sequence 8, s, t, u, 18, ..., answer the following questions.
    (a) What is s?
    (b) What is t?
    (c) What is u?
    (d) What is the sum of the first 20 terms?

Set B–Applications of Arithmetic Sequences

Gross income

19. Suppose that a bookkeeper obtains a job with a starting salary of $15,000 and that he receives an $800 raise each year thereafter.
    (a) What would his salary be during the 15th year?

(b) What would his total gross income be for the 15-year period?

Gross income

20. Suppose that a college professor obtains a job with a starting salary of $16,000 and that she receives a $900 raise each year thereafter.
    (a) How much would her salary be during the 10th year?
    (b) What would her total gross income be for the 10-year period?

Salary comparison

21. Which of the following salary schedules would produce the greater total gross income over a 20-year period?
    (a) A starting salary of $12,000 with a raise of $900 per year, or
    (b) A starting salary of $15,000 with a raise of $600 per year.

Daily savings

22. A student begins the year by saving 5 cents the first day, 10 cents the second, 15 cents the third, 20 cents the fourth, etc. How many days will it take to save a total of $91.50?

Daily savings

23. If the savings pattern described in Exercise 22 could be continued for one year (365 days), how much would the student save?

Monthly savings

24. A person starts the year by saving $5.00 the first month, $10.00 the second month, $15.00 the third month, etc.
    (a) How many months will it take for the monthly savings to reach $300?
    (b) At the end of that time, what would be the total amount of savings? (*Note:* We do not assume that the money is deposited in an interest bearing account; such problems are discussed in Section 8.5.)

Grocery display

25. A grocery store manager has 25 cases of canned peaches, each containing 12 cans. He decides to display the cans by stacking them in a "pyramid" where each row after the bottom row contains one less can. Is it possible for him to use all the cans and end up with a top row having only one can? If so, how many cans should be put in the bottom row?

Grocery display

26. A grocery store manager has 5 cases of canned pears, each containing 24 cans. She decides to display the cans by stacking them in a "pyramid" where each row after the bottom row contains one less can. Is it possible for her to use all the cans and end up with a top row having only one can? If so, how many cans should be put in the bottom row?

Dental payments

27. The recipient of a dental bill for $250 is offered the option of paying $12.50 at the end of each month plus a finance charge of $1\frac{1}{2}\%$ on the balance due before the payment is deducted. If she accepts this option, what will be the total amount paid?

Furniture payments

28. A young couple receives a furniture bill for $1400 and is offered the option of paying $200 down and $100 at the end of each month plus a finance charge of $1\frac{1}{2}\%$ on the balance due before the payment is deducted. If this option is accepted, what will the total amount paid be?

Television payments

29. A television set is on sale for $900. It can be bought by paying $200 down and then paying $50 at the end of each month plus a finance charge of $1\frac{1}{2}\%$ on the balance due before the payment is deducted. If this payment plan is used, what will be the total cost of the set?

8.3 GEOMETRIC SEQUENCES

Consider the following salary problem. If Smith were paid a salary of $10,000 a day for 31 days and if Jones were paid 1 cent the first day, 2 cents the second day, 4 cents the third day, 8 cents the fourth day, 16 cents the fifth day, 32 cents the sixth day, and 2^{n-1} cents the *n*th day, which 31-day total salary is greater? Since we bother to ask the question, one may suspect that Jones has the better salary. How much do you think Smith would need to make each day in order to get the same total salary as Jones? (Make a guess.) To find out, let us determine Jones's total salary for 31 days.

If S is Jones's salary in cents for the 31 days, then

$$S = 1 + 2 + 2^2 + 2^3 + \cdots + 2^{29} + 2^{30}. \tag{1}$$

Although this sum could be found by adding the 31 given numbers, it is much easier to use the following procedure:

First multiply each side of Eq. (1) by 2 to obtain

$$2S = 2 + 2^2 + 2^3 + 2^4 + \cdots + 2^{30} + 2^{31}. \tag{2}$$

Next subtract Eq. (1) from Eq. (2) to get

$$S = 2^{31} - 1;$$

that is, $S = 2{,}147{,}483{,}647$ cents. Therefore,

$$S = \$21{,}474{,}836.47$$

is Jones's total salary in dollars. Since $21,474,836 divided by 31 is approximately $696,737, Smith would have to make more than

$\frac{2}{3}$ of a million dollars a day for his 31-day total salary to equal that of Jones.

The salary schedule for Jones forms what is called a *geometric sequence*. In general, if after the first term of a sequence each term can be obtained by multiplying the preceding term by the same number r, then the sequence is called a **geometric sequence;** the number r is called the **common ratio.** In a geometric sequence, if a is the first term and r is the common ratio, then the second term a_2 is ar, the third term a_3 is ar^2, the fourth term a_4 is ar^3, and the nth term a_n is given by

geometric sequence
common ratio

nth term of a geometric sequence

$$a_n = ar^{n-1}.$$

For example,

$$1, 2, 2^2, 2^3, \ldots, 2^{29}, 2^{30}$$

is a geometric sequence where $a = 1$, $r = 2$ and $n = 31$. Note that for any geometric sequence the ratio r can be found by dividing any term after the first by the preceding term in the sequence; that is,

$$r = \frac{a_n}{a_{n-1}}.$$

Geometric sequence

EXAMPLE 1

(a) If $6, 2, \frac{2}{3}, \ldots$ is a geometric sequence, find its common ratio.

(b) Find the seventh term of the geometric sequence $18, -12, 8, \ldots$.

Solution

(a) Since in any geometric sequence the common ratio r can be found by dividing any term after the first by the preceding term,

$$r = 2 \div 6 = \frac{1}{3}.$$

8.3 GEOMETRIC SEQUENCES

(b) Since $a = 18$ and $r = -12 \div 18 = -\frac{2}{3}$,

$$a_7 = 18\left(-\frac{2}{3}\right)^6 = 18\left(\frac{64}{729}\right) = \frac{128}{81}.$$

The procedure used to find the sum of the geometric sequence with $a = 1$, $r = 2$, and $n = 31$ can be employed to find the sum of any geometric sequence containing a finite number of terms. Let us derive the general formula. Consider the geometric sequence

$$a, ar, ar^2, ar^3, \ldots, ar^{n-2}, ar^{n-1}$$

with n terms and let

$$s_n = a + ar + ar^2 + ar^3 + \cdots + ar^{n-2} + ar^{n-1}. \quad (3)$$

Multiplying each side of Eq. (3) by the common ratio r, we obtain

$$rs_n = ar + ar^2 + ar^3 + ar^4 + \cdots + ar^{n-1} + ar^n. \quad (4)$$

By subtracting Eq. (3) from Eq. (4), all the terms except the first of Eq. (3) and the last of Eq. (4) drop out and we get

$$rs_n - s_n = ar^n - a,$$
$$s_n(r - 1) = ar^n - a.$$

Therefore,

sum of a geometric sequence

$$s_n = \frac{ar^n - a}{r - 1} \quad \text{provided} \quad r \neq 1.$$

Notes

(1) If $r = 1$, each term of the sequence is a and the sum is the product na.

(2) If the numerator and denominator of the fraction in the formula are multiplied by -1, then s_n is also given by

sum of a geometric sequence

$$s_n = \frac{a - ar^n}{1 - r} = a\left(\frac{1 - r^n}{1 - r}\right).$$

Geometric sequence

EXAMPLE 2 Let $a = 64$ be the first term of a geometric sequence and let $r = -\frac{1}{2}$ be the common ratio.

(a) Find the eighth term.

(b) Find the sum of the first eight terms.

Solution

(a) The eighth term is $64(-\frac{1}{2})^7 = -\frac{1}{2}$.

(b) $$s_8 = \frac{(64)(-\frac{1}{2})^8 - 64}{(-\frac{1}{2}) - 1} = \frac{\frac{1}{4} - 64}{-\frac{3}{2}} = \frac{85}{2}.$$

PRACTICE PROBLEM

(a) For the geometric sequence with 54 as first term and $-\frac{1}{3}$ as common ratio, find the sixth term.

(b) Find the sum of the first six terms.

Answer: (a) $a_6 = -\frac{2}{9}$. (b) $s_6 = \frac{364}{9}$. ∎

Geometric sequence

EXAMPLE 3 If $27, u, v, 8$ are the first four terms of a geometric sequence, what are the common ratio r and the terms u and v?

Solution: Since the fourth term of a geometric sequence is ar^3,

$$8 = 27r^3,$$

$$r^3 = \frac{8}{27},$$

$$r = \frac{2}{3}.$$

Hence, $u = (27)(\frac{2}{3}) = 18$ and $v = (18)(\frac{2}{3}) = 12$.

PRACTICE PROBLEM Let $125, u, v,$ and 8 be the first four terms of a geometric sequence. Find the common ratio r and the terms u and v.

Answer: $r = \frac{2}{5}$, $u = 50$, and $v = 20$. ∎

Gross income

EXAMPLE 4 Suppose that a physician assistant's starting salary is $15,000 a year and that he receives an 8% yearly salary increase for 10 years.

(a) What salary would he earn in the 10th year?

(b) What would be his gross income for the 10-year period?

Solution

(a) The salary for each year after the first would be the preceding year's salary plus an 8% increase. Thus

Second year salary: $15,000 + 0.08($15,000)$
                       $= \mathbf{\$15{,}000(1.08)}$

Third year salary: $15,000(1.08) + 0.08[$15,000(1.08)]$
                       $= 15,000(1.08)[1 + 0.08]$
                       $= \mathbf{\$15{,}000(1.08)^2}$

Fourth year salary: $15,000(1.08)^2 + 0.08[$15,000(1.08)^2]$
                       $= 15,000(1.08)^2[1 + 0.08]$
                       $= \mathbf{\$15{,}000(1.08)^3}$

$\vdots \qquad \vdots \qquad \vdots$

Tenth year salary: $\mathbf{\$15{,}000(1.08)^9}$

Note that the annual salaries form a geometric sequence. By using a calculator, logarithms, or Table 1 in Appendix A, we find that the salary for the 10th year is

$$\$15{,}000(1.08)^9 \doteq \$15{,}000(1.999) = \$29{,}985.$$

Note: The annual salary would nearly double in 10 years.

(b) The total 10-year salary S_{10} is the sum of the geometric sequence where $a = 15{,}000$, $r = 1.08$, and $n = 10$. Hence

$$S_{10} = 15{,}000 \frac{(1.08)^{10} - 1}{1.08 - 1}$$

$$= \frac{15{,}000}{0.08}[(1.08)^{10} - 1]$$

$$= 187{,}500[2.158925 - 1]$$

$$= 217{,}298.44.$$

Consequently, the total salary (before taxes, etc.) would be $217,298.44.

The solution to Example 4 can be used as a pattern to develop the following two general formulas. If B represents the starting (beginning) salary and i represents the percent of yearly increase, then the **salary** a_n for the nth year is given by

salary for nth year

$$a_n = B(1 + i)^{n-1}$$

Furthermore, the total **gross income** S_n for n years is given by

gross income for n years

$$S_n = \frac{B}{i}[(1 + i)^n - 1].$$

EXERCISES

Set A—Geometric Sequences

For each of the geometric sequences in Exercises 1 through 9, find the common ratio and the seventh term.

1. $3, -3, 3, \ldots$
2. $2, 6, 18, \ldots$
3. $4, -2, 1, \ldots$
4. $3, 6, 12, \ldots$
5. $81, -27, 9, \ldots$
6. $60, 30, 15, \ldots$
7. $\frac{2}{3}, \frac{1}{3}, \frac{1}{6}, \ldots$
8. $8, -2, \frac{1}{2}, \ldots$
9. $3, 3.6, 4.32, \ldots$

Find the sum of the first 6 terms of the geometric sequences in Exercises 10, 11, and 12.

10. $5, -5, 5, \ldots$
11. $8, -2, \frac{1}{2}, \ldots$
12. $81, -27, 9, \ldots$

13. For each of the geometric sequences, answer the given questions.
    (a) $2, x, 9, \ldots$ What is x if x is positive?
    (b) $16, x, y, 54, \ldots$ What are x and y?

14. For each of the geometric sequences, answer the given questions.
    (a) $49, x, 16, \ldots$ What is x if x is positive?
    (b) $125, x, y, 8, \ldots$ What are x and y?

8.3 GEOMETRIC SEQUENCES

Set B—Annual Salary and Gross Income

Gross income 15. Consider the following salary schedule: a $13,200 starting salary with a 5% yearly increase.
(a) What would be the salary during the 10th year?
(b) What would be the gross income during the first 10 years?

Gross income 16. Consider the following salary schedule: $12,000 starting salary with a 7% yearly increase.
(a) What would be the salary during the 10th year?
(b) What would be the gross income during the first ten years?

Gross income 17. A job pays a starting salary of $20,000 and calls for a 6% yearly increase thereafter.
(a) What would it pay in the 10th year?
(b) What would be the total gross income for the first 10 years?

Gross income 18. A job pays a starting salary of $18,000 and calls for an 8% yearly increase thereafter.
(a) What would be the salary for the 15th year?
(b) What would be the total gross income for the first 15 years?

Salary comparison 19. Consider the following salary schedules: (i) a $15,000 salary the first year and a $500 annual raise thereafter; (ii) a $12,000 salary the first year and a 6% annual raise thereafter.
(a) In each case, what would be the salary for the 10th year?
(b) In the first 10 years, which salary would provide the greater gross income? How much greater?
(c) In the first 20 years, which salary would provide the greater gross income? How much greater?

Effect of beginning salary 20. For a given position a person is offered a starting salary of $15,000 and an 8% yearly raise thereafter. However, the person insists on a starting salary of $15,800 and gets it. In 20 years, what will be the difference in the gross incomes obtained from the two starting salaries?

Effect of beginning salary 21. For a given position a person is offered a starting salary of $14,000 and a 6% yearly raise thereafter. However, the person insists on a starting salary of $14,800 and gets it. In 20 years, what will be the total difference in the gross incomes obtained from the two starting salaries?

Beginning salary 22. A manager's salary for her 15th year on a job is $36,715. If she received an 8% raise each year, what was her starting salary to the nearest dollar?

8.4 INTEREST: SIMPLE AND COMPOUND

interest

interest rate

As we know, money paid for the use of money is called **interest.** If a sum of money, called **principal,** is invested and interest is earned, then the amount of interest received depends upon the **interest rate,** a number expressed as a percentage for a given time period. For example, a principal of $100 invested at an annual interest rate of 6% earns

$$0.06 \times 100 = \$6.00$$

interest in one year (that is, 6 cents on each dollar invested).

Simple interest is computed on principal only. For example, if $100 were invested for two years at an annual simple interest rate of 6%, then the interest earned would be $6.00 each year, a total of $12.00 in interest. In general, if P represents principal, r denotes rate of interest per unit of time, t denotes the number of time units in which the interest rate is expressed, and I represents interest, then

simple interest

$$I = Prt$$

is the formula for *simple interest*. Since almost nobody invests money at simple interest (banks are forbidden by law to pay simple interest), our attention will be focused on what is called *compound interest.*

compound interest

Compound interest is interest paid not only on principal but also on interest. For example, if $100 is invested for two years at a 6% annual interest rate compounded yearly, then the amount (balance) on deposit at the end of the first year would be the initial principal plus the first year's interest; that is,

$$P_1 = \$100 + (0.06)(\$100) = \$106,$$

where P_1 is the amount at the end of the first year. Although this is the same amount that would result from simple interest, things change in the second year. During the second year, interest is paid on $106 and the second year's interest is

$$(0.06)(\$106) = \$6.36.$$

Therefore, the amount P_2 on deposit at the end of two years is

$$P_2 = \$106 + \$6.36 = \$112.36.$$

8.4 INTEREST: SIMPLE AND COMPOUND

Hence, the compounding of interest yields $12.36 in interest for the two years as opposed to $12 if interest were not compounded. The difference in the amounts of interest may not seem very much in this case, but over longer periods of time it can be dramatic. For example, it would take more than 16 years to earn $100 interest on $100 invested at 6% simple interest while the time required to earn $100 interest on $100 invested at 6% interest compounded annually is less than 12 years. (We shall prove this fact soon.)

Suppose a person deposits P dollars where the annual interest rate i is 6% compounded annually. At the end of 1 year the total amount P_1 on deposit is the initial principal P plus the interest earned; hence,

$$P_1 = P + 0.06P = (1.06)P.$$

If the money is left on deposit, then at the end of the second year the amount of money P_2 on deposit is the sum of the amount on deposit at the end of the first year plus the second year's interest; that is,

$$P_2 = (1.06)P + (0.06)[(1.06)P]$$
$$= (1.06)P(1 + 0.06)$$
$$= P(1.06)^2.$$

In a similar way we find that the amount P_3 at the end of 3 years is

$$P_3 = (1.06)^2 P + (0.06)[(1.06)^2 P] = P(1.06)^3.$$

Continuing, we find that $P_1, P_2, P_3, P_4, \ldots$ is a geometric sequence, and the amount P_n on deposit at the end of n years starting from an initial deposit P at a 6% interest rate compounded annually is

$$P_n = P(1.06)^n.$$

(Note the similarity to the solution of Example 4, Section 8.3.)

In general, if P dollars is the initial principal invested at an interest rate of i percent **compounded annually,** then the amount P_n at the end of n years is given by

annual compound interest

$$P_n = P(1 + i)^n.$$

PRACTICE PROBLEM Suppose $1000 is deposited in a savings institution which pays 8% interest compounded annually. How much would be on deposit at the end of 4 years?

Answer: $1360.49. ∎

To find how long it takes to double an original deposit P at 6% interest compounded annually involves finding the number n such that

$$2P = P(1.06)^n$$

or

$$2 = (1.06)^n.$$

One way to find n is by using interest tables. (See Appendix A, Table 1.) We find that

$$(1.06)^{12} = 2.01219647.$$

Since $(1.06)^{11} < 2$, it would take 12 years to double the amount.

Another way to solve the equation $(1.06)^n = 2$ for n is by using logarithms. If

$$(1.06)^n = 2,$$

then

$$\log_{10}(1.06)^n = \log_{10}2,$$

and

$$n(\log_{10}1.06) = \log_{10}2.$$

Using logarithm tables (see Appendix A, Table 4), we obtain

$$0.0253n = 0.3010$$

$$n = \frac{3010}{253} \doteq 11.9.$$

Consequently, 12 years would more than double the original deposit.

Doubling a deposit at 7%

EXAMPLE 1 How many years does it take to double a given amount placed on deposit at 7% compounded annually?

8.4 INTEREST: SIMPLE AND COMPOUND

Solution: We need to solve $(1 + 0.07)^n = 2$ for n. Using logarithms, interest tables, or a calculator, we find that if $n = 11$, then

$$(1.07)^{11} = 2.1048.$$

Furthermore, since $(1.07)^{10} < 2$, 11 years are required.

Suppose $100 is deposited at an annual interest rate of 6%, and that the interest is compounded quarterly. How much will be on deposit at the end of the second quarter? At the end of the first quarter, the interest earned will be

$$\frac{1}{4}(0.06)(\$100) = (0.015)(\$100);$$

that is, $1.50, which is one-fourth the yearly interest at 6%. The amount A_1 at the end of the first quarter on which interest will be computed during the second quarter is

$$A_1 = \$100 + (0.015)(\$100) = \$100(1.015).$$

At the end of the second quarter the amount A_2 will be $100(1.015)$ plus the interest on it; that is,

$$A_2 = \$100(1.015) + \$100(1.015)\frac{0.06}{4}$$

$$= \$100(1.015)(1 + 0.015)$$

$$= \$100(1.015)^2.$$

Since $(1.015)^2 = 1.030\,225$, the amount at the end of two quarters will be $103.02.

Let P be the principal invested and let i *percent be the annual interest rate.* If the interest is compounded four times per year (quarterly), then the amount on deposit at the end of the first quarter is

$$A_1 = P + P\left(\frac{i}{4}\right) = P\left(1 + \frac{i}{4}\right).$$

The amount at the end of the second quarter is

$$A_2 = P\left(1 + \frac{i}{4}\right) + P\left(1 + \frac{i}{4}\right)\left(\frac{i}{4}\right) = P\left(1 + \frac{i}{4}\right)^2,$$

and the amount at the end of the third quarter is

$$A_3 = P\left(1 + \frac{i}{4}\right)^2 + P\left(1 + \frac{i}{4}\right)^2\left(\frac{i}{4}\right) = P\left(1 + \frac{i}{4}\right)^3.$$

We see that $A_1, A_2, A_3, \ldots$ is a geometric sequence, and the amount at the end of m quarters is

$$A_m = P\left(1 + \frac{i}{4}\right)^m.$$

In fact, if there were q compoundings in a year, the amount A_m at the end of m compounding periods would be

$$A_m = P\left(1 + \frac{i}{q}\right)^m.$$

Note: In n years, $m = nq$.

interest rate per period

If we let $j = i/q$ be the **interest rate per period** and let m be the **number of compoundings,** we have the following general formula where P is the initial principal and A_m is the future value of the initial principal after m periods:

compound interest

$$A_m = P(1 + j)^m.$$

Quarterly compounding

EXAMPLE 2 Suppose $100 is deposited in a savings institution where the annual interest rate is 6% compounded quarterly. How much would be on deposit at the end of two years?

Solution: Using the formula $A_m = P(1 + j)^m$ where $P = \$100$, $j = i/q = 0.06/4 = 0.015$, $m = nq = 8$, we get

$$\$100(1.015)^8 \doteq \$100(1.126\ 49) \doteq \$112.64.$$

Important Notes

(1) The amount really is $112.65 to the *nearest cent,* but most institutions ignore tenths of a cent when "rounding off" since this "omission" is in their favor.

8.4 INTEREST: SIMPLE AND COMPOUND

(2) The phrase "i percent interest" means an annual interest rate of i percent. Since

$$\frac{0.06}{4} = 0.015,$$

6% interest is equivalent to a *quarterly interest* rate of 1.5%.

PRACTICE PROBLEM Suppose $500 is deposited in a savings institution where the annual interest rate is 6% compounded monthly. How much will be on deposit at the end of two years?

Answer: $563.58. ■

Future value of a principal

EXAMPLE 3 If $5000 is deposited at 8% interest compounded quarterly, what will be the amount on deposit after 10 years?

Solution: Since the interest is compounded quarterly for 10 years,

$$m = qn = 4(10) = 40.$$

Since $i = 0.08$, $j = i/q = 0.08/4 = 0.02$ is the interest rate per period. Hence, we need to find A_{40} where $P = \$5000$. Using the formula and a calculator (or Table 1, Appendix A), we find

$$A_{40} = \$5000(1 + 0.02)^{40}$$
$$= \$5000(1.02)^{40}$$
$$= \$5000(2.208\ 039\ 66)$$
$$= \$11{,}040.19.$$

The future value of $5000 deposited for 10 years at 8% compounded quarterly is $11,040.19.

The amount on deposit after 1 year starting with a principal of 1 dollar invested, for example, at an interest rate of 6% compounded quarterly is

$$\left(1 + \frac{0.06}{4}\right)^4 = (1.015)^4 \doteq 1.061\ 36.$$

The interest rate which would yield the same amount in one year without compounding (simple interest) is

$$(1.06136 - 1)(100) = 6.136 \text{ percent};$$

this is called the *effective interest rate* of 6% interest compounded quarterly.

When interest is compounded daily, it is generally assumed that the year contains 360 days. The *effective interest rate for daily compounding* of 6% interest is

$$\left[\left(1 + \frac{0.06}{360}\right)^{360} - 1\right]100\%.$$

which, with the aid of a calculator, we find to be 6.18312% correct to five decimal places. In general, if i is the annual interest rate and the compounding is done q times per year then the **effective interest rate** E_q is given by

effective interest rate

$$E_q = \left[\left(1 + \frac{i}{q}\right)^q - 1\right] \times 100.$$

Effective interest rate

EXAMPLE 4 What is the effective interest rate of 8% compounded monthly?

Solution

$$E_{12} = \left[\left(1 + \frac{0.08}{12}\right)^{12} - 1\right]100$$
$$\doteq [(1.066\,666\,6)^{12} - 1]100$$
$$\doteq [1.082\,999 - 1]100$$
$$= 8.2999\%.$$

Thus, it is 8.3% to the nearest hundredth.

PRACTICE PROBLEM What is the effective interest rate of 9% compounded monthly?

Answer: 9.38%. ■

8.4 INTEREST: SIMPLE AND COMPOUND

Present value

EXAMPLE 5 How much money should be deposited at an interest rate of 8% compounded quarterly for 4 years so that the amount on deposit at the end of that time is $5000? (The problem is to find the present value of $5000 at the given interest rate for the given time period.)

Solution: Since $q = 4$, $i = 0.08$, and $n = 4$, $m = nq = 16$ and $j = i/q = 0.08/4 = 0.02$. We want to find P when $A_{16} = \$5000$. Since

$$A_{16} = P(1 + j)^{16},$$

it follows that

$$\$5000 = P(1.02)^{16},$$

$$P = \frac{\$5000}{1.372\ 7857}$$

$$= \$3642.23.$$

Therefore the present value of $5000 at 8% compounded quarterly for 4 years is $3642.23; it is the principal needed to yield $5000.

Note: Since $P = \$5000(1.02)^{-16}$, if a calculator is not available, we can use Table 2, Appendix A, to avoid dividing by a decimal.

PRACTICE PROBLEM How much money should be deposited at 12% interest compounded quarterly for 4 years so that the amount on deposit at the end of that time is $5000?

Answer: $3115.84. ∎

EXERCISES

Set A—Interest Rate per Period

In Exercises 1 through 6, find the interest rate per period j and the number of periods m where the annual interest rate is i, the number of compoundings per year is q, and the number of years is n.

1. $i = 6\%$, $q = 12$, $n = 3$
2. $i = 5\%$, $q = 4$, $n = 5$
3. $i = 10\%$, $q = 4$, $n = 2$
4. $i = 12\%$, $q = 12$, $n = 4$
5. $i = 12\%$, $q = 2$, $n = 5$
6. $i = 16\%$, $q = 4$, $n = 5$

Set B–Compound Interest Problems

Future value
7. If $1000 is invested at 5% interest compounded quarterly, what is the amount of the investment at the end of 1 year?

Future value
8. If $3000 is invested at 6% interest compounded monthly, what is the amount of the investment at the end of 3 years?

Future value
9. What is the future value of $3000 deposited for 3 years at 8% compounded quarterly?

Future value
10. What is the future value of $10,000 deposited for 4 years at 6% compounded monthly?

Effective interest rate
11. What is the effective interest rate of 5% interest compounded quarterly?

Effective interest rate
12. What is the effective interest rate of 8% interest compounded quarterly?

Effective interest rate
13. What is the effective interest rate of 12% interest compounded monthly?

Effective interest rate
14. What is the effective interest rate of 16% interest compounded daily?

Interest
15. What is the interest obtained from $10,000 deposited for one year at 6% compounded daily? (Use 360 days.)

Interest
16. What is the interest obtained from $8000 deposited for one year at 8% compounded daily?

Doubling amount
17. How long will it take to double an amount left on deposit at 6% compounded quarterly?

Doubling amount
18. How long will it take to double an amount left on deposit at 5% compounded quarterly?

Doubling amount
19. How long will it take to double an amount left on deposit at 12% compounded monthly?

Doubling amount
20. How long will it take to double an amount left on deposit at 16% compounded daily?

Present value
21. What is the present value of $5000 if the interest rate is 8% compounded quarterly for 4 years?

Present value
22. What is the present value of $4000 if the interest rate is 10% compounded quarterly for 4 years?

Present value
23. What is the present value of $10,000 if the interest rate is 12% compounded quarterly for 4 years?

Present value
24. How much should be placed in a 4-year savings certificate earning 8% compounded quarterly in order to have $20,000 on deposit at maturity of the certificate?

Present value
25. How much should be placed in a 4-year savings certificate earning 12% compounded monthly in order to have $20,000 on deposit at maturity of the certificate?

8.5 CONTINUOUS INTEREST

Let us now reconsider the geometric sequence

$$P\left(1 + \frac{i}{q}\right)^q,$$

which gives the amount obtained in one year from an initial deposit P where the annual interest rate is i percent compounded q times in one year. If we let $u = i/q$, then $q = i/u$ and

$$P\left(1 + \frac{i}{q}\right)^q = P(1 + u)^{i/u} = P[(1 + u)^{1/u}]^i.$$

Since $u = i/q$, it follows for fixed i that if q increases, then u approaches 0, and

$$(1 + u)^{1/u}$$

approaches $e \doteq 2.718\,28$. (See Section 2.2.) As the number of compoundings q increases, u increases, and

$$[(1 + u)^{1/u}]^i \quad \text{approaches} \quad e^i \doteq (2.718\,28)^i.$$

With an initial principal P, the amount obtained in one year from what is *defined* as i percent *annual interest with* **continuous compounding** is denoted by A_C and given by the formula

Continuous interest

$$A_C = Pe^i.$$

Continuous interest

EXAMPLE 1 If $5000 is put into an account drawing 8% annual interest with continuous compounding, what is the amount on deposit at the end of 1 year?

Solution: Using the formula $A_C = Pe^i$ where $P = \$5000$ and $i = 0.08$, we get

$$A_C = \$5000 e^{0.08}.$$

From Table 3 in Appendix A (or by using the e^x key on a calculator), we find that

$$A_C = \$5000(1.083\ 287)$$
$$= \$5416.43.$$

Note: The interest from the deposit is $416.43.

PRACTICE PROBLEM If $4000 is deposited into an account earning 5% annual interest with continuous compounding, what is the amount on deposit at the end of one year?

Answer: $4205.08. ■

If we let $P = \$1.00$ and $i = 0.06$, the amount obtained from 6% interest with continuous compounding is

$$A_C = e^{0.06} \doteq 1.0618.$$

Consequently, the interest on $1.00 in one year at 6% interest with continuous compounding is 6.18 cents. Since this is equivalent to the interest on $1.00 at 6.18% simple interest, 6.18% is the *effective interest rate of 6% interest with continuous compounding*. In general, the **effective interest rate** for an annual interest rate of *i* percent compounded continuously is given by

effective interest rate

$$E_C = (e^i - 1)(100).$$

Effective interest rate

EXAMPLE 2 What is the effective interest rate for an annual interest rate of 8% with continuous compounding?

Solution: Using the formula $E_C = (e^i - 1)(100)$ where $i = 0.08$, we get

$$E_C = (e^{0.08} - 1)(100) \doteq 8.328\ 7.$$

Therefore, the effective interest rate is approximately 8.33%.

8.5 CONTINUOUS INTEREST

Daily and continuous interest

EXAMPLE 3 What would be the difference in the amount of interest obtained from $8000 deposited for 1 year at 6% interest compounded daily and 6% compounded continuously?

Solution: Using the formula derived in the last section, we find that the amount obtained from 6% interest compounded daily is

$$A_{360} = \$8000\left(1 + \frac{0.06}{360}\right)^{360} \doteq \$8000(1.061\ 831\ 2) \doteq \$8494.64.$$

The amount obtained from continuous compounding at 6% interest is

$$A_C = \$8000 e^{0.06} \doteq \$8000(1.061\ 836\ 5) \doteq \$8494.69.$$

Therefore, the difference in the amount of interest for 1 year is 5 cents.

Continuous compounding

EXAMPLE 4 What is the present value of a deposit invested at an interest rate of 7.5% compounded continuously for 1 year when the amount in the account at the end of the year is $4000?

Solution: In the equation $A_C = Pe^i$, the problem is to find P if $A_C = \$4000$ and $i = 0.075$.

$$\$4000 = Pe^{0.075},$$
$$(1.077\ 88)P = \$4000,$$
$$P = \$3710.97.$$

Hence, at 7.5% interest compounded continuously, your deposit must be $3710.97 in order for you to have $4000 at the end of 1 year.

PRACTICE PROBLEM Find the value of the deposit which, if the interest rate is 5% compounded continuously for one year grows to $2000.

Answer: $1902.46. ■

Annual interest rate

EXAMPLE 5 Suppose the interest earned in one year on a deposit of $4000 at a given interest rate compounded continuously is $298.62. What is the annual interest rate?

Solution: In the equation $A_C = Pe^i$, the problem is to find i if $A_C = \$4298.62$ and $P = \$4000$.

$$\$4298.62 = \$4000e^i,$$

$$e^i = 1.074\ 655.$$

From Table 3, Appendix A (or by taking the logarithm of each side in order to solve the exponential equation), we find that

$$i = 0.072; \quad \text{that is,} \quad i = 7.2\%.$$

In order to obtain the formula for the future value of a principal P deposited at an interest rate of i percent compounded continuously for n years, we use the fact that as q increases,

$$P\left(1 + \frac{i}{q}\right)^{qn} = P\left[\left(1 + \frac{i}{q}\right)^q\right]^n \quad \text{approaches} \quad P(e^i)^n.$$

We conclude that the future value after n years of continuous compounding at i percent is given by

$$Pe^{in}.$$

Continuous compounding

EXAMPLE 6 Find the amount resulting from a principal of $5000 deposited for 10 years at 6% compounded continuously.

Solution
$$Pe^{in} = \$5000(e^{0.06})^{10}$$
$$= \$5000e^{0.6}$$
$$= \$5000(1.82212)$$
$$= \$9110.60.$$

PRACTICE PROBLEM Find the amount resulting from a principal of $4000 deposited for 4 years at 8% compounded continuously.

Answer: $5508.51. ∎

Before turning to the exercises, let us consider the amounts obtained from $100 deposited at an interest rate of 12% for 1 year compounded at different time intervals.

8.5 CONTINUOUS INTEREST

$100 at 12% Interest Rate

Compounded	Periods per year	Amount	Effective rate
Annually	1	$112.00	12%
Quarterly	4	$112.55	12.55%
Monthly	12	$112.68	12.68%
Daily	360	$112.74	12.747%
Continuously	~	$112.75	12.749%

EXERCISES

Set A–Effective Interest Rate

For the given interest rate in Exercises 1 through 5, find the effective interest rate with continuous compounding.

1. 5%   2. 9%   3. 10%   4. 14%   5. 15%

Set B–Problems on Continuous Compounding of Interest

Future value

6. Find the amount resulting from a principal of $2000 deposited for 4 years at 6% interest compounded continuously.

Future value

7. Find the amount resulting from a principal of $5000 deposited for 4 years at 7% interest compounded continuously.

Future value

8. Find the amount resulting from a principal of $5000 deposited for 4 years at 8% compounded continuously.

Difference in interest

9. How much difference is there between the interest on $5000 deposited at 6% interest compounded continuously for 10 years and at 6% interest compounded quarterly for 10 years?

Future value

10. At the birth of a child suppose $2000 were placed on deposit at 6% interest compounded continuously. Assuming no money is added or withdrawn from the account, how much money will be in the account when the person reaches age 60?

Present value

11. What is the present value of a deposit invested at 8% interest compounded continuously for 1 year when the amount in the account at the end of the year is $6000?

Present value

12. What is the present value of a deposit invested at 10% interest compounded continuously for 1 year when the amount in the account at the end of the year is $8000?

Present value

13. What is the present value of a deposit invested at 8% interest compounded continuously for 4 years if the amount in the account at the end of the 4 years is $10,000?

Present value

14. What is the present value of a deposit invested at 7.5% interest compounded continuously for 4 years if the amount in the account at the end of the 4-year period is $20,000?

Interest rate

15. Suppose the interest earned in one year on $2000 deposited at an interest rate compounded continuously is $188.34. What is the annual interest rate?

Interest rate

16. Suppose the interest earned in one year on $3000 deposited at an interest rate compounded continuously is $153.81. What is the annual interest rate?

Interest rate

17. Suppose the interest earned in 6 years on $4000 deposited at an interest rate compounded continuously is $1399.43. What is the annual interest rate?

Interest rate

18. Suppose the interest earned in 8 years on $6000 deposited at an interest rate compounded continuously is $4932.71. What is the annual interest rate?

8.6 INTEREST ON PERIODIC SAVINGS

In the preceding two sections our attention was focused on interest obtained from a given principal. Although such savings as money market certificates and certificates of deposit are of this nature, many types of savings are of a periodic nature. Let us look at some examples.

Quarterly savings

EXAMPLE 1 Suppose at the beginning of a given year a person deposits $300 at 6% interest compounded quarterly. Suppose further that at the beginning of each quarter thereafter another $300 is deposited and that the interest is left on deposit. How much would be on deposit at the end of 2 years? (*Note:* We do not assume that there is a final deposit at the end of 2 years. More about this later.)

Solution: The first deposit of $300 will be on deposit for 8 quarters (2 years) and the amount A_8 of this deposit at that time will be

$$A_8 = \$300\left(1 + \frac{0.06}{4}\right)^8 = \$300(1.015)^8.$$

The second deposit of $300 will be on deposit for 7 quarters and the amount A_7 for this deposit will be

$$A_7 = \$300(1.015)^7.$$

8.6 INTEREST ON PERIODIC SAVINGS

Continuing, we see that the last deposit of $300 will draw interest for one quarter, and its amount A_1 will be

$$A_1 = \$300(1.015).$$

The total amount A of all the deposits is

$$A = A_1 + A_2 + A_3 + \cdots + A_7 + A_8;$$

that is,

$$A = \$300(1.015) + \$300(1.015)^2 + \cdots + \$300(1.015)^8.$$

This is the sum of a geometric sequence with $a_1 = 300(1.015)$, $r = 1.015$, and $n = 8$. Using the formula for the sum of a geometric sequence, we obtain

$$A = 300(1.015)\frac{1.015^8 - 1}{1.015 - 1}$$

$$= \frac{300(1.015)}{0.015}(1.015^8 - 1)$$

$$= (20,300)(0.12649)$$

$$- \$2567.80.$$

Note: Since the total deposits amount to 8($300) = $2400, the interest accrued is $167.80.

In general, let i be the annual interest rate, q be the number of deposits per year, and m be the total number of deposits. It should be clear from examining Example 1 that the **amount** A at the end of m periods resulting from m regular deposits of D dollars at the *beginning of each period* where interest is compounded at $j = i/q$ percent per period is given by

Periodic deposits

$$A = \frac{D}{j}[(1 + j)^{m+1} - (1 + j)].$$

Monthly savings

EXAMPLE 2 Suppose deposits of $100 drawing 6% interest compounded monthly are made on the first of each month. What is the amount on deposit at the end of 2 years?

Solution: In this case, $D = \$100$, $j = 0.06/12 = 0.005$, and $m = 2(12) = 24$. Therefore,

$$A = \frac{\$100}{0.005}[(1.005)^{25} - 1.005]$$

$$= \$20,000[0.127\ 795\ 6]$$

$$= \$2555.91$$

PRACTICE PROBLEM Suppose deposits of $50 earning 6% interest compounded monthly are made on the first of each month. (a) What is the amount on deposit at the end of 4 years? (b) How much interest was earned?

Answer: (a) $2718.41. (b) $318.41. ∎

Suppose a person can obtain 6% interest compounded quarterly on periodic savings for an extended period of time. How much should the person deposit at the beginning of each quarter over a 10-year period in order to save $30,000? In this case, $A = \$30,000$, $j = 0.06/4 = 0.015$, $m = 40$, and D is to be determined.

$$\$30,000 = \frac{D}{0.015}[(1.015)^{41} - 1.015],$$

$$\$450 = D(0.826\ 228\ 7),$$

$$D = \$544.64$$

Therefore $544.64 should be deposited at the beginning of each quarter.

Let us return to Example 1. Suppose that for some reason the person making the $300 quarterly deposits had to miss the third deposit. What would be the total savings under these circumstances? To answer the question, we can still use the general formula so long as we realize that the solution to Example 1 will be decreased by just A_6, the amount and interest resulting from the

8.6 INTEREST ON PERIODIC SAVINGS

third deposit. Therefore, the total amount at the end of 2 years would be

$$\$2567.80 - A_6 = \$2567.80 - \$300(1.015)^6$$
$$= \$2567.80 - \$328.03$$
$$= \$2239.77$$

instead of the $2567.80 obtained in Example 1.

annuity

simple annuity

annuity due
ordinary annuity

A periodic savings plan is an example of what is called an *annuity*. An **annuity** involves equal payments (deposits) made at equal intervals. Apart from equal deposits at regular intervals, installment loan payments on automobiles, house payments, life insurance premiums, etc., are examples of annuities. An annuity is called **simple** if the conversion period for interest coincides with the payment periods. (We consider only simple annuities.) A simple annuity where the payments are made at the *beginning of each period* is called an **annuity due.** If the payments are made at the *end of each period*, it is called an **ordinary annuity.** The savings problems we have been considering are examples of annuity due problems; life insurance premiums are often in the same category. Car payments are usually ordinary annuities.

Let us reconsider Example 1 but as an ordinary annuity. That is, suppose $300 is deposited for 8 quarters *at the end of each quarter* where $j = 0.06/4 = 0.015$ is the interest rate per quarter. The total amount $\overline{A}$ at the end of 8 quarters from the deposits would be

$$\overline{A} = \$300(1.015)^7 + \$300(1.015)^6 + \cdots + \$300(1.015) + \$300$$
$$= \$300[1 + 1.015 + (1.015)^2 + \cdots + (1.015)^6 + (1.015)^7].$$

Observe that the last deposit would earn no interest and the first deposit would earn interest for only 7 quarters. Using the formula for the sum of a geometric sequence, we get

$$\overline{A} = \$300 \frac{(1.015)^8 - 1}{1.015 - 1}$$
$$= \frac{\$300}{0.015}[(1.015)^8 - 1]$$
$$= \$20{,}000(0.126\ 492\ 6)$$
$$= \$2529.85.$$

Note: The annuity due yielded $2567.80.

CHAPTER 8 SEQUENCES AND MATHEMATICS OF FINANCE

The general formula for determining the amount for an **ordinary annuity** is

ordinary annuity

$$\overline{A} = \frac{D}{j}[(1 + j)^m - 1],$$

where j is the interest rate per period, m is the number of periods, and D is the amount of each of the equal deposits (payments).

Future value of an annuity

EXAMPLE 3 Find the future value of an annuity where $1000 is invested at the end of each quarter for 5 years if the interest rate is 8% compounded quarterly.

Solution: This is an ordinary annuity where $D = \$1000$, $j = 0.08/4 = 0.02$, and $m = 5(4) = 20$. Hence,

$$\overline{A} = \frac{\$1000}{0.02}[(1.02)^{20} - 1]$$

$$= \$50,000(0.485\ 947)$$

$$= \$24,297.35$$

The following is a table listing the amount of $1.00 in m periods of an ordinary annuity with j as interest rate per period.

Amount of $1.00 for an Ordinary Annuity

Periods	2%	5%	6%
1	1.000 000	1.000 000	1.000 000
2	2.020 000	2.050 000	2.060 000
3	3.060 400	3.152 500	3.183 600
4	4.121 608	4.310 125	4.373 616
5	5.204 040	5.525 631	5.637 093
6	6.308 121	6.801 913	6.975 319
7	7.434 283	8.142 009	8.393 838
8	8.582 969	9.549 108	9.897 468
9	9.754 628	11.026 564	11.491 316
10	10.949 721	12.577 893	13.180 795

8.6 INTEREST ON PERIODIC SAVINGS

Ordinary annuity

EXAMPLE 4 Verify the entries in the preceding table for $m = 8$.

Solution

(a) If $j = 0.02$ and $m = 8$, then

$$\bar{A} = \frac{\$1.00}{0.02}[(1.02)^8 - 1]$$
$$= 50(0.171\ 659\ 38)$$
$$= 8.582\ 969.$$

(b) If $j = 0.05$ and $m = 8$, then

$$\bar{A} = \frac{\$1.00}{0.05}[(1.05)^8 - 1]$$
$$= 20(0.477\ 455\ 44)$$
$$= 9.549\ 108.$$

(c) If $j = 0.06$ and $m = 8$, then

$$\bar{A} = \frac{\$1.00}{0.06}[(1.06)^8 - 1]$$
$$= \frac{50}{3}(0.593\ 848\ 1)$$
$$= 9.897\ 468.$$

EXERCISES

Set A—Periodic Savings Problems

In Exercises 1 through 14, we assume that all deposits are made at the beginning of the interest period.

Total savings
1. If $100 a month is deposited at 6% compounded monthly, how much will be on deposit after 5 years?

Total savings
2. If $100 a month is deposited at 12% compounded monthly, how much will be on deposit after 5 years?

Total savings
3. If $150 is deposited every 3 months at 8% compounded quarterly, how much will be on deposit after 10 years?

Total savings

4. If $200 is deposited every 3 months at 12% compounded quarterly, how much will be on deposit after 10 years?

Total savings

5. If $200 is deposited every month at 9% compounded monthly, how much will be on deposit after 5 years?

Total savings

6. If $100 is deposited every month at 9% compounded monthly, how much will be on deposit after 5 years?

Omitted deposit

7. In Example 2, what would the final amount be if the third deposit had been missed?

Total savings

8. Suppose $50 a month is deposited each month at 8% compounded monthly. What will be the amount on deposit after 40 years?

Monthly deposit

9. How much would a person need to deposit each month at 6% compounded monthly for 2 years in order to save $800 for the purchase of a television set?

Monthly deposit

10. How much would a person need to save each month at 6% compounded monthly in order to save $4000 in 4 years?

Monthly deposit

11. How much would a person need to save each month at 6% compounded monthly in order to save $200,000 in 40 years?

Monthly deposit

12. How much would a person need to save each month at 6% compounded monthly in order to save $100,000 in 30 years?

Savings and interest

13. At the beginning of each month, $200 is deposited in a savings account paying interest of 6% compounded monthly.
    (a) How much will be in the account after 5 years?
    (b) How many months will it take to save $24,000?
    (c) How much of the $24,000 is interest?

Savings and interest

14. At the beginning of each month $300 is deposited in a savings account paying interest of 6% compounded monthly.
    (a) How much will be in the account after 5 years?
    (b) How many months will it take to save $30,000?
    (c) How much of the $30,000 is interest?

15. Suppose an individual 30 years old can invest money at 12% compounded monthly for an indefinite period of time. How much must he/she invest each month to be a millionaire by the age of 60?

Set B—Ordinary Annuity and Annuity Due Problems

16. Find the amount for each of the indicated annuities:
    (a) $1000 invested annually at the end of the year at 6% compounded annually for 4 years.
    (b) $1000 invested annually at the beginning of the year at 6% compounded annually for 4 years.

17. Find the amount for each of the indicated annuities:
    (a) $2000 invested annually at the end of the year at 8% compounded annually for 4 years.
    (b) $2000 invested annually at the beginning of the year at 8% compounded annually for 4 years.
18. Determine the amount at the end of 6 years of an annuity of $4000 per year at 6% compounded annually provided it is
    (a) an ordinary annuity.
    (b) an annuity due.
19. Determine the amount at the end of 10 years of an annuity of $2000 per year at 5% compounded annually provided it is
    (a) an ordinary annuity.
    (b) an annuity due.
20. Verify the entries in the ordinary annuity table for $m = 4$.
21. Verify the entries in the ordinary annuity table for $m = 20$.

8.7 INSTALLMENT PAYMENTS AND RETIREMENT INCOME

Let us now investigate another type of savings. Suppose two parents want to set up a trust fund for their child. If they are able to obtain 6% interest compounded quarterly and want the child to receive $1500 quarterly for the next 20 years, how much is needed to set up the fund? The problem here is to find what is called the **present value** *of the given annuity.*

Let us keep this example in mind while deriving the general formula where V is the present value of the annuity, $j = 0.06/4 = 0.015$ is the interest rate for each of the $m = (4)(20) = 80$ periods, and $P = \$1500$ is the periodic payment made to the child. At the end of the first period before the first payment P is withdrawn, the amount on deposit is the present value V (amount deposited to set up the fund) plus Vj, the interest accumulated on V during the first period; that is, it is

$$V + Vj = V(1 + j).$$

After withdrawal of the first payment P, the amount on deposit *at the end of the first period is*

$$V(1 + j) - P.$$

Now, this amount will earn interest during the next period.

Hence, the amount on deposit at the end of the second period before P is withdrawn is

$$[V(1 + j) - P](1 + j) = V(1 + j)^2 - P(1 + j).$$

Therefore, the amount on deposit *at the end of the second period* (after withdrawal of P) *is*

$$V(1 + j)^2 - P(1 + j) - P.$$

Similarly, we find that the amount on deposit after *three periods* is

$$V(1 + j)^3 - P(1 + j)^2 - P(1 + j) - P = \\ V(1 + j)^3 - P[(1 + j)^2 + (1 + j) + 1].$$

Finally, after m periods it is

$$V(1 + j)^m - P[1 + (1 + j) + (1 + j)^2 + \cdots + (1 + j)^{m-1}].$$

Using the formula for the sum of a geometric sequence, we find that the amount after m periods is given by

$$V(1 + j)^m - P\frac{(1 + j)^m - 1}{j}.$$

Since the withdrawals are to exhaust the funds,

$$V(1 + j)^m - P\frac{(1 + j)^m - 1}{j} = 0$$

and

$$V(1 + j)^m = \frac{P}{j}[(1 + j)^m - 1].$$

By dividing each side of the preceding equation by $(1 + j)^m$, we obtain the following formula for the **present value** V of an annuity for m periodic payments P at the end of each period where the interest rate is j percent for each period. The formula is

present value

$$V = \frac{P}{j}[1 - (1 + j)^{-m}].$$

8.7 INSTALLMENT PAYMENTS AND RETIREMENT INCOME

In our original problem, $P = \$1500$, $j = 0.015$, and $m = 80$. Therefore,

$$V = \frac{\$1500}{0.015}[1 - (1.015)^{-80}]$$

$$= \$100{,}000(0.696\ 109\ 85)$$

$$= \$69{,}610.99.$$

It is worthwhile to observe that the child will eventually receive

$$(\$1500)(80) = \$120{,}000$$

from the approximately $69,611 set aside in trust.

30-year annuity

EXAMPLE 1 At the age of 40 a person inherits $150,000 (after taxes) and decides to buy a 30-year annuity earning 8% annually.

(a) How much can the person expect to receive each year from annual payments?

(b) What is the total amount the person will receive during the 30-year period?

Solution

(a) The present value V is known; it is $150,000. The problem is to find P where $j = 0.08$ and $m = 30$. Thus

$$\$150{,}000 = \frac{P}{0.08}[1 - (1.08)^{-30}],$$

$$\$12{,}000 = P(0.900\ 622\ 67),$$

$$P = \$13{,}324.12.$$

(b) The total amount received is $(30)(\$13{,}324.12) = \$399{,}723.60$. (Note that the total amount received is about $2\frac{2}{3}$ times the cost of the annuity.)

PRACTICE PROBLEM A person inherits $100,000 (after taxes) and decides to buy a 20-year annuity earning 6% annually. How much can the person expect to receive each year from annual payments?

Answer: $8718.45. ∎

Suppose a person borrows $4000 to buy a car; the annual interest rate is 12%. How much should the equal monthly payments be if the loan is to be paid off in 36 months? Essentially, this is an annuity problem from the standpoint of the *lender* where the present value V is $4000. Since $j = 0.012/12 = 0.01$ and $m = 36$,

$$\$4000 = \frac{P}{0.01}[1 - (1.01)^{-36}],$$

$$\$40 = P(0.301\ 075),$$

$$P = \$132.86.$$

Note that the total amount paid is $(36)(132.86) = \$4782.96$; the amount of interest paid is $\$4782.96 - \$4000 = \$782.96$.

PRACTICE PROBLEM A person borrows $3000 to buy a car; the annual interest rate is 9%. How much should the equal monthly payments be if the loan is to be paid off in 24 months?

Answer: $137.06. ∎

Mortgage payments

EXAMPLE 2 The figures given in Section 8.1 from a newspaper clipping indicated that a 30-year mortgage of $50,000 at 15% required monthly payments of $623.23. Verify this amount.

Solution: Essentially, $V = \$50,000$ is the present value of an annuity on the part of the lender. The problem is to find P where $j = 0.15/12 = 0.0125$ and $m = 360$. Hence

$$\$50,000 = \frac{P}{0.0125}[1 - (1.0125)^{-360}],$$

$$\$625 = P(0.988\ 576\ 78),$$

$$P = \$632.22^{+}.$$

The lender will "round up" to $623.23.

Savings and retirement payments

EXAMPLE 3 A couple decides at age 20 that they would like to be assured of an income of $4000 every 3 months for 15 years after they reach the age of 55. If they are able to arrange for an 8% annual interest rate compounded quarterly, how much per quarter do they need to save to meet their goal?

8.7 INSTALLMENT PAYMENTS AND RETIREMENT INCOME

Solution: First we need to find the present value (at age 55) of their annuity where $P = \$4000$, $j = 0.08/4 = 0.02$, and $m = (4)(15) = 60$.

$$V = \frac{\$4000}{0.02}[1 - (1.02)^{-60}]$$

$$= \$200,000(0.695\ 217\ 7)$$

$$= \$139,043.54.$$

Now we resort to the savings plan formula using $A = \$139,043.54$, $j = 0.02$, and $m = (3)(35) = 140$ to find out how much they need to deposit at the beginning of each quarter for 35 years in order to obtain $139,043.54. Since

$$A = \frac{D}{j}[(1 + j)^{n+1} - (1 + j)],$$

it follows that

$$\$139,043.54 = \frac{D}{0.02}[(1.02)^{141} - 1.02],$$

$$\$2780.87 = D(15.296\ 395),$$

$$D = \$181.80.$$

They would need to save $181.80 every three months to meet their goal.

---

add-on interest

Suppose a person borrows $6000 to buy a car and plans to pay off the loan in 48 equal monthly payments. The lender states that there is an **add-on interest rate** of 8.5% per year. The payments would be figured as follows. The interest for 1 year is

$$0.085 \times 6000 = \$510.$$

Therefore for 4 years (48 months) the interest is

$$4 \times \$510 = \$2040.$$

Since the loan plus the interest is $6000 + $2040 = $8040, the monthly payments are

$$\frac{\$8040}{48} = \$167.60.$$

APR

Federal law requires that the **annual percentage rate,** denoted by APR, must be disclosed. Although solving the equation necessary to find the APR for the 4-year 8.5% annual add-on interest rate is not easy algebraically, it is easy to verify that it is 15.17%. To do this, note that from the lender's viewpoint $6000 is the present value of an annuity where $m = 48$, $j = 0.1517/12 = 0.012\,641\,6\overline{6}$, and P is the monthly payments. Using the formula

$$V = \frac{P}{j}[1 - (1 + j)^{-m}],$$

we find that

$$\$6000(0.012\,641\,667) = P(0.452\,830\,4),$$
$$P = \$167.50.$$

Since $167.50 was the computed monthly payments, the APR is 15.17%.

Add-on interest and APR

EXAMPLE 4

(a) Suppose a $1200 loan is obtained for 1 year with a 6% add-on interest rate. What would be the monthly payments?

(b) Verify that the APR for this loan is approximately 10.9%.

Solution

(a) Since $0.06 \times \$1200 = \$72$ is the interest, the monthly payments are $(1200 + \$72)/12 = \$106$.

(b) We use the present-value formula for an annuity with $V = \$1200$, $j = 0.109/12 = 0.009\,083\,3\overline{3}$, and $m = 12$:

$$\$1200 = \frac{P}{j}[1 - (1 + j)^{-12}],$$
$$\$10.90 = P(0.102\,828),$$
$$P = \$106.$$

Before turning to the exercises, consider the following problem. Suppose you could deposit $100 at the beginning of each month for 20 years at an interest rate of 6% compounded monthly. For how many years thereafter could you withdraw $300 a month from the account? (Make a guess.)

8.7 INSTALLMENT PAYMENTS AND RETIREMENT INCOME

First we need to find the amount on deposit after 20 years of making monthly deposits where $D = \$100$, $j = 0.06/12 = 0.005$, and $m = (12)(20) = 240$.

$$A = \frac{\$100}{0.005}[(1.005)^{241} - 1.005]$$

$$= \$20,000(2.321\ 755\ 5)$$

$$= \$46,435.20.$$

Now, $46,435.20 should be considered as the present value V of an annuity where $P = \$300$, $j = 0.005$, and m is unknown. From

$$V = \frac{P}{j}[1 - (1 + j)^{-m}],$$

we get

$$\$46,435.20 = \frac{\$300}{0.005}[1 - (1.005)^{-m}],$$

$$0.077\ 392 = 1 - (1.005)^{-m},$$

$$-0.226\ 08 = -(1.005)^{-m}$$

$$0.226\ 08 = (1.005)^{-m}$$

$$= \frac{1}{(1.005)^m},$$

$$(1.005)^m = \frac{1}{0.226\ 08}$$

$$= 4.423\ 213,$$

$$m[\log(1.005)] = \log(4.423\ 213)$$

$$(0.002\ 166)m = 0.645\ 737\ 85$$

$$m = 298^+.$$

Therefore, you could get $300 a month for 298 months, that is, for 24 years and 10 months. (Note that you can withdraw $300 a month for 298 months as a result of only 240 monthly deposits of $100.)

EXERCISES

Set A–Installment Payments and Retirement Income Problems

Trust fund

1. Two parents want to set up a trust fund for their child. They can obtain 8% interest compounded quarterly and want the

child to receive $300 quarterly for the next 10 years. How much is needed to set up the fund?

Trust fund

2. Two parents want to set up a trust fund for their child. They can obtain 8% interest compounded quarterly and want the child to receive $1000 quarterly for the next 10 years. How much is needed to set up the fund?

20-year annuity

3. Upon graduation, a college graduate is given a 20-year annuity at a cost of $20,000 paying 8% interest compounded annually.
   (a) How much will she receive each year?
   (b) What is the total amount that she will receive?

20-year annuity

4. Upon graduation, a college graduate is given a 20-year annuity at a cost of $30,000 paying 6% compounded annually.
   (a) How much will he receive each year?
   (b) What is the total amount he will receive?

Car payments

5. Suppose a person borrows $5000 to buy a car. The interest rate is 1% per month. How much should the equal monthly payments be if the loan is to be paid off in 36 months?

Car payments

6. Suppose a person borrows $6000 to buy a car. The interest rate is $1\frac{1}{2}$% per month. How much should the equal monthly payments be if the loan is to be paid off in 36 months?

Mortgage payments

In Exercises 7 through 10, verify the monthly payment given in Section 8.1 for the 30-year mortgage of $50,000 at:

7. 12%   8. 8%   9. 10%   10. 18%

Add-on interest

11. Suppose a person borrows $4000 for 3 years with an annual add-on interest rate of 9%. What are the monthly payments?

Add-on interest

12. Suppose a person borrows $4000 for 4 years with an annual add-on interest rate of 9%. What are the monthly payments?

APR

13. A 3-year $4000 loan is secured from a bank. The monthly payments to repay the loan are $141.11. Verify that the APR is 16.24%. (See Exercise 11.)

APR

14. A 4-year $4000 loan is secured from a bank. The monthly payments to repay the loan are $113.33. Verify that the APR is 15.98%. (See Exercise 12.)

Savings and retirement payments

15. Suppose a person were to deposit $200 a month for 20 years at 6% compounded monthly. For how many years thereafter could the person withdraw $1000 a month from the account?

Savings and retirement payments

16. Suppose a person were to deposit $100 a month for 30 years at 8% compounded monthly. For how many years thereafter could the person withdraw $500 a month from the account?

Savings and retirement payments

17. A couple decide that in 30 years they would like to receive an income of $3000 every three months for 10 years. If they can

obtain 6% compounded quarterly, how much per quarter must they save to meet their goal?

Savings and retirement payments

18. A couple decide that in 30 years they would like to receive an income of $5000 every three months for 10 years. If they can obtain 8% compounded quarterly, how much per quarter do they need to save to meet their goal?

8.8 SUMMARY AND REVIEW EXERCISES

1. **ARITHMETIC SEQUENCES**

    (a) If a_1 is the first term of an arithmetic sequence then $a_1 + d$ is the second term where d is the **common difference.** The nth term is
    $$a_n = a_1 + (n-1)d.$$

    (b) The **sum** s_n of the first n terms of an arithmetic sequence is given by
    $$s_n = \frac{n}{2}[2a_1 + (n-1)d] = \frac{n}{2}(a_1 + a_n).$$

2. **GEOMETRIC SEQUENCES**

    (a) If a is the first term of a geometric sequence then ar is the second term where r is the **common ratio.** The nth term is
    $$a_n = ar^{n-1}.$$

    (b) The **sum** s_n of the first n terms of a geometric sequence is given by
    $$s_n = \frac{ar^n - a}{r - 1} = \frac{a - ar^n}{1 - r} = a\left(\frac{1 - r^n}{1 - r}\right)$$

    provided $r \neq 1$.

3. **COMPOUND INTEREST**

    (a) If i is the annual interest rate, q is the number of compoundings per year for n years, then $j = i/q$ is the interest rate per period, $m = nq$ is the number of periods, and the **amount** A_m (future value) of an initial principal P after m periods is given by
    $$A_m = P(1 + j)^m.$$

(b) If i is the annual interest rate and q is the number of compoundings per year, then the **effective interest rate** E_q is given by

$$E_q = \left[\left(1 + \frac{i}{q}\right)^q - 1\right] \times 100.$$

(c) If i is the annual interest rate and P is the initial principal, then the amount in n years resulting from **continuous compounding** is

$$A_C = Pe^{in}.$$

(d) The **effective interest rate** for an annual interest rate of i percent compounded continuously is given by

$$E_C = (e^i - 1)(100).$$

4. INTEREST ON PERIODIC SAVINGS

(a) The **amount** A at the end of m periods resulting from m regular deposits of D dollars *at the beginning of each period* at $j = i/q$ percent interest rate per period is given by

$$A = \frac{D}{j}[(1 + j)^{m+1} - (1 + j)].$$

(b) The **amount** $\bar{A}$ at the end of m periods resulting from m regular deposits of D dollars *at the end of each period* at $j = i/q$ percent interest rate per period is given by

$$\bar{A} = \frac{D}{j}[(1 + j)^m - 1].$$

5. INSTALLMENT PAYMENTS AND RETIREMENT INCOME

(a) The **present value** V of an annuity for m periodic payments P at the end of each period where the interest rate is j percent for each period is given by

$$V = \frac{P}{j}[1 - (1 + j)^{-m}].$$

6. REVIEW EXERCISES

For each arithmetic sequence in Exercises 1 through 4, do each of the following.
(a) Find the common difference. (b) List the next 3 terms.

(c) Find the sum of the first 30 terms of the sequence.

1. 40, 32, 24, ...
2. 1.5, 2.7, 3.9, ...
3. $\frac{2}{3}, \frac{7}{6}, \frac{5}{3}, \ldots$
4. $-20, -17, -14, \ldots$

For each geometric sequence in Exercises 5 and 6, do each of the following.
(a) Find the common ratio. (b) List the next 3 terms.

5. 64, −16, 4, ...
6. 9, 6, 4, ...

For each geometric sequence in Exercises 7 and 8, do each of the following.
(a) Find the common ratio. (b) Find x and y.

7. 2.5, x, y, 6.86, ...
8. 5, x, y, 8.64, ...

9. The sum of an arithmetic sequence with 20 terms is 130 and the 20th term is 54.
   (a) Find the first term.
   (b) What is the common difference?

10. The sum of an arithmetic sequence with 30 terms is 180 and the 30th term is 62.
    (a) Find the first term.
    (b) What is the common difference?

11. Suppose that a person obtains a job with a starting salary of $18,000 and receives a $1200 raise each year thereafter.
    (a) How much would the person's salary be during the 10th year?
    (b) What would the person's total gross income be for the 10-year period?

12. Suppose that a person obtains a job with a starting salary of $18,000 and that he receives a 6% raise each year thereafter.
    (a) How much would the person's salary be during the 10th year?
    (b) What would the person's total gross income be for the 10-year period?

13. What is the total interest from a fixed deposit of $300 for 4 years earning an annual interest rate of 12% compounded quarterly?

14. What is the present value of $8000 if the interest rate is 8% compounded quarterly for 4 years?

15. How much should an investor put in a 4-year savings certificate earning 8% interest compounded quarterly in order to have $10,000 on deposit at maturity of the certificate?

16. What is the total interest from a fixed deposit of $1500 for 1 year at 6% interest compounded continuously?

17. Suppose $155.76 is the interest earned in 1 year from a $2000 deposit with interest compounded continuously. What is the annual interest rate?

18. Suppose $100 is deposited each month and earns 8% compounded monthly. What will be the amount on deposit after 20 years?

19. How much would a person need to save each month earning 9% compounded monthly in order to save $100,000 in 20 years?

20. What should be the monthly payment for a 30-year home mortgage of $60,000 at 12% annual interest?

John von Neumann

The Bettmann Archive

John von Neumann, a twentieth-century mathematical genius who exhibited truly exceptional mathematical creativity and versatility, was born at Budapest, Hungary, on December 28, 1903. He received a doctorate in mathematics from the University of Budapest at the age of 23. From 1926 until 1929 he taught at the University of Hamburg. In 1930 he came to the United States with a professorship in mathematical physics at Princeton University. In 1933 when the Institute for Advanced Study at Princeton was founded, he became a professor of mathematics at this important research institution and remained there until his death.

He directed the development of the first electronic computers at the Institute for Advanced Study. He created important new avenues of investigation in the field of mathematical economics and was one of the founders of game theory. He made important and basic contributions in developing such diverse fields as operator theory, quantum theory, atomic energy, and long-range weather forecasting. Von Neumann was both a member of the Atomic Energy Commission and a consultant on the Atomic Bomb Project at Los Alamos. In addition, he was also a consultant to both the Army and Navy during World War II. In 1956 the Atomic Energy Commission honored him with the $50,000 Enrico Fermi award for his relevant basic scientific contributions. Interestingly von Neumann and Fermi, along with J. Robert Oppenheimer, had all been students of David Hilbert. Also, like some of his great mathematical predecessors, von Neumann had a reputation for being able to perform phenomenal computations and solve complicated problems mentally.

Besides his significant contributions to applied mathematics, von Neumann also left his mark on pure mathematics. Some of his research efforts were devoted to group theory, set theory, mathematical logic, and the foundations of mathematics. It was von Neumann who first gave the name to, and a formal axiomatic description of, what today is called Hilbert space.

APPENDIX

A

TABLES

TABLE 1 Compound Interest $(1 + i)^n$

n	1%	1½%	2%	2½%	3%	4%	5%	6%
1	1.0100	1.0150	1.0200	1.0250	1.0300	1.0400	1.0500	1.0600
2	1.0201	1.0302	1.0404	1.0506	1.0609	1.0816	1.1025	1.1236
3	1.0303	1.0457	1.0612	1.0769	1.0927	1.1249	1.1576	1.1910
4	1.0406	1.0614	1.0824	1.1038	1.1255	1.1699	1.2155	1.2625
5	1.0510	1.0773	1.1041	1.1314	1.1593	1.2167	1.2763	1.3382
6	1.0615	1.0934	1.1262	1.1597	1.1941	1.2653	1.3401	1.4185
7	1.0721	1.1098	1.1487	1.1887	1.2299	1.3159	1.4071	1.5036
8	1.0829	1.1265	1.1717	1.2184	1.2668	1.3686	1.4775	1.5938
9	1.0937	1.1434	1.1951	1.2489	1.3048	1.4233	1.5513	1.6895
10	1.1046	1.1605	1.2190	1.2801	1.3439	1.4802	1.6289	1.7908
11	1.1157	1.1779	1.2434	1.3121	1.3842	1.5395	1.7103	1.8983
12	1.1268	1.1956	1.2682	1.3449	1.4258	1.6010	1.7959	2.0122
13	1.1381	1.2136	1.2936	1.3785	1.4685	1.6651	1.8856	2.1329
14	1.1495	1.2318	1.3195	1.4130	1.5126	1.7317	1.9799	2.2609
15	1.1610	1.2502	1.3459	1.4483	1.5580	1.8009	2.0789	2.3966
16	1.1726	1.2690	1.3728	1.4845	1.6047	1.8730	2.1829	2.5404
17	1.1843	1.2880	1.4002	1.5216	1.6528	1.9479	2.2920	2.6928
18	1.1961	1.3073	1.4282	1.5597	1.7024	2.0258	2.4066	2.8543
19	1.2081	1.3270	1.4568	1.5987	1.7535	2.1068	2.5270	3.0256
20	1.2202	1.3469	1.4859	1.6386	1.8061	2.1911	2.6533	3.2071
21	1.2324	1.3671	1.5157	1.6796	1.8603	2.2788	2.7860	3.3996
22	1.2447	1.3876	1.5460	1.7216	1.9161	2.3699	2.9253	3.6035
23	1.2572	1.4084	1.5769	1.7646	1.9736	2.4647	3.0715	3.8197
24	1.2697	1.4295	1.6084	1.8087	2.0328	2.5633	3.2251	4.0489
25	1.2824	1.4509	1.6406	1.8539	2.0938	2.6658	3.3864	4.2919
26	1.2953	1.4727	1.6734	1.9003	2.1566	2.7725	3.5557	4.5494
27	1.3082	1.4948	1.7069	1.9478	2.2213	2.8834	3.7335	4.8223
28	1.3213	1.5172	1.7410	1.9965	2.2879	2.9987	3.9201	5.1117
29	1.3345	1.5400	1.7758	2.0464	2.3566	3.1187	4.1161	5.4184
30	1.3478	1.5631	1.8114	2.0976	2.4273	3.2434	4.3219	5.7435
31	1.3613	1.5865	1.8476	2.1500	2.5001	3.3731	4.5380	6.0881
32	1.3749	1.6103	1.8845	2.2038	2.5751	3.5081	4.7649	6.4534
33	1.3887	1.6345	1.9222	2.2589	2.6523	3.6484	5.0032	6.8406
34	1.4026	1.6590	1.9607	2.3153	2.7319	3.7943	5.2533	7.2510
35	1.4166	1.6839	1.9999	2.3732	2.8139	3.9461	5.5160	7.6861
36	1.4308	1.7091	2.0399	2.4325	2.8983	4.1039	5.7918	8.1473
37	1.4451	1.7348	2.0807	2.4933	2.9852	4.2681	6.0814	8.6361
38	1.4595	1.7608	2.1223	2.5557	3.0748	4.4388	6.3855	9.1543
39	1.4741	1.7872	2.1647	2.6196	3.1670	4.6164	6.7048	9.7035
40	1.4889	1.8140	2.2080	2.6851	3.2620	4.8010	7.0400	10.2857

APPENDIX A

TABLE 1 Compound Interest $(1 + i)^n$ *(cont.)*

n	7%	7½%	8%	8½%	9%	10%	11%	12%
1	1.0700	1.0750	1.0800	1.0850	1.0900	1.1000	1.1100	1.1200
2	1.1449	1.1556	1.1664	1.1772	1.1881	1.2100	1.2321	1.2544
3	1.2250	1.2423	1.2597	1.2773	1.2950	1.3310	1.3676	1.4049
4	1.3108	1.3355	1.3605	1.3859	1.4116	1.4641	1.5181	1.5735
5	1.4026	1.4356	1.4693	1.5037	1.5386	1.6105	1.6851	1.7623
6	1.5007	1.5433	1.5869	1.6315	1.6771	1.7716	1.8704	1.9738
7	1.6058	1.6590	1.7138	1.7701	1.8280	1.9487	2.0762	2.2107
8	1.7182	1.7835	1.8509	1.9206	1.9926	2.1436	2.3045	2.4760
9	1.8385	1.9172	1.9990	2.0839	2.1719	2.3579	2.5580	2.7731
10	1.9672	2.0610	2.1589	2.2610	2.3674	2.5937	2.8394	3.1058
11	2.1049	2.2156	2.3316	2.4532	2.5804	2.8531	3.1518	3.4785
12	2.2522	2.3818	2.5182	2.6617	2.8127	3.1384	3.4985	3.8960
13	2.4098	2.5604	2.7196	2.8879	3.0658	3.4523	3.8833	4.3635
14	2.5785	2.7524	2.9372	3.1334	3.3417	3.7975	4.3104	4.8871
15	2.7590	2.9589	3.1722	3.3997	3.6425	4.1772	4.7846	5.4736
16	2.9522	3.1808	3.4259	3.6887	3.9703	4.5950	5.3109	6.1304
17	3.1588	3.4194	3.7000	4.0023	4.3276	5.0545	5.8951	6.8660
18	3.3799	3.6758	3.9960	4.3425	4.7171	5.5599	6.5436	7.6900
19	3.6165	3.9515	4.3157	4.7116	5.1417	6.1159	7.2633	8.6128
20	3.8697	4.2479	4.6610	5.1120	5.6044	6.7275	8.0623	9.6463
21	4.1406	4.5664	5.0338	5.5466	6.1088	7.4002	8.9492	10.8038
22	4.4304	4.9089	5.4365	6.0180	6.6586	8.1403	9.9336	12.1003
23	4.7405	5.2771	5.8715	6.5296	7.2579	8.9543	11.0263	13.5523
24	5.0724	5.6729	6.3412	7.0846	7.9111	9.8497	12.2392	15.1786
25	5.4274	6.0983	6.8485	7.6868	8.6231	10.8347	13.5855	17.0001
26	5.8074	6.5557	7.3964	8.3401	9.3992	11.9182	15.0799	19.0401
27	6.2139	7.0474	7.9881	9.0490	10.2451	13.1100	16.7386	21.3249
28	6.6488	7.5759	8.6271	9.8182	11.1671	14.4210	18.5799	23.8839
29	7.1143	8.1441	9.3173	10.6528	12.1722	15.8631	20.6237	26.7499
30	7.6123	8.7550	10.0627	11.5583	13.2677	17.4494	22.8923	29.9599
31	8.1451	9.4116	10.8677	12.5407	14.4618	19.1943	25.4104	33.5551
32	8.7153	10.1174	11.7371	13.6067	15.7633	21.1138	28.2056	37.5817
33	9.3253	10.8763	12.6760	14.7632	17.1820	23.2252	31.3082	42.0915
34	9.9781	11.6920	13.6901	16.0181	18.7284	25.5477	34.7521	47.1425
35	10.6766	12.5689	14.7853	17.3796	20.4140	28.1024	38.5749	52.7996
36	11.4239	13.5115	15.9682	18.8569	22.2512	30.9127	42.8181	59.1356
37	12.2236	14.5249	17.2456	20.4597	24.2538	34.0039	47.5281	66.2318
38	13.0793	15.6143	18.6253	22.1988	26.4367	37.4043	52.7562	74.1797
39	13.9948	16.7853	20.1153	24.0857	28.8160	41.1448	58.5593	83.0812
40	14.9745	18.0442	21.7245	26.1330	31.4094	45.2593	65.0009	93.0510

TABLE 2 Present Value of a Dollar $(1 + i)^{-n}$

n	1%	1½%	2%	2½%	3%	4%	5%	6%
1	.99010	.98522	.98039	.97561	.97087	.96154	.95238	.94340
2	.98030	.97066	.96117	.95181	.94260	.92456	.90703	.89000
3	.97059	.95632	.94232	.92860	.91514	.88900	.86384	.83962
4	.96098	.94218	.92385	.90595	.88849	.85480	.82270	.79209
5	.95147	.92826	.90573	.88385	.86261	.82193	.78353	.74726
6	.94205	.91454	.88797	.86230	.83748	.79031	.74622	.70496
7	.92372	.90103	.87056	.84127	.81309	.75992	.71068	.66506
8	.92348	.88771	.85349	.82075	.78941	.73069	.67684	.62741
9	.91434	.87459	.83676	.80073	.76642	.70259	.64461	.59190
10	.90529	.86167	.82035	.78120	.74409	.67556	.61391	.55839
11	.89632	.84893	.80426	.76214	.72242	.64958	.58468	.52679
12	.88745	.83639	.78849	.74356	.70138	.62460	.55684	.49697
13	.87866	.82403	.77303	.72542	.68095	.60057	.53032	.46884
14	.86996	.81185	.75788	.70773	.66112	.57748	.50507	.44230
15	.86135	.79985	.74301	.69047	.64186	.55526	.48102	.41727
16	.85282	.78803	.72845	.67362	.62317	.53391	.45811	.39365
17	.84438	.77639	.71416	.65720	.60502	.51337	.43630	.37136
18	.83602	.76491	.70016	.64117	.58739	.49363	.41552	.35034
19	.82774	.75361	.68643	.62553	.57029	.47464	.39573	.33051
20	.81954	.74247	.67297	.61027	.55368	.45639	.37689	.31180
21	.81143	.73150	.65978	.59539	.53755	.43883	.35894	.29416
22	.80340	.72069	.64684	.58086	.52189	.42196	.34185	.27751
23	.79544	.71004	.63416	.56670	.50669	.40573	.32557	.26180
24	.78757	.69954	.62172	.55288	.49193	.39012	.31007	.24698
25	.77977	.68921	.60953	.53939	.47761	.37512	.29530	.23300
26	.77205	.67902	.59758	.52623	.46369	.36069	.28124	.21981
27	.76440	.66899	.58586	.51340	.45019	.34682	.26785	.20737
28	.75684	.65910	.57437	.50088	.43708	.33348	.25509	.19563
29	.74934	.64936	.56311	.48866	.42435	.32065	.24295	.18456
30	.74192	.63976	.55207	.47674	.41199	.30832	.23138	.17411
31	.73458	.63031	.54125	.46511	.39999	.29646	.22036	.16425
32	.72730	.62099	.53063	.45377	.38834	.28506	.20987	.15496
33	.72010	.61182	.52023	.44270	.37703	.27409	.19987	.14619
34	.71297	.60277	.51003	.43191	.36604	.26355	.19035	.13791
35	.70591	.59387	.50003	.42137	.35538	.25342	.18129	.13011
36	.69892	.58509	.49022	.41109	.34503	.24367	.17266	.12274
37	.69200	.57644	.48061	.40107	.33498	.23430	.16444	.11579
38	.68515	.56792	.47119	.39128	.32523	.22529	.15661	.10924
39	.67837	.55953	.46195	.38174	.31575	.21662	.14915	.10306
40	.67165	.55126	.45289	.37243	.30656	.20829	.14205	.09722

APPENDIX A

TABLE 2 Present Value of a Dollar $(1 + i)^{-n}$ (cont.)

n	7%	7½%	8%	8½%	9%	10%	11%	12%
1	0.9346	0.9302	0.9259	0.9217	0.9174	0.9091	0.9009	0.8929
2	0.8734	0.8653	0.8573	0.8495	0.8417	0.8264	0.8116	0.7972
3	0.8163	0.8050	0.7938	0.7829	0.7722	0.7513	0.7312	0.7118
4	0.7629	0.7488	0.7350	0.7216	0.7084	0.6830	0.6587	0.6355
5	0.7130	0.6966	0.6806	0.6650	0.6499	0.6209	0.5935	0.5674
6	0.6663	0.6480	0.6302	0.6129	0.5963	0.5645	0.5346	0.5066
7	0.6227	0.6028	0.5835	0.5649	0.5470	0.5132	0.4817	0.4523
8	0.5820	0.5607	0.5403	0.5207	0.5019	0.4665	0.4339	0.4039
9	0.5439	0.5216	0.5002	0.4799	0.4604	0.4241	0.3909	0.3606
10	0.5083	0.4852	0.4632	0.4423	0.4224	0.3855	0.3522	0.3220
11	0.4751	0.4513	0.4289	0.4076	0.3875	0.3505	0.3173	0.2875
12	0.4440	0.4199	0.3971	0.3757	0.3555	0.3186	0.2858	0.2567
13	0.4150	0.3906	0.3677	0.3463	0.3262	0.2897	0.2575	0.2292
14	0.3878	0.3633	0.3405	0.3191	0.2992	0.2633	0.2320	0.2046
15	0.3624	0.3380	0.3152	0.2941	0.2745	0.2394	0.2090	0.1827
16	0.3387	0.3144	0.2919	0.2711	0.2519	0.2176	0.1883	0.1631
17	0.3166	0.2925	0.2703	0.2499	0.2311	0.1978	0.1696	0.1456
18	0.2959	0.2720	0.2502	0.2303	0.2120	0.1799	0.1528	0.1300
19	0.2765	0.2531	0.2317	0.2122	0.1945	0.1635	0.1377	0.1161
20	0.2584	0.2354	0.2145	0.1956	0.1784	0.1486	0.1240	0.1037
21	0.2415	0.2190	0.1987	0.1803	0.1637	0.1351	0.1117	0.0926
22	0.2257	0.2037	0.1839	0.1662	0.1502	0.1228	0.1007	0.0826
23	0.2109	0.1895	0.1703	0.1531	0.1378	0.1117	0.0907	0.0738
24	0.1971	0.1763	0.1577	0.1412	0.1264	0.1015	0.0817	0.0659
25	0.1842	0.1640	0.1460	0.1301	0.1160	0.0923	0.0736	0.0588
26	0.1722	0.1525	0.1352	0.1199	0.1064	0.0839	0.0663	0.0525
27	0.1609	0.1419	0.1252	0.1105	0.0976	0.0763	0.0597	0.0469
28	0.1504	0.1320	0.1159	0.1019	0.0895	0.0693	0.0538	0.0419
29	0.1406	0.1228	0.1073	0.0939	0.0822	0.0630	0.0485	0.0374
30	0.1314	0.1142	0.0994	0.0865	0.0754	0.0573	0.0437	0.0334
31	0.1228	0.1063	0.0920	0.0797	0.0691	0.0521	0.0394	0.0298
32	0.1147	0.0988	0.0852	0.0735	0.0634	0.0474	0.0355	0.0266
33	0.1072	0.0919	0.0789	0.0677	0.0582	0.0431	0.0319	0.0238
34	0.1002	0.0855	0.0730	0.0624	0.0534	0.0391	0.0288	0.0212
35	0.0937	0.0796	0.0676	0.0575	0.0490	0.0356	0.0259	0.0189
36	0.0875	0.0740	0.0626	0.0530	0.0449	0.0323	0.0234	0.0169
37	0.0818	0.0688	0.0580	0.0489	0.0412	0.0294	0.0210	0.0151
38	0.0765	0.0640	0.0537	0.0450	0.0378	0.0267	0.0190	0.0135
39	0.0715	0.0596	0.0497	0.0415	0.0347	0.0243	0.0171	0.0120
40	0.0668	0.0554	0.0460	0.0383	0.0318	0.0221	0.0154	0.0107

TABLE 3 Exponential Functions (e^x)

x	e^x	x	e^x	x	e^x	x	e^x
0.00	1.0000	**0.50**	1.6487	**1.00**	2.7183	**1.50**	4.4817
0.01	1.0101	0.51	1.6653	1.01	2.7456	1.51	4.5267
0.02	1.0202	0.52	1.6820	1.02	2.7732	1.52	4.5722
0.03	1.0305	0.53	1.6989	1.03	2.8011	1.53	4.6182
0.04	1.0408	0.54	1.7160	1.04	2.8292	1.54	4.6646
0.05	1.0513	**0.55**	1.7333	**1.05**	2.8577	**1.55**	4.7115
0.06	1.0618	0.56	1.7507	1.06	2.8864	1.56	4.7588
0.07	1.0725	0.57	1.7683	1.07	2.9154	1.57	4.8066
0.08	1.0833	0.58	1.7860	1.08	2.9447	1.58	4.8550
0.09	1.0942	0.59	1.8040	1.09	2.9743	1.59	4.9037
0.10	1.1052	**0.60**	1.8221	**1.10**	3.0042	**1.60**	4.9530
0.11	1.1163	0.61	1.8404	1.11	3.0344	1.61	5.0028
0.12	1.1275	0.62	1.8589	1.12	3.0649	1.62	5.0531
0.13	1.1388	0.63	1.8776	1.13	3.0957	1.63	5.1039
0.14	1.1503	0.64	1.8965	1.14	3.1268	1.64	5.1552
0.15	1.1618	**0.65**	1.9155	**1.15**	3.1582	**1.65**	5.2070
0.16	1.1735	0.66	1.9348	1.16	3.1899	1.66	5.2593
0.17	1.1853	0.67	1.9542	1.17	3.2220	1.67	5.3122
0.18	1.1972	0.68	1.9739	1.18	3.2544	1.68	5.3656
0.19	1.2092	0.69	1.9937	1.19	3.2871	1.69	5.4195
0.20	1.2214	**0.70**	2.0138	**1.20**	3.3201	**1.70**	5.4739
0.21	1.2337	0.71	2.0340	1.21	3.3535	1.71	5.5290
0.22	1.2461	0.72	2.0544	1.22	3.3872	1.72	5.5845
0.23	1.2586	0.73	2.0751	1.23	3.4212	1.73	5.6407
0.24	1.2712	0.74	2.0959	1.24	3.4556	1.74	5.6973
0.25	1.2840	**0.75**	2.1170	**1.25**	3.4903	**1.75**	5.7546
0.26	1.2969	0.76	2.1383	1.26	3.5254	1.76	5.8124
0.27	1.3100	0.77	2.1598	1.27	3.5609	1.77	5.8709
0.28	1.3231	0.78	2.1815	1.28	3.5966	1.78	5.9299
0.29	1.3364	0.79	2.2034	1.29	3.6328	1.79	5.9895
0.30	1.3499	**0.80**	2.2255	**1.30**	3.6693	**1.80**	6.0496
0.31	1.3634	0.81	2.2479	1.31	3.7062	1.81	6.1104
0.32	1.3771	0.82	2.2705	1.32	3.7434	1.82	6.1719
0.33	1.3910	0.83	2.2933	1.33	3.7810	1.83	6.2339
0.34	1.4049	0.84	2.3164	1.34	3.8190	1.84	6.2965
0.35	1.4191	**0.85**	2.3396	**1.35**	3.8574	**1.85**	6.3598
0.36	1.4333	0.86	2.3632	1.36	3.8962	1.86	6.4237
0.37	1.4477	0.87	2.3869	1.37	3.9354	1.87	6.4883
0.38	1.4623	0.88	2.4109	1.38	3.9749	1.88	6.5535
0.39	1.4770	0.89	2.4351	1.39	4.0149	1.89	6.6194
0.40	1.4918	**0.90**	2.4596	**1.40**	4.0552	**1.90**	6.6859
0.41	1.5068	0.91	2.4843	1.41	4.0960	1.91	6.7531
0.42	1.5220	0.92	2.5093	1.42	4.1371	1.92	6.8210
0.43	1.5373	0.93	2.5345	1.43	4.1787	1.93	6.8895
0.44	1.5527	0.94	2.5600	1.44	4.2207	1.94	6.9588
0.45	1.5683	**0.95**	2.5857	**1.45**	4.2631	**1.95**	7.0287
0.46	1.5841	0.96	2.6117	1.46	4.3060	1.96	7.0993
0.47	1.6000	0.97	2.6379	1.47	4.3492	1.97	7.1707
0.48	1.6161	0.98	2.6645	1.48	4.3929	1.98	7.2427
0.49	1.6323	0.99	2.6912	1.49	4.4371	1.99	7.3155
0.50	1.6487	**1.00**	2.7183	**1.50**	4.4817	**2.00**	7.3891

TABLE 3 Exponential Functions (e^x) (cont.)

x	e^x	x	e^x	x	e^x	x	e^x
2.00	7.3891	**2.50**	12.182	**3.00**	20.086	**3.50**	33.115
2.01	7.4633	2.51	12.305	3.01	20.287	3.51	33.448
2.02	7.5383	2.52	12.429	3.02	20.491	3.52	33.784
2.03	7.6141	2.53	12.554	3.03	20.697	3.53	34.124
2.04	7.6906	2.54	12.680	3.04	20.905	3.54	34.467
2.05	7.7679	**2.55**	12.807	**3.05**	21.115	**3.55**	34.813
2.06	7.8460	2.56	12.936	3.06	21.328	3.56	35.163
2.07	7.9248	2.57	13.066	3.07	21.542	3.57	35.517
2.08	8.0045	2.58	13.197	3.08	21.758	3.58	35.874
2.09	8.0849	2.59	13.330	3.09	21.977	3.59	36.234
2.10	8.1662	**2.60**	13.464	**3.10**	22.198	**3.60**	36.598
2.11	8.2482	2.61	13.599	3.11	22.421	3.61	36.966
2.12	8.3311	2.62	13.736	3.12	22.646	3.62	37.338
2.13	8.4149	2.63	13.874	3.13	22.874	3.63	37.713
2.14	8.4994	2.64	14.013	3.14	23.104	3.64	38.092
2.15	8.5849	**2.65**	14.154	**3.15**	23.336	**3.65**	38.475
2.16	8.6711	2.66	14.296	3.16	23.571	3.66	38.861
2.17	8.7583	2.67	14.440	3.17	23.807	3.67	39.252
2.18	8.8463	2.68	14.585	3.18	24.047	3.68	39.646
2.19	8.9352	2.69	14.732	3.19	24.288	3.69	40.045
2.20	9.0250	**2.70**	14.880	**3.20**	24.533	**3.70**	40.447
2.21	9.1157	2.71	15.029	3.21	24.779	3.71	40.854
2.22	9.2073	2.72	15.180	3.22	25.028	3.72	41.264
2.23	9.2999	2.73	15.333	3.23	25.280	3.73	41.679
2.24	9.3933	2.74	15.487	3.24	25.534	3.74	42.098
2.25	9.4877	**2.75**	15.643	**3.25**	25.790	**3.75**	42.521
2.26	9.5831	2.76	15.800	3.26	26.050	3.76	42.948
2.27	9.6794	2.77	15.959	3.27	26.311	3.77	43.380
2.28	9.7767	2.78	16.119	3.28	26.576	3.78	43.816
2.29	9.8749	2.79	16.281	3.29	26.843	3.79	44.256
2.30	9.9742	**2.80**	16.445	**3.30**	27.113	**3.80**	44.701
2.31	10.074	2.81	16.610	3.31	27.385	3.81	45.150
2.32	10.176	2.82	16.777	3.32	27.660	3.82	45.604
2.33	10.278	2.83	16.945	3.33	27.938	3.83	46.063
2.34	10.381	2.84	17.116	3.34	28.219	3.84	46.525
2.35	10.486	**2.85**	17.288	**3.35**	28.503	**3.85**	46.993
2.36	10.591	2.86	17.462	3.36	28.789	3.86	47.465
2.37	10.697	2.87	17.637	3.37	29.079	3.87	47.942
2.38	10.805	2.88	17.814	3.38	29.371	3.88	48.424
2.39	10.913	2.89	17.993	3.39	29.666	3.89	48.911
2.40	11.023	**2.90**	18.174	**3.40**	29.964	**3.90**	49.402
2.41	11.134	2.91	18.357	3.41	30.265	3.91	49.899
2.42	11.246	2.92	18.541	3.42	30.569	3.92	50.400
2.43	11.359	2.93	18.728	3.43	30.877	3.93	50.907
2.44	11.473	2.94	18.916	3.44	31.187	3.94	51.419
2.45	11.588	**2.95**	19.106	**3.45**	31.500	**3.95**	51.935
2.46	11.705	2.96	19.298	3.46	31.817	3.96	52.457
2.47	11.822	2.97	19.492	3.47	32.137	3.97	52.985
2.48	11.941	2.98	19.688	3.48	32.460	3.98	53.517
2.49	12.061	2.99	19.886	3.49	32.786	3.99	54.055
2.50	12.182	**3.00**	20.086	**3.50**	33.115	**4.00**	54.598

TABLE 3 Exponential Functions (e^x) (*cont.*)

x	e^x	x	e^x	x	e^x	x	e^x
4.00	54.598	**4.50**	90.017	**5.00**	148.41	**5.50**	244.69
4.01	55.147	4.51	90.922	5.01	149.90	5.55	257.24
4.02	55.701	4.52	91.836	5.02	151.41	5.60	270.43
4.03	56.261	4.53	92.759	5.03	152.93	5.65	284.29
4.04	56.826	4.54	93.691	5.04	154.47	5.70	298.87
4.05	57.397	**4.55**	94.632	**5.05**	156.02	**5.75**	314.19
4.06	57.974	4.56	95.583	5.06	157.59	5.80	330.30
4.07	58.557	4.57	96.544	5.07	159.17	5.85	347.23
4.08	59.145	4.58	97.514	5.08	160.77	5.90	365.04
4.09	59.740	4.59	98.494	5.09	162.39	5.95	383.75
4.10	60.340	**4.60**	99.484	**5.10**	164.02	**6.00**	403.43
4.11	60.947	4.61	100.48	5.11	165.67	6.05	424.11
4.12	61.559	4.62	101.49	5.12	167.34	6.10	445.86
4.13	62.178	4.63	102.51	5.13	169.02	6.15	468.72
4.14	62.803	4.64	103.54	5.14	170.72	6.20	492.75
4.15	63.434	**4.65**	104.58	**5.15**	172.43	**6.25**	518.01
4.16	64.072	4.66	105.64	5.16	174.16	6.30	544.57
4.17	64.715	4.67	106.70	5.17	175.91	6.35	572.49
4.18	65.366	4.68	107.77	5.18	177.68	6.40	601.85
4.19	66.023	4.69	108.85	5.19	179.47	6.45	632.70
4.20	66.686	**4.70**	109.95	**5.20**	181.27	**6.50**	655.14
4.21	67.357	4.71	111.05	5.21	183.09	6.55	699.24
4.22	68.033	4.72	112.17	5.22	184.93	6.60	735.10
4.23	68.717	4.73	113.30	5.23	186.79	6.65	772.78
4.24	69.408	4.74	114.43	5.24	188.67	6.70	812.41
4.25	70.105	**4.75**	115.58	**5.25**	190.57	**6.75**	854.06
4.26	70.810	4.76	116.75	5.26	192.48	6.80	897.85
4.27	71.522	4.77	117.92	5.27	194.42	6.85	943.88
4.28	72.240	4.78	119.10	5.28	196.37	6.90	992.27
4.29	72.966	4.79	120.30	5.29	198.34	6.95	1043.1
4.30	73.700	**4.80**	121.51	**5.30**	200.34	**7.00**	1096.6
4.31	74.440	4.81	122.73	5.31	202.35	7.05	1152.9
4.32	75.189	4.82	123.97	5.32	204.38	7.10	1212.0
4.33	75.944	4.83	125.21	5.33	206.44	7.15	1274.1
4.34	76.708	4.84	126.47	5.34	208.51	7.20	1339.4
4.35	77.478	**4.85**	127.74	**5.35**	210.61	**7.25**	1408.1
4.36	78.257	4.86	129.02	5.36	212.72	7.30	1480.3
4.37	79.044	4.87	130.32	5.37	214.86	7.35	1556.2
4.38	79.838	4.88	131.63	5.38	217.02	7.40	1636.0
4.39	80.640	4.89	132.95	5.39	219.20	7.45	1719.9
4.40	81.451	**4.90**	134.29	**5.40**	221.41	**7.50**	1808.0
4.41	82.269	4.91	135.64	5.41	223.63	7.55	1900.7
4.42	83.096	4.92	137.00	5.42	225.88	7.60	1998.2
4.43	83.931	4.93	138.38	5.43	228.15	7.65	2100.6
4.44	84.775	4.94	139.77	5.44	230.44	7.70	2208.3
4.45	85.627	**4.95**	141.17	**5.45**	232.76	**7.75**	2321.6
4.46	86.488	4.96	142.59	5.46	235.10	7.80	2440.6
4.47	87.357	4.97	144.03	5.47	237.46	7.85	2565.7
4.48	88.235	4.98	145.47	5.48	239.85	7.90	2697.3
4.49	89.121	4.99	146.94	5.49	242.26	7.95	2835.6
4.50	90.017	**5.00**	148.41	**5.50**	244.69	**8.00**	2981.0

APPENDIX A

TABLE 4 Four-Place Logarithms (Base Ten)

t	0	1	2	3	4	5	6	7	8	9
1.0	0.0000	0.0043	0.0086	0.0128	0.0170	0.0212	0.0253	0.0294	0.0334	0.0374
1.1	0.0414	0.0453	0.0492	0.0531	0.0569	0.0607	0.0645	0.0682	0.0719	0.0755
1.2	0.0792	0.0828	0.0864	0.0899	0.0934	0.0969	0.1004	0.1038	0.1072	0.1106
1.3	0.1139	0.1173	0.1206	0.1239	0.1271	0.1303	0.1335	0.1367	0.1399	0.1430
1.4	0.1461	0.1492	0.1523	0.1553	0.1584	0.1614	0.1644	0.1673	0.1703	0.1732
1.5	0.1761	0.1790	0.1818	0.1847	0.1875	0.1903	0.1931	0.1959	0.1987	0.2014
1.6	0.2041	0.2068	0.2095	0.2122	0.2148	0.2175	0.2201	0.2227	0.2253	0.2279
1.7	0.2304	0.2330	0.2355	0.2380	0.2405	0.2430	0.2455	0.2480	0.2504	0.2529
1.8	0.2553	0.2577	0.2601	0.2625	0.2648	0.2672	0.2695	0.2718	0.2742	0.2765
1.9	0.2788	0.2810	0.2833	0.2856	0.2878	0.2900	0.2923	0.2945	0.2967	0.2989
2.0	0.3010	0.3032	0.3054	0.3075	0.3096	0.3118	0.3139	0.3160	0.3181	0.3201
2.1	0.3222	0.3243	0.3263	0.3284	0.3304	0.3324	0.3345	0.3365	0.3385	0.3404
2.2	0.3424	0.3444	0.3464	0.3483	0.3502	0.3522	0.3541	0.3560	0.3579	0.3598
2.3	0.3617	0.3636	0.3655	0.3674	0.3692	0.3711	0.3729	0.3747	0.3766	0.3784
2.4	0.3802	0.3820	0.3838	0.3856	0.3874	0.3892	0.3909	0.3927	0.3945	0.3962
2.5	0.3979	0.3997	0.4014	0.4031	0.4048	0.4065	0.4082	0.4099	0.4116	0.4133
2.6	0.4150	0.4166	0.4183	0.4200	0.4216	0.4232	0.4249	0.4265	0.4281	0.4298
2.7	0.4314	0.4330	0.4346	0.4362	0.4378	0.4393	0.4409	0.4425	0.4440	0.4456
2.8	0.4472	0.4487	0.4502	0.4518	0.4533	0.4548	0.4564	0.4579	0.4594	0.4609
2.9	0.4624	0.4639	0.4654	0.4669	0.4683	0.4698	0.4713	0.4728	0.4742	0.4757
3.0	0.4771	0.4786	0.4800	0.4814	0.4829	0.4843	0.4857	0.4871	0.4886	0.4900
3.1	0.4914	0.4928	0.4942	0.4955	0.4969	0.4983	0.4997	0.5011	0.5024	0.5038
3.2	0.5051	0.5065	0.5079	0.5092	0.5105	0.5119	0.5132	0.5145	0.5159	0.5172
3.3	0.5185	0.5198	0.5211	0.5224	0.5237	0.5250	0.5263	0.5276	0.5289	0.5302
3.4	0.5315	0.5328	0.5340	0.5353	0.5366	0.5378	0.5391	0.5403	0.5416	0.5428
3.5	0.5441	0.5453	0.5465	0.5478	0.5490	0.5502	0.5514	0.5527	0.5539	0.5551
3.6	0.5563	0.5575	0.5587	0.5599	0.5611	0.5623	0.5635	0.5647	0.5658	0.5670
3.7	0.5682	0.5694	0.5705	0.5717	0.5729	0.5740	0.5752	0.5763	0.5775	0.5786
3.8	0.5798	0.5809	0.5821	0.5832	0.5843	0.5855	0.5866	0.5877	0.5888	0.5899
3.9	0.5911	0.5922	0.5933	0.5944	0.5955	0.5966	0.5977	0.5988	0.5999	0.6010
4.0	0.6021	0.6031	0.6042	0.6053	0.6064	0.6075	0.6085	0.6096	0.6107	0.6117
4.1	0.6128	0.6138	0.6149	0.6160	0.6170	0.6180	0.6191	0.6201	0.6212	0.6222
4.2	0.6232	0.6243	0.6253	0.6263	0.6274	0.6284	0.6294	0.6304	0.6314	0.6325
4.3	0.6335	0.6345	0.6355	0.6365	0.6375	0.6385	0.6395	0.6405	0.6415	0.6425
4.4	0.6435	0.6444	0.6454	0.6464	0.6474	0.6484	0.6493	0.6503	0.6513	0.6522
4.5	0.6532	0.6542	0.6551	0.6561	0.6571	0.6580	0.6590	0.6599	0.6609	0.6618
4.6	0.6628	0.6637	0.6646	0.6656	0.6665	0.6675	0.6684	0.6693	0.6702	0.6712
4.7	0.6721	0.6730	0.6739	0.6749	0.6758	0.6767	0.6776	0.6785	0.6794	0.6803
4.8	0.6812	0.6821	0.6830	0.6839	0.6848	0.6857	0.6866	0.6875	0.6884	0.6893
4.9	0.6902	0.6911	0.6920	0.6928	0.6937	0.6946	0.6955	0.6964	0.6972	0.6981
5.0	0.6990	0.6998	0.7007	0.7016	0.7024	0.7033	0.7042	0.7050	0.7059	0.7067
5.1	0.7076	0.7084	0.7093	0.7101	0.7110	0.7118	0.7126	0.7135	0.7143	0.7152
5.2	0.7160	0.7168	0.7177	0.7185	0.7193	0.7202	0.7210	0.7218	0.7226	0.7235
5.3	0.7243	0.7251	0.7259	0.7267	0.7275	0.7284	0.7292	0.7300	0.7308	0.7316
5.4	0.7324	0.7332	0.7340	0.7348	0.7356	0.7364	0.7372	0.7380	0.7388	0.7396
t	0	1	2	3	4	5	6	7	8	9

TABLE 4 Four-Place Logarithms (Base Ten) (*cont.*)

t	0	1	2	3	4	5	6	7	8	9
5.5	0.7404	0.7412	0.7419	0.7427	0.7435	0.7443	0.7451	0.7459	0.7466	0.7474
5.6	0.7482	0.7490	0.7497	0.7505	0.7513	0.7520	0.7528	0.7536	0.7543	0.7551
5.7	0.7559	0.7566	0.7574	0.7582	0.7589	0.7597	0.7604	0.7612	0.7619	0.7627
5.8	0.7634	0.7642	0.7649	0.7657	0.7664	0.7672	0.7679	0.7686	0.7694	0.7701
5.9	0.7709	0.7716	0.7723	0.7731	0.7738	0.7745	0.7752	0.7760	0.7767	0.7774
6.0	0.7782	0.7789	0.7796	0.7803	0.7810	0.7818	0.7825	0.7832	0.7839	0.7846
6.1	0.7853	0.7860	0.7868	0.7875	0.7882	0.7889	0.7896	0.7903	0.7910	0.7917
6.2	0.7924	0.7931	0.7938	0.7945	0.7952	0.7959	0.7966	0.7973	0.7980	0.7987
6.3	0.7993	0.8000	0.8007	0.8014	0.8021	0.8028	0.8035	0.8041	0.8048	0.8055
6.4	0.8062	0.8069	0.8075	0.8082	0.8089	0.8096	0.8102	0.8109	0.8116	0.8122
6.5	0.8129	0.8136	0.8142	0.8149	0.8156	0.8162	0.8169	0.8176	0.8182	0.8189
6.6	0.8195	0.8202	0.8209	0.8215	0.8222	0.8228	0.8235	0.8241	0.8248	0.8254
6.7	0.8261	0.8267	0.8274	0.8280	0.8287	0.8293	0.8299	0.8306	0.8312	0.8319
6.8	0.8325	0.8331	0.8338	0.8344	0.8351	0.8357	0.8363	0.8370	0.8376	0.8382
6.9	0.8388	0.8395	0.8401	0.8407	0.8414	0.8420	0.8426	0.8432	0.8439	0.8445
7.0	0.8451	0.8457	0.8463	0.8470	0.8476	0.8482	0.8488	0.8494	0.8500	0.8506
7.1	0.8513	0.8519	0.8525	0.8531	0.8537	0.8543	0.8549	0.8555	0.8561	0.8567
7.2	0.8573	0.8579	0.8585	0.8591	0.8597	0.8603	0.8609	0.8615	0.8621	0.8627
7.3	0.8633	0.8639	0.8645	0.8651	0.8657	0.8663	0.8669	0.8675	0.8681	0.8686
7.4	0.8692	0.8698	0.8704	0.8710	0.8716	0.8722	0.8727	0.8733	0.8739	0.8745
7.5	0.8751	0.8756	0.8762	0.8768	0.8774	0.8779	0.8785	0.8791	0.8797	0.8802
7.6	0.8808	0.8814	0.8820	0.8825	0.8831	0.8837	0.8842	0.8848	0.8854	0.8859
7.7	0.8865	0.8871	0.8876	0.8882	0.8887	0.8893	0.8899	0.8904	0.8910	0.8915
7.8	0.8921	0.8927	0.8932	0.8938	0.8943	0.8949	0.8954	0.8960	0.8965	0.8971
7.9	0.8976	0.8982	0.8987	0.8993	0.8998	0.9004	0.9009	0.9015	0.9020	0.9025
8.0	0.9031	0.9036	0.9042	0.9047	0.9053	0.9058	0.9063	0.9069	0.9074	0.9079
8.1	0.9085	0.9090	0.9096	0.9101	0.9106	0.9112	0.9117	0.9122	0.9128	0.9133
8.2	0.9138	0.9143	0.9149	0.9154	0.9159	0.9165	0.9170	0.9175	0.9180	0.9186
8.3	0.9191	0.9196	0.9201	0.9206	0.9212	0.9217	0.9222	0.9227	0.9232	0.9238
8.4	0.9243	0.9248	0.9253	0.9258	0.9263	0.9269	0.9274	0.9279	0.9284	0.9289
8.5	0.9294	0.9299	0.9304	0.9309	0.9315	0.9320	0.9325	0.9330	0.9335	0.9340
8.6	0.9345	0.9350	0.9355	0.9360	0.9365	0.9370	0.9375	0.9380	0.9385	0.9390
8.7	0.9395	0.9400	0.9405	0.9410	0.9415	0.9420	0.9425	0.9430	0.9435	0.9440
8.8	0.9445	0.9450	0.9455	0.9460	0.9465	0.9469	0.9474	0.9479	0.9484	0.9489
8.9	0.9494	0.9499	0.9504	0.9509	0.9513	0.9518	0.9523	0.9528	0.9533	0.9538
9.0	0.9542	0.9547	0.9552	0.9557	0.9562	0.9566	0.9571	0.9576	0.9581	0.9586
9.1	0.9590	0.9595	0.9600	0.9605	0.9609	0.9614	0.9619	0.9624	0.9628	0.9633
9.2	0.9638	0.9643	0.9647	0.9652	0.9657	0.9661	0.9666	0.9671	0.9675	0.9680
9.3	0.9685	0.9689	0.9694	0.9699	0.9703	0.9708	0.9713	0.9717	0.9722	0.9727
9.4	0.9731	0.9736	0.9741	0.9745	0.9750	0.9754	0.9759	0.9763	0.9768	0.9773
9.5	0.9777	0.9782	0.9786	0.9791	0.9795	0.9800	0.9805	0.9809	0.9814	0.9818
9.6	0.9823	0.9827	0.9832	0.9836	0.9841	0.9845	0.9850	0.9854	0.9859	0.9863
9.7	0.9868	0.9872	0.9877	0.9881	0.9886	0.9890	0.9894	0.9899	0.9903	0.9908
9.8	0.9912	0.9917	0.9921	0.9926	0.9930	0.9934	0.9939	0.9943	0.9948	0.9952
9.9	0.9956	0.9961	0.9965	0.9969	0.9974	0.9978	0.9983	0.9987	0.9991	0.9996
t	0	1	2	3	4	5	6	7	8	9

APPENDIX

ALGEBRA REVIEW

B

B.1 EXPONENTS

B.2 POLYNOMIALS, ALGEBRAIC EXPRESSIONS, AND FACTORING

B.3 QUADRATIC EQUATIONS

B.1 EXPONENTS

If x is a number, then the product $x \cdot x$ may be written x^2. Such an expression is called an *exponential expression*, x is called the *base*, and 2 is the *exponent*. Another example of an exponential expression is $y^4 = y \cdot y \cdot y \cdot y$. In general, we define integer exponents as follows.

> **DEFINITION OF INTEGER EXPONENTS**
> (a) $x^1 = x$ and $x^0 = 1$.
> (b) If n is a positive integer, $x^n = x \cdot x \cdot x \cdots x$ (n factors).
> (c) If n is a positive integer, $x^{-n} = 1/x^n$, $x \neq 0$.

Consider the expression $(xy)^4$. From the definition, we have

$$(xy)^4 = (xy)(xy)(xy)(xy).$$

The factors on the right-hand side of the preceding equation can be rearranged by the commutative and associative laws of multiplication to give

$$(xy)^4 = (x \cdot x \cdot x \cdot x)(y \cdot y \cdot y \cdot y).$$

Now the exponent definition can be used to obtain

$$(xy)^4 = x^4 y^4.$$

The definitions for exponents lead immediately to the following laws of exponents.

> **LAWS OF EXPONENTS** If x and y are any real numbers and m and n are integers, then
> (a) $x^m x^n = x^{m+n}$.
> (b) $(xy)^m = x^m y^m$.
> (c) $(x^m)^n = x^{mn}$.
> (d) $\left(\dfrac{x}{y}\right)^n = \dfrac{x^n}{y^n}$ provided $y \neq 0$.
> (e) $\dfrac{x^m}{x^n} = x^{m-n}$ provided $x \neq 0$.

EXAMPLE 1 Use the laws of exponents to simplify the following:
(a) $u^8 u^{-3}$. (b) $(2u)^3$. (c) $(u^2 v^{-2})^3$.

Solution

(a) $\qquad u^8 u^{-3} = u^{8-3} = u^5.$

(b) $\qquad (2u)^3 = 2^3 u^3 = 8u^3.$

(c) $\qquad (u^2 v^{-2})^3 = (u^2)^3 (v^{-2})^3 = u^6 v^{-6}.$

Note: Henceforth, answers to problems such as in part (c) will be given in terms of positive exponents. Thus, $u^6 v^{-6}$ may be written as u^6/v^6 or as $(u/v)^6$.

PRACTICE PROBLEM Use the laws of exponents to simplify the following expressions. (a) $w^{-5} w^{14}$. (b) $(3w)^2$. (c) $(w^3 x^{-1})^2$.

Answers: (a) w^9. (b) $9w^2$. (c) w^3/x^2. ∎

EXAMPLE 2 Use the laws of exponents to simplify the following:
(a) $(5u^{-2})(2u^7)$. (b) $(u^{-2}/v^{-3})^2$. (c) $(2u^5 v^4)^2 / (u^{-1} v^2)^{-1}$.

Solution

(a) $\qquad (5u^{-2})(2u^7) = (5)(2) u^{-2+7} = 10 u^5.$

(b) $\qquad \left(\dfrac{u^{-2}}{v^{-3}}\right)^2 = \dfrac{(u^{-2})^2}{(v^{-3})^2} = \dfrac{u^{-4}}{v^{-6}} = \dfrac{v^6}{u^4}.$

(c) $\qquad \dfrac{(2u^5 v^4)^2}{(u^{-1} v^2)^{-1}} = \dfrac{2^2 (v^5)^2 (v^4)^2}{(u^{-1})^{-1}(v^2)^{-1}} = \dfrac{4 u^{10} v^8}{u^1 v^{-2}} = 4 u^9 v^{10}.$

PRACTICE PROBLEM Use the laws of exponents to simplify the following expressions. (a) $(3w^5)(-2w^{-10})$. (b) $(2w^3/x^2)^{-2}$. (c) $(w^{-2} x^2)^{-1} / (w^{-1} x^{-1})^{-2}$.

Answers: (a) $-6/w^5$. (b) $x^4/4w^6$. (c) $1/x^4$. ∎

Let us now turn our attention to defining *radicals* and *rational exponents* of the form p/q where p and q are integers, $q \neq 0$. First let us consider the equation

$$x^n = y \qquad (1)$$

B.1 EXPONENTS

where n is an integer greater than or equal to 2 and y is a real number. Development of the definitions of radicals and rational exponents depends upon the following facts about Eq. (1).

(a) If n is an odd integer, then Eq. (1) has exactly one solution (in the set of real numbers). The solution has the same sign as y.

(b) If n is an even integer and $y > 0$, then Eq. (1) has exactly two solutions, one positive and the other negative.

(c) If n is even and $y = 0$, then Eq. (1) has only the solution $x = 0$.

(d) If n is even and $y < 0$, then Eq. (1) has no real number solution.

Now we are prepared to define what is called the *principal nth root* of a real number y.

DEFINITION OF PRINCIPAL ROOT Given an integer $n \geq 2$ and a real number y.

(a) If n is odd, the single real solution of the equation

$$x^n = y$$

is called the **principal nth root** of y.

(b) If n is even and $y > 0$, the positive solution of $x^n = y$ is called the **principal nth root** of y.

(c) The *principal* nth *root* of 0 is 0.

(d) If n is even and $y < 0$, y has no principal nth root.

EXAMPLE 3 (a) The principal third root of 1 is 1. (b) The principal second root (square root) of 16 is 4. (c) The principal fifth root of 0 is 0. (d) The principal fifth root of -32 is -2.

PRACTICE PROBLEM Find the principal (a) eighth root of 1, (b) third root (cube root) of $-\frac{1}{8}$, and (c) fourth root of 16.

Answers: (a) 1. (b) $-\frac{1}{2}$. (c) 2. ■

DEFINITION OF RADICALS The nth principal root of a real number y, if it exists, is denoted by the *radical* expression

$$\sqrt[n]{y}.$$

Note: Generally, $\sqrt{y}$ is the notation used for $\sqrt[2]{y}$.

EXAMPLE 4 Calculate the value of each of the following radical expressions:

(a) $\sqrt{4}$. (b) $\sqrt[3]{8}$. (c) $\sqrt[5]{-1}$. (d) $\sqrt[8]{0}$. (e) $\sqrt[4]{16}$. (f) $\sqrt[3]{-\frac{1}{27}}$.

Solution

(a) 2   (b) 2   (c) -1
(d) 0   (e) 2   (f) $-\frac{1}{3}$

PRACTICE PROBLEM Calculate (a) $\sqrt[5]{-32}$, (b) $\sqrt[4]{81}$, and (c) $\sqrt[4]{\frac{1}{16}}$.

Answers: (a) -2 (b) 3   (c) $\frac{1}{2}$ ∎

We are now prepared to define rational exponents.

DEFINITION OF RATIONAL EXPONENTS Given an integer $n \geq 2$ and a real number x for which $\sqrt[n]{x}$ is defined. Then

$$x^{1/n} = \sqrt[n]{x}.$$

If m is an integer, then

$$x^{m/n} = (x^{1/n})^m$$

The definition just given extends the concept of exponents to all rational numbers. It also allows us to convert from exponential notation to radical notation and *vice versa*. It can be shown that

$$x^{m/n} = (x^m)^{1/n};$$

furthermore, the laws of exponents for integers hold for rational numbers. Finally, we have

$$x^{m/n} = (\sqrt[n]{x})^m = \sqrt[n]{x^m}.$$

B.1 EXPONENTS

EXAMPLE 5 Convert the following exponential expressions to radicals and evaluate: (a) $16^{3/2}$. (b) $8^{-4/3}$.

Solution

(a) $\quad 16^{3/2} = (\sqrt{16})^3 = 4^3 = 64.$

(b) $\quad 8^{-4/3} = (8^{1/3})^{-4} = (\sqrt[3]{8})^{-4} = 2^{-4} = \dfrac{1}{2^4} = \dfrac{1}{16}.$

PRACTICE PROBLEM Convert the exponential expression $32^{2/5}$ to a radical expression and evaluate.

Answer: 4. ∎

EXAMPLE 6 Convert the following radicals to exponential notation: (a) $\sqrt[5]{x^2}$. (b) $\sqrt[4]{a^2b^2}$. (c) $\sqrt{7x+1}$.

Solution

(a) $\quad \sqrt[5]{x^2} = (x^2)^{1/5} = x^{2/5}.$

(b) $\quad \sqrt[4]{a^2b^2} = (a^2b^2)^{1/4} = [(ab)^2]^{1/4} = (ab)^{1/2}.$

(c) $\quad \sqrt{7x+1} = (7x+1)^{1/2}.$

PRACTICE PROBLEM Convert the radical expression $\sqrt[3]{(2x+1)^2}$ to an exponential expression.

Answer: $(2x+1)^{2/3}$. ∎

EXAMPLE 7 Use the laws of exponents to simplify the following expressions: (a) $x^{1/2}x^{3/2}$. (b) $(x^{5/2}y^{3/4})^4$. (c) $(a^{-1/3}b^{-4/3}/c^{-8/3})^6$.

Solution

(a) $\quad x^{1/2}x^{3/2} = x^{4/2} = x^2.$

(b) $\quad (x^{5/2}y^{3/4})^4 = (x^{5/2})^4(y^{3/4})^4 = x^{10}y^3.$

(c) $\quad \left(\dfrac{a^{-1/3}b^{-4/3}}{c^{-8/3}}\right)^6 = \dfrac{(a^{-1/3})^6(b^{-4/3})^6}{(c^{-8/3})^6} = \dfrac{a^{-2}b^{-8}}{c^{-16}} = \dfrac{c^{16}}{a^2b^8}.$

PRACTICE PROBLEM Use the laws of exponents to simplify $(x^2y^3)^{1/6}(x^4y^{-3})^{1/6}$.

Answer: x. ∎

B.2 POLYNOMIALS, ALGEBRAIC EXPRESSIONS, AND FACTORING

The expression $3x + 1$ is called a polynomial of degree 1, or *linear polynomial*. Furthermore, $5x^2 - 2.1x + 3$ is a polynomial of degree 2, or *quadratic polynomial*, and $x^3 + 7x$ is a polynomial of degree 3, or *cubic polynomial*. In general, a **polynomial** *of degree n* is an expression of the form

$$a_0x^n + a_1x^{n-1} + a_2x^{n-2} + \cdots + a_{n-1}x + a_n$$

where $a_0, a_1, a_2, \ldots, a_n$ represent real numbers and $a_n \neq 0$; these numbers are called the *coefficients*. For the cubic polynomial $2x^3 - 3x^2 + 7x + 5$, the *terms* of the polynomial are $2x^3$, $3x^2$, $7x$, and 5 and, for example, 7 is the *coefficient* of the linear term $7x$. An expression in x, such as

$$6x + 3x^{1/2} + x^{-2},$$

which involves the algebraic operations of addition, subtraction, multiplication, division, or taking roots is called an *algebraic expression*.

Let us now consider how one adds, subtracts, and multiplies algebraic expressions. To illustrate addition, consider the expressions

$$\sqrt{xy} + 2 \quad \text{and} \quad 3\sqrt{xy} + 5.$$

Note: Each could be written $(xy)^{1/2} + 2$ and $3(xy)^{1/2} + 5$. Their sum is found as follows:

$$(\sqrt{xy} + 2) + (3\sqrt{xy} + 5) = (\sqrt{xy} + 3\sqrt{xy}) + (2 + 5)$$
$$= 4\sqrt{xy} + 7.$$

To perform the addition we grouped similar terms with like powers and added. Polynomials are added as in the following example.

B.2 POLYNOMIALS, ALGEBRAIC EXPRESSIONS, AND FACTORING

EXAMPLE 8 Add the following:

(a) $(-x^3 + 5x^2 - 7x + 1) + (x^4 + 3x^3 - 2x^2 - 4x - 8)$.

(b) $(5x^2y - 8xy + 15xy^2) + (-2x^2y + 29xy - 8xy^2)$.

Solution

(a) $(-x^3 + 5x^2 - 7x + 1) + (x^4 + 3x^3 - 2x^2 - 4x - 8)$

$= x^4 + (-x^3 + 3x^3) + (5x^2 - 2x^2) + (-7x - 4x) + (1 - 8)$

$= x^4 + 2x^3 + 3x^2 - 11x - 7$.

(b) $(5x^2y - 8xy + 15xy^2) + (-2x^2y + 29xy - 8xy^2)$

$= (5x^2y - 2x^2y) + (-8xy + 29xy) + (15xy^2 - 8xy^2)$

$= 3x^2y + 21xy + 7xy^2$.

Multiplication of two algebraic expressions, such as $(x^{1/2} + x + 3)$ and $(2x^3 - 3x + 1)$, is carried out by using the *distributive property*

$$P(Q + R) = PQ + PR,$$

where P, Q, and R are algebraic expressions, and by the other techniques already discussed. For example, consider the following product:

$$2x^3(3x^2 - 7x + 1) = (2x^3)(3x^2) + (2x^3)(-7x) + (2x^3)(1)$$
$$= 6x^5 - 14x^4 + 2x^3.$$

EXAMPLE 9 Find the following products:

(a) $(\sqrt{xy} + 1)(\sqrt{xy} - 1)$.

(b) $(x^2 + 1)(3x^2 - 8x - 1)$.

Solution

(a) $(\sqrt{xy} + 1)(\sqrt{xy} - 1)$

$= (\sqrt{xy} + 1)(\sqrt{xy}) + (\sqrt{xy} + 1)(-1)$

$= (\sqrt{xy})^2 + (1)(\sqrt{xy}) + (\sqrt{xy})(-1) + (1)(-1)$

$= xy + \sqrt{xy} - \sqrt{xy} - 1$

$= xy - 1$.

(b) $(x^2 + 1)(3x^2 - 8x - 1)$
$$= (x^2 + 1)(3x^2) + (x^2 + 1)(-8x) + (x^2 + 1)(-1)$$
$$= 3x^4 + 3x^2 - 8x^3 - 8x - x^2 - 1$$
$$= 3x^4 - 8x^3 + 2x^2 - 8x - 1.$$

PRACTICE PROBLEM Multiply: $(3x^2 - 5)(x^2 - 2x - 5)$.

Answer: $3x^4 - 6x^3 - 20x^2 + 10x + 25$. ∎

Let us now consider some "special" product formulas which occur frequently in applications. It is helpful to be able to recognize these patterns and to obtain the resulting products quickly.

SPECIAL PRODUCTS

1. $(a + b)^2 = a^2 + 2ab + b^2.$
2. $(a - b)^2 = a^2 - 2ab + b^2.$
3. $(a + b)(a - b) = a^2 - b^2.$
4. $(a + b)(a^2 - ab + b^2) = a^3 + b^3.$
5. $(a - b)(a^2 + ab + b^2) = a^3 - b^3.$

EXAMPLE 10 Use the special product formulas to obtain the following: (a) $(2x - 3)^2$. (b) $(ab^2 + c^2)^2$. (c) $(\sqrt{x} + \sqrt{y})(\sqrt{x} - \sqrt{y})$. (d) $(2x + 1)(4x^2 - 2x + 1)$.

Solution

(a) $(2x - 3)^2 = 4x^2 - 12x + 9.$
(b) $(ab^2 + c^2)^2 = a^2b^4 + 2ab^2c^2 + c^4.$
(c) $(\sqrt{x} + \sqrt{y})(\sqrt{x} - \sqrt{y}) = (\sqrt{x})^2 - (\sqrt{y})^2 = x - y.$
(d) $(2x + 1)(4x^2 - 2x + 1) = (2x)^3 + (1)^3 = 8x^3 + 1.$

PRACTICE PROBLEM Use a special product formula to multiply
$$(2xy^2 - 5)^2.$$

Answer: $4x^2y^4 - 20xy^2 + 25$. ∎

B.2 POLYNOMIALS, ALGEBRAIC EXPRESSIONS, AND FACTORING

If a polynomial is expressed as a product of two or more polynomials, it is said to be *factored*. For example,

$$x^2 - x - 6$$

is factored if written as

$$(x - 3)(x + 2).$$

Each one of the special products listed earlier gives rise to *special factors*. For example, the sum of two cubes is factored as follows:

$$a^3 + b^3 = (a + b)(a^2 - ab + b^2).$$

The distributive property can be used to factor an expression where each term contains the same common factor. For example,

$$2ax^2 + 3ay^2 + 4az^3 = a(2x^2 + 3y^2 + 4z^3).$$

EXAMPLE 11 Factor the following polynomials completely: (a) $7x^3 - 9x^2$. (b) $xa - ya + xb - yb$. (c) $4x^2 - 4x + 1$.

Solution

(a) $\quad 7x^3 - 9x^2 = x^2(7x - 9).$

(b) $\quad xa - ya + xb - yb = a(x - y) + b(x - y)$
$\quad\quad\quad\quad\quad\quad\quad\quad\quad = (x - y)(a + b).$

(c) $\quad 4x^2 - 4x + 1 = (2x - 1)^2.$

PRACTICE PROBLEM Factor $w^2 - 4x^2$.

Answer: $(w + 2x)(w - 2x)$. ■

EXAMPLE 12 Factor (a) $x^3 + 1$ and (b) $8a^3 - 27$.

Solution

(a) $\quad\quad\quad\quad x^3 + 1 = x^3 + 1^3$
$\quad\quad\quad\quad\quad\quad = (x + 1)(x^2 - x(1) + 1^2)$
$\quad\quad\quad\quad\quad\quad = (x + 1)(x^2 - x + 1).$

(b) $\quad 8a^3 - 27 = (2a)^3 - (3)^3$
$= (2a - 3)((2a)^2 + (2a)(3) + 3^2)$
$= (2a - 3)(4a^2 + 6a + 9).$

PRACTICE PROBLEM Factor $8w^3 + z^3v^3$.

Answer: $(2w + zv)(4w^2 - 2wzv + z^2v^2)$. ■

B.3 QUADRATIC EQUATIONS

A **quadratic equation** *in one variable* is an equation which can be expressed as

$$ax^2 + bx + c = 0$$

where a, b, and c are real numbers and $a \neq 0$. Let us now review techniques for solving such equations.

If the linear term bx is missing from a quadratic equation, then it is easy to solve. For example,

$$x^2 = 4$$

is such an equation. Both 2 and -2 are solutions to the equation since, when substituted for x, they make the equality a true statement. In general, $ax^2 + c = 0$ can be written as $ax^2 = -c$ or $x^2 = -c/a$. So any quadratic equation with no linear term can be written in the form

$$x^2 = k.$$

(1) If $k < 0$, then $x^2 = k$ has no (real) solution.

(2) If $k = 0$, then $x^2 = k$ has $x = 0$ as the only solution.

(3) If $k > 0$, then $x^2 = k$ has two solutions; they are $x = \sqrt{k}$ and $x = -\sqrt{k}$.

PRACTICE PROBLEM Solve (a) $x^2 = 16$, (b) $x^2 = 7$, and (c) $x^2 = 0$.

Answers: (a) $x = 4$, $x = -4$. (b) $x = \sqrt{7}$, $x = -\sqrt{7}$. (c) $x = 0$. ■

B.3 QUADRATIC EQUATIONS

Some quadratic equations can be easily solved as a result of factoring. For example, we can solve

$$x^2 - x - 6 = 0$$

in the following way:

$$x^2 - x - 6 = 0,$$
$$(x - 3)(x + 2) = 0,$$
$$x - 3 = 0 \quad \text{or} \quad x + 2 = 0,$$
$$x = 3 \quad \text{or} \quad x = -2.$$

If a quadratic equation is not easily factored, the *quadratic formula* can be used. It is given below.

THE QUADRATIC FORMULA The solutions to the quadratic equation

$$ax^2 + bx + c = 0$$

may be computed by the formula

$$x = \frac{-b \pm \sqrt{b^2 - 4ac}}{2a}.$$

(1) If $b^2 - 4ac > 0$, the quadratic has two real solutions.

(2) If $b^2 - 4ac = 0$, the quadratic has exactly one real solution; it is $-b/2a$.

(3) If $b^2 - 4ac < 0$, the quadratic has no real solutions.

EXAMPLE 13 Solve by the quadratic formula: (a) $x^2 - 2x - 1 = 0$. (b) $2x^2 + x - 1 = 0$.

Solution

(a) For $x^2 - 2x - 1 = 0$, $a = 1$, $b = -2$, and $c = -1$. Thus

$$x = \frac{-(-2) \pm \sqrt{(-2)^2 - 4(1)(-1)}}{2(1)}$$

$$= \frac{2 \pm \sqrt{8}}{2}$$

$$= \frac{2 \pm 2\sqrt{2}}{2}.$$

The two solutions are $x = 1 + \sqrt{2}$ and $x = 1 - \sqrt{2}$.

(b) For $2x^2 + x - 1 = 0$, $a = 2$, $b = 1$, and $c = -1$. Thus

$$x = \frac{-1 \pm \sqrt{(1)^2 - 4(2)(-1)}}{2(2)}$$

$$= \frac{-1 \pm \sqrt{9}}{4}$$

$$= \frac{-1 \pm 3}{4}.$$

The two solutions are

$$x = (-1 + 3)/4 = \tfrac{1}{2} \quad \text{and} \quad x = (-1 - 3)/4 = -1.$$

Note: Since the solutions are rational numbers, this means that the original quadratic polynomial was factorable. That is,

$$2x^2 + x - 1 = (2x - 1)(x + 1).$$

PRACTICE PROBLEM Solve each of the following equations by the quadratic formula. (a) $x^2 - 2x - 2 = 0$. (b) $12x^2 + 18 = 35x$.

Answers: (a) $x = -1 + \sqrt{3}$, $x = -1 - \sqrt{3}$. (b) $x = \tfrac{2}{3}$, $x = \tfrac{9}{4}$. ∎

APPENDIX C

ANSWERS TO ODD-NUMBERED EXERCISES AND SOLUTIONS TO SELECTED EXERCISES

CHAPTER 1. FUNCTIONS AND LINEAR MODELS
CHAPTER 2. DIFFERENTIAL CALCULUS AND ITS APPLICATIONS
CHAPTER 3. FURTHER APPLICATIONS OF THE DERIVATIVE
CHAPTER 4. EXPONENTIAL AND LOGARITHMIC FUNCTIONS
CHAPTER 5. THE ANTIDERIVATIVE AND ITS APPLICATIONS
CHAPTER 6. THE DEFINITE INTEGRAL AND ITS APPLICATIONS
CHAPTER 7. MULTIVARIABLE CALCULUS AND ITS APPLICATIONS
CHAPTER 8. SEQUENCES AND MATHEMATICS OF FINANCE

CHAPTER 1. FUNCTIONS AND LINEAR MODELS

1.2 THE REAL NUMBER LINE AND LINEAR INEQUALITIES

1. $x > -\frac{9}{2}$ 3. $x \leq -\frac{6}{13}$ 5. $x \geq 7$ 7. $x < -\frac{9}{26}$
9. $x \leq \frac{2}{3}$ 11. $x < \frac{27}{13}$ 13. $x \leq \frac{29}{21}$ 15. $x > -40$
17. $x < 41$ 19. $x > -34.8$ 21. $324 23. 120 miles
25. (a) $212 (b) $364 27. More than 50,000 units

1.3 THE RECTANGULAR COORDINATE SYSTEM AND SLOPE OF A LINE

1. $\frac{3}{5}$ 3. $\frac{3}{4}$ 5. $-\frac{7}{12}$ 7. $\frac{4}{3}$
9. -2 11. $\left(\frac{11}{2}, \frac{17}{2}\right)$ 13. $\left(0, \frac{5}{2}\right)$ 15. $\left(-1, -\frac{1}{2}\right)$
17. $(0, -5)$ 19. $\left(\frac{1}{2}, 1\right)$ 21. $\sqrt{34}$ 23. 5
25. $\sqrt{193}$ 27. 10 29. $6\sqrt{5}$

31.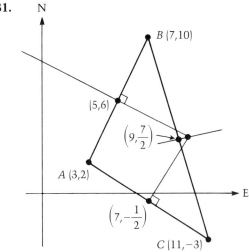

33. $A(-40, 60)$, $B(100, 40)$, and $C(20, 140)$. $|AC| = 100$ miles and $|BC| \doteq 128.1$ miles.

1.4 EQUATIONS OF LINES

1. $y = -7$ 3. $x = 2$ 5. $2x + y = -2$
7. $4x - 3y = 34$ 9. $11x - 4y = 46$ 11. $x + 3y = 20$
13. $x = -3$ 15. $10x + 7y = 42$ 17. $3x - 3y = -1$
19. $2x + 5y = 7$ 21. $3x - 5y = -23$ (see p. C4)
23. Midpoint: $\left(-1, \frac{7}{2}\right)$. Slope: $\frac{1}{2}$. Slope of perpendicular bisector: -2. $2x + y = \frac{3}{2}$ or $4x + 2y = 3$ (see p. C4)
25. Midpoint: $(1, 2)$; slope: -1; equation of line: $x + y = 3$
27. $6x + 10y = -2$

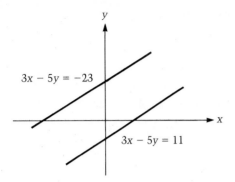

Figure for answer to Exercise 21

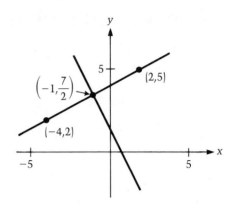

Figure for answer to Exercise 23

29. Midpoint: $(\frac{1}{6}, \frac{5}{24})$; slope of line: $-\frac{13}{16}$; slope of perpendicular bisector: $\frac{16}{13}$; equation of perpendicular bisector: $384x - 312y = -1$

31. (a) Slope of line containing $(0, 32)$ and $(100, 212)$ is $\frac{9}{5}$. An equation of the line is $9C - 5F = -160$, or $F = \frac{9}{5}C + 32$.
 (b) $37\frac{7}{9}°C$ (c) $104°F$

1.5 LINEAR SYSTEMS AND THEIR APPLICATIONS

1. $x = 2, y = -3$ 3. $x = \frac{1}{2}, y = 1$ 5. $x = \frac{8}{41}, y = \frac{76}{41}$
7. $x = 8, y = -12$ 9. $x = \frac{4}{3}, y = -\frac{5}{4}$

11. Let x be the number of $6.00 shirts and y be the number of $9.00 shirts. Then $6x + 9y = 7680$ and $6.25x + 9.50y = 8050$, which are equivalent to $2x + 3y = 2560$ and $x + 1.52y = 1288$. $x = 680$ and $y = 400$

13. The perpendicular bisector of AB is $x + 2y = 17$. The perpendicular bisector of BC is $8x - 26y = -19$. The solution is $x = \frac{202}{21}$ and $y = \frac{155}{42}$; that is, approximately 9.6 miles east and 3.7 miles north of City Hall.

15. Let x be the number of shares at $20 a share and let y be the number of shares at $15 a share. Then $20x + 15y = 16{,}250$ and $1.60x + 1.00y = 1300$. $x = 750$ and $y = 100$

17. Let x be the number of gallons of 9% alcohol and let y be the number of gallons of 13% alcohol. Then $x + y = 2000$ and $0.09x + 0.13y = 200$. $x = 1500$ gallons and $y = 500$ gallons

19. Let x be the number of Chewey bars in the shipment and let y be the number of Crispy bars. Then $0.75x + 0.35y = 171.25$ and $0.50x + 0.22y = 111.50$. $x = 135$ and $y = 200$

21. Let x be the number of gallons of 25% solution and let y be the number of gallons of 14% solution. Then

CHAPTER 1. FUNCTIONS AND LINEAR MODELS C5

$x + y = 500$ and $0.25x + 0.14y = 0.18(500)$. $x = 181.82$ gallons and $y = 318.18$ gallons

23. Let x be the amount invested in stocks and let y be the amount invested in bonds. Then $x + y = 85,000$ and $0.12x + 0.08y = 8680$. $x = \$47,000$

25. (a) 0.82 ounces of Product A and 6.12 ounces of Product B
(b) Yes

1.6 FUNCTIONS AND THEIR GRAPHS

1. (a) 19   (b) 27   (c) No   (d) No
3. (a) 12   (b) 27   (c) No   (d) No
5. (a) 8   (b) 13   (c) No   (d) No
7. (a) 8   (b) 27   (c) No   (d) Yes
9. $x < \frac{22}{3}$ **11.** $x > \frac{1}{5}$ **13.** $x < \frac{7}{2}$ **15.** $x < 1$
17. (a)   (b) 11   (c) -7

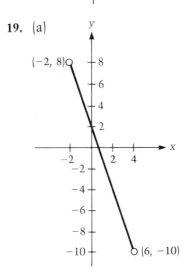

19. (a)   (b) None   (c) None

21. (a) 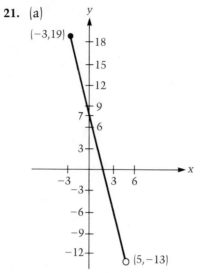 (b) 19 (c) None

23. (a) 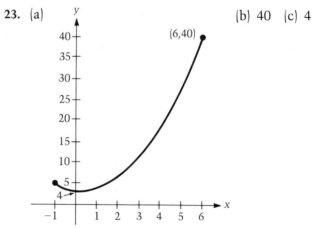 (b) 40 (c) 4

25. (a) 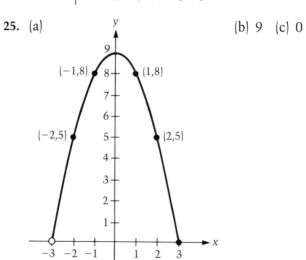 (b) 9 (c) 0

CHAPTER 1. FUNCTIONS AND LINEAR MODELS

27. (a) 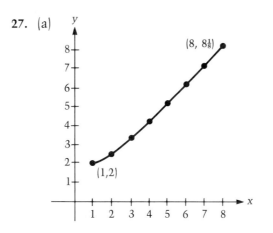 (b) $\frac{65}{8}$ (c) 2

29. (a) 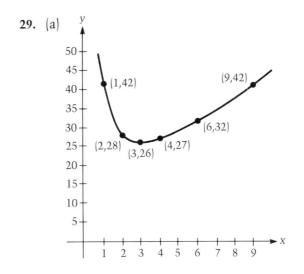 (b) None
(c) $f(3) = 26$

1.7 APPLICATIONS OF LINEAR FUNCTIONS

1. (a) $V(t) = 8000 - 1200t$ (b) $V(2) = \$5600$
   (c) $8000 - 1200t = 0$ when $t = 6\frac{2}{3}$ years = 6 years 8 months (d) $V(2.5) = \$5000$

3. (a) $D(3.50) = 48{,}000$ units (b) $D(6) = 28{,}000$ units
   (c) $100(76 - 8p) = 0$ when $p = \$9.50$ (d) The change (increase or decrease) in demand when the price is increased from p to $p + 1$ dollars. It is decreased by $\$8000$.

5. (a) $S(3.50) = 28{,}000$ units (b) $S(6) = 48{,}000$ units
   (c) $1000(76 - 8p) = 8000p$ when $p = \$4.75$
   (d) $S(4.75) = 38{,}000$ units

7. (a) $C(t) = kt$, $43.50 = k(\frac{3}{2})$, and $k = 29$ (b) The hourly charge (c) $C(t) = 29t$ (d) $C(4.75) = \$137.75$

9. (a) $3.50 + 20(0.15) = \$6.50$ (b) $C(x) = 0.15x + 3.50$
   (c) $C(35) = \$8.75$ (d) $\overline{C}(x) = 0.15 + \left(\dfrac{3.50}{x}\right)$
   (e) $\overline{C}(35) = \$0.25$
   (f)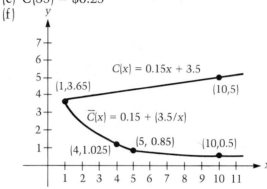

11. $0.15x + 3.50 = 0.10x + 3.00$ when $x = 30$ checks
13. (a) $W(65) \doteq 128.33$ pounds (b) 84 inches
15. (a) $R(x) = 600N(x) = 24{,}000x + 7{,}200{,}000$
    (b) $R(90) = \$9{,}360{,}000$
17. (a) $P(x) = R(x) - C(x) = 5000x - 1{,}700{,}000$
    (b) $P(90) = -\$1{,}250{,}000$, a loss of $\$1{,}250{,}000$
19. $3,300,000 profit
21. (a) $R(x) = 500N(x) = 15{,}000x + 4{,}000{,}000$
    (b) $R(100) = \$5{,}500{,}000$
23. (a) $P(x) = R(x) - C(x) = 2000x - 1{,}700{,}000$ (b) $1,500,000 loss
25. $300,000 profit

1.8 SUMMARY AND REVIEW EXERCISES

1. Multiplying each side of the given inequality by 20 yields
$$4(4x - 3) - 10(2x + 1) \geq 5(5x + 6),$$
$$16x - 12 - 20x - 10 \geq 25x + 30,$$
$$16x - 20x - 25x \geq 30 + 10 + 12,$$
$$-29x \geq 52, \quad x \leq -\tfrac{52}{29}$$

2. Multiplying each side of the given inequality by 100 yields
$$135x + 400 < 545x - 830, \quad -410x < -1230, \quad x > 3$$

3. Slope of line is $\tfrac{2}{3}$. Equation of line is $2x - 3y = -13$.

CHAPTER 1. FUNCTIONS AND LINEAR MODELS   C9

4. Slope of line is $-\frac{5}{6}$. Equation of line is $5x + 6y = -13$.
5. Midpoint is $(2, 6)$; slope of line is $-\frac{1}{6}$; slope of perpendicular bisector is 6; equation of line is $6x - y = 6$.
6. Midpoint is $(1, 2)$; slope of line is $\frac{3}{2}$; slope of perpendicular bisector is $-\frac{2}{3}$; equation of line is $2x + 3y = 8$.
7. $x = \frac{3}{2}$ and $y = -2$.   8. $x = -3$ and $y = 7$.
9. Maximum value is $f(1) = -4$ and the minimum value is $f(5) = -28$.
10. Since $f(x) = (x + 1)^2 + 4 \geq 4$, $f(-1) = 4$ is the minimum value of the function; $f(6) = 53$ is the maximum value.
11. Let x be the number of \$35 radios in the shipment and let y be the number of \$62 radios. Therefore $x + y = 225$ and $35x + 62y = 11{,}088$. Since $y = 225 - x$,

$$35x + 62(225 - x) = 11{,}088,$$
$$35x + 13{,}950 - 62x = 11{,}088,$$
$$-27x = -2862, \quad \text{and} \quad x = 106$$

12. Let x be the number of dimes in the meter and let y be the number of nickels. Then $x + y = 419$ and $10x + 5y = 2825$. Since $y = 419 - x$,

$$10x + 5(419 - x) = 2825, \quad 10x + 2095 - 5x = 2825,$$
$$5x = 730, \quad x = 146$$

13. Let x be the number of gallons of regular gasoline. Then $76 - x$ is the number of gallons of unleaded gas. Then

$$117x + 123(76 - x) = 9144, \quad 117x + 9348 - 123x = 9144,$$
$$-6x = -204, \quad x = 34$$

14. Let x be the number of miles truck A travels. Then $600 - x$ is the number of miles truck B travels. Therefore,

$$\frac{x}{12} + \frac{600 - x}{9} \leq 56, \quad 3x + 2400 - 4x \leq 2016,$$
$$-x \leq -384, \quad x \geq 384$$

15. $A(300, -400)$, $B(-500, 200)$, and $C(100, 400)$ are the locations of the three outlets from headquarters. Midpoint of AB is $(-100, -100)$; slope of AB is $-\frac{3}{4}$; slope of perpendicular bisector of AB is $\frac{4}{3}$; equation of perpendicular bisector of AB is $4x - 3y = -100$ (Eq. 1).

Midpoint of BC is $(-200, 300)$; slope of BC is $\frac{1}{3}$; slope of perpendicular bisector is -3; equation of perpendicular bisector is $3x + y = -300$ (Eq. 2). From Eq. (2), $y = -3x - 300$. Substituting in Eq. (1), we get

$$4x - 3(-3x - 300) = -100, \qquad 4x + 9x + 900 = -100,$$
$$13x = -1000, \qquad x \doteq -76.92.$$

We find that $y \doteq -69.23$. Therefore, the warehouse should be 76.92 miles west and 69.23 miles south of the company headquarters.

16. Let x be the number of nut bars and y be the number of chocolate bars. Therefore, $45x + 30y = 6750$ and $65x + 45y = 9950$. Multiplying the first equation by 3 and the second equation by 2 yields $135x + 90y = 20,250$ (Eq. 1) and $130x + 90y = 19,900$ (Eq. 2). Subtracting Eq. (2) from Eq. (1) yields $5x = 350$, $x = 70$. By substituting $x = 70$ into either of the original equations we get $y = 120$.

17. Let x be the number of \$328 motors sold and y be the number of \$450 motors sold. Therefore, $(0.05)328x + (0.07)450y = 1216.50$ and $328x + 450y = 20,190$. Multiplying the second equation by 0.05 yields

$$(0.05)328x + (0.05)450y = 1009.5.$$

Subtracting this from the first equation yields $(0.02)(450)y = 207$, $y = 23$ and $x = 30$.

18. $3x^2 + 2x + 15 = 48x$, $3x^2 - 46x + 15 = 0$, $(3x - 1)(x - 15) = 0$, $x = 15$. Note: $x = \frac{1}{3}$ is an untenable solution.

19. $C(t) = kt$, $62.50 = k(\frac{5}{3})$, $k = 37.5$. Hence, $C(t) = 37.5t$.

20. $\overline{C}(x) = 3x + 2 + (27/x)$. Construct table of values:

$$C(1) = 32 \qquad C(2) = 21.5 \qquad C(3) = 20$$
$$C(4) = 20.75 \qquad C(5) = 22.4 \qquad C(6) = 24.5$$

It appears that $x = 3$ minimizes $\overline{C}(x)$. The minimum average cost is \$20.

CHAPTER 2. DIFFERENTIAL CALCULUS AND ITS APPLICATIONS

2.2 THE LIMIT OF A FUNCTION AND CONTINUITY

1. -2 3. -3 5. 7 7. 24 9. $\frac{10}{7}$ 11. $\frac{10}{13}$ 13. 6 15. 6 17. 5 19. 12 21. All reals 23. All reals 25. $x \neq \frac{7}{5}$ 27. All reals 29. $x \neq 3 \pm \sqrt{29}$

CHAPTER 2. DIFFERENTIAL CALCULUS AND ITS APPLICATIONS

2.3 THE NEWTON QUOTIENT

1. 7
3. 8
5. $2x + \Delta x$
7. $-4x - 2(\Delta x)$
9. $8x + 4(\Delta x) - 3$
11. $2x + \Delta x + 5$
13. $3x^2 + 3x(\Delta x) + (\Delta x)^2$
15. $20 + 12(\Delta x) + 2(\Delta x)^2$
17. $\dfrac{-1}{3(3 + \Delta x)}$
19. $\dfrac{-7}{13(13 + \Delta x)}$
21. (a) $600 is the initial (fixed) costs. (b) The cost of the 15th unit; it is $38. (c) $38 (d) $38
23. (a) $54 (b) $4x + 2(\Delta x) + 4$ (c) $54 (d) $44 (e) $44.40
25. (a) $2100 (b) $100[4x + 2(\Delta x) + 4]$ (c) $2100
27. (a) $2x + \Delta x$ (b) 6.2 (c) 6.1 (d) 6
29. (a) $\dfrac{-1}{(x + \Delta x)x}$ (b) $\dfrac{-1}{15.96}$ (c) $-\dfrac{1}{16}$
31. (a) $96 - 32x - 16(\Delta x)$. It represents average velocity in the time interval from x to $x + \Delta x$ seconds. (b) 30.4 ft/sec (c) 1.6 ft/sec (d) -1.6 ft/sec (e) The bullet is going up in the time interval from 2.9 to 3 seconds since $\Delta x > 0$ and $\dfrac{\Delta h}{\Delta x} > 0$ implies $\Delta h > 0$ and the height above the ground is increasing. The bullet is going down in the time interval from 3 to 3.1 seconds since $\Delta x > 0$ and $\dfrac{\Delta h}{\Delta x} < 0$ implies $\Delta h < 0$ and the height above the ground is decreasing.
33. (a) $-0.32x - 0.16(\Delta x) + 4.0$ (b) $320 (c) 0 (d) $-$320 (e) Profit increases from 11,000 to 12,000 units. Profit is the same at 12,000 and 13,000 units. Profit decreases from 13,000 to 14,000 units.

2.4 THE DERIVATIVE OF A FUNCTION

1. $f'(x) = 6$ and $f''(x) = 0$
3. $g'(x) = 12x^3 - 10x + 7$ and $g''(x) = 36x^2 - 10$
5. $G'(x) = 80x^9 - 12x^2 + 5$ and $G''(x) = 720x^8 - 24x$
7. $f'(x) = 4x^{1/3} + x^{-1/2}$ and $f''(x) = \tfrac{4}{3}x^{-2/3} - \tfrac{1}{2}x^{-3/2}$
9. $g'(x) = -2x^{-4/3} + \tfrac{21}{5}x^{-8/5}$ and $g''(x) = \tfrac{8}{3}x^{-7/3} - \tfrac{168}{25}x^{-13/5}$
11. $f'(x) = 3x^2 - 3 - \dfrac{1}{x^2}$ and $f''(x) = 6x + \dfrac{2}{x^3}$
13. $f'(1) = 10$ and $f''(1) = 10$
15. $f'(8) = -1$ and $f''(8) = -\tfrac{1}{12}$
17. $m = 2$
19. $x - y = 1$
21. $15x - 4y = 12$
23. $8x - 4y = 3$

25. (a) 0 (b) All $x \neq 0$ (c) None (d) $x > 0$ (e) $x < 0$
27. (a) $\frac{3}{2}$ (b) $x > \frac{3}{2}$ (c) $x < \frac{3}{2}$ (d) All x (e) None
29. (a) $x = 0$, $x = \frac{2}{3}$ (b) $x < 0$ or $x > \frac{2}{3}$ (c) $0 < x < \frac{2}{3}$
    (d) $x > \frac{1}{3}$ (e) $x < \frac{1}{3}$

2.5 PRODUCT, QUOTIENT, AND GENERAL POWER RULES FOR DIFFERENTIATION

1. $f'(x) = \frac{15}{2}(x^3 + 6x)^{3/2}(x^2 + 2)$ 3. $f'(x) = 3(2x + 1)^{1/2}$
5. $F'(x) = \frac{6}{7}(2x^3 - 3x)^{-5/7}(2x^2 - 1)$ 7. $g'(x) = -\frac{4}{5}(3 - x)^{-1/5}$
9. $f'(x) = 3(x + 1)(2x + 3)^{-1/2}$ 11. $f'(x) = \frac{x}{2}(x + 1)^{1/2}(7x + 4)$
13. $f'(x) = \dfrac{-11}{(3x - 4)^2}$ 15. $f'(x) = \dfrac{-x^2 + 1}{(x^2 + 1)^2}$
17. (a) $f'(x) = -\dfrac{1}{x^2}$ (b) $f'(x) = -\dfrac{1}{x^2}$
19. (a) $f'(x) = x(2)(x^2 + 3)(2x) + (x^2 + 3)^2 = (x^2 + 3)(5x^2 + 3)$
    (b) $f(x) = x^5 + 6x^3 + 9x$, $f'(x) = 5x^4 + 18x^2 + 9$,
    $5x^4 + 18x^2 + 9 = (x^2 + 3)(5x^2 + 3)$
21. $f'(x) = \dfrac{(x^2 + 1)(4) - (4x - 3)(2x)}{(x^2 + 1)^2}$ and
    $m = f'(2) = \dfrac{20 - 20}{25} = 0$
23. $f'(x) = \dfrac{x}{2}(x + 3)^{-1/2} + (x + 3)^{1/2}$ and $m = f'(1) = \frac{9}{4}$
25. $f'(x) = 2x(4x + 1)^{3/2}(9x + 1)$ and $m = f'(2) = 2052$
27. $4x - 2y = 5$ 29. $10x - 3y = 1$
31. (a) $h'(x) = 80 - 32x$; $h'(0) = 80$ ft/sec is the initial velocity. (b) $h'(2) = 16$ ft/sec (c) $80 - 32x = 0$ at $x = 2.5$ seconds. This is the time it takes to reach its maximum height. (d) 5 seconds
33. $v(x) = s'(x) = 32x$ and $v'(x) = 32$ ft/sec^2

2.6 THE DIFFERENTIAL AND MARGINAL ANALYSIS

1. $\dfrac{dy}{dx} = 8x - 6 - \dfrac{1}{x^2}$; $\dfrac{d^2y}{dx^2} = 8 + \dfrac{2}{x^3}$
3. $\dfrac{dy}{dx} = -2x^{-3/2} + 4x^{-1/3}$; $\dfrac{d^2y}{dx^2} = 3x^{-5/2} - \frac{4}{3}x^{-4/3}$
5. $\dfrac{dC}{dp} = -3 + 2p - 6p^2$; $\dfrac{d^2C}{dp^2} = 2 - 12p$

CHAPTER 2. DIFFERENTIAL CALCULUS AND ITS APPLICATIONS

7. $\dfrac{dq}{dp} = 15p^2 - 2 - \dfrac{3}{p^2}$; $\dfrac{d^2q}{dp^2} = 30p + \dfrac{6}{p^3}$

9. $\dfrac{dV}{dr} = 4\pi r^2$; $\dfrac{d^2V}{dr^2} = 8\pi r$

11. (a) $C'(x) = 0.04x + 6$ and $C'(50) = \$8$
    (b) $C(51) - C(50) = \$8.02$ is the cost of the 51st unit.
    (c) $E = \tfrac{1}{2}C''(t)(\Delta x)^2 = \tfrac{1}{2}(0.04)(1)^2 = 0.02$ (d) $C'(100) = \$10$
    (e) No. Since $C'(x) > 10$ for $x > 100$, the cost per unit would exceed the selling price per unit.

13. (a) $C'(x) = 1.6x + 1.2$ and $dC = C'(5)(0.001) = 0.0092$ thousand. Hence, $C(5.001) - C(5) \doteq .0092$, or $\$9.20$.
    (b) $C(5.001) - C(5) = 0.0092008$, or $\$9.20$ to the nearest cent. (c) $E = \tfrac{1}{2}C''(t)(\Delta x)^2 = \tfrac{1}{2}(1.6)(0.001)^2 = 0.0000008$

15. (a) $R'(x) = -0.16x + 2.8$ and $R'(11) = 1.04$
    (b) $R(12) - R(11) = 0.96$

17. Since $P(x) = R(x) - C(x)$ and since $P'(x) = R'(x) - C'(x)$, $P'(x) = 0$ if $R'(x) = C'(x)$.

19. $C'(x) = 0.04x + 6 = 22$ when $0.04x = 16$, $x = 400$ units

21. (a) $f'(x) = \tfrac{1}{5}x^{-4/5}$ and $f'(32) = \tfrac{1}{80}$. Consequently,

$$\sqrt[5]{34} \doteq f(32) + f'(32)(2) = 2 + \tfrac{1}{40} = 2.025.$$

(b) $f''(x) = -(\tfrac{4}{25})x^{-9/5}$. Since $|f''(t)| = \dfrac{4}{25t^{9/5}}$ where $32 < t < 34$ is less than $\dfrac{4}{25(32)^{9/5}} = \dfrac{1}{3200}$,

$$E < \dfrac{1}{2}\left(\dfrac{1}{3200}\right)(2)^2 = 6.25 \times 10^{-4}.$$

Note: Since $f''(t)$ is negative, 2.025 exceeds $\sqrt[5]{34}$ and we conclude that

$$2.024375 < \sqrt[5]{34} < 2.025.$$

23. $f'(x) = -\dfrac{1}{x^2}$ and $f'(4) = -\tfrac{1}{16}$. Therefore

$$f(4.01) \doteq f(4) + f'(4)(.01) = 0.25 - 0.000625 = 0.249375.$$

Note: $1/(4.01) = 0.249376$, correct to six decimal places.

2.7 SUMMARY AND REVIEW EXERCISES

1. $\frac{7}{26}$ 2. $\frac{8}{7}$ 3. All x. 4. $x \ne 3$, $x \ne -2$.

5. $\dfrac{\dfrac{x + \Delta x}{3(x + \Delta x) + 4} - \dfrac{x}{3x + 4}}{\Delta x} = \dfrac{4}{(3 + 3\Delta x + 4)(3x + 4)}$

6. $C'(x) = 18x^2 - 10x + 3$ and $C''(x) = 36x - 10$

7. $\dfrac{dy}{dx} = -5x^{3/2} - \dfrac{6}{x^3}$ and $\dfrac{d^2y}{dx^2} = -\dfrac{15}{2}x^{1/2} + \dfrac{18}{x^4}$

8. $f'(x) = \frac{2}{3}(3x^2 + 5)^{-1/3}(6x) = 4x(3x^2 + 5)^{-1/3}$

9. $\dfrac{dy}{dx} = \dfrac{(x^2 + 4)(3) - (3x + 1)(2x)}{(x^2 + 4)^2} = \dfrac{-3x^2 - 2x + 12}{(x^2 + 4)^2}$

10. $\dfrac{du}{dv} = \dfrac{(3v + 1) - 3v}{(3v + 1)^2} = \dfrac{1}{(3v + 1)^2} = (3v + 1)^{-2}$,

    $\dfrac{d^2u}{dv^2} = -2(3v + 1)^{-3}(3) = -6(3v + 1)^{-3}$

11. (a) $\dfrac{C(14) - C(10)}{4} = \dfrac{17.84}{4} = 4.46$

    (b) $\dfrac{\Delta C}{\Delta x} = 0.08x + 0.04(\Delta x) + 3.5$

    (c) $\dfrac{\Delta C}{\Delta x} = 0.08(10) + 0.04(4) + 3.5 = 4.46$

12. $\dfrac{4}{(15 + 0.03 + 4)(15 + 4)} = \dfrac{4}{(19.03)(19)} \doteq 0.011$

13. $f'(x) = 3x^{1/2} + \dfrac{4}{x^2}$, $m = f'(4) = \frac{25}{4}$, $f(4) = 15$, and an equation of the tangent line is $25x - 4y = 40$.

14. (a) $C(11) - C(10) = 79.34 - 75 = \$4.34$
    (b) $C'(x) = 0.08x + 3.5$ and $C'(10) = \$4.30$

15. (a) $P'(x) = 0.24x + 5.2$ and $P'(20) = 10$
    (b) $P(21) - P(20) = 152.12 - 142 = 10.02$

16. $C'(x) = 0.08x + 3.5$ and $0.08x + 3.5 = 6.3$ where $0.08x = 2.8$, $x = 35$

17. (a) $C'(x) = 20 + 2x^{-1/2}$ and
    $C(104) - C(100) \doteq C'(100)(4) = 80.8$
    (b) $C''(x) = -x^{-3/2}$. Thus, $E < |\frac{1}{2}C''(100)(4)^2| = 0.008$. Since $C''(x) < 0$, the error is negative and

    $$80.792 \le C(104) - C(100) \le 80.8.$$

    (c) $C(104) - C(100) \doteq 80.792156$

CHAPTER 3. FURTHER APPLICATIONS OF THE DERIVATIVE

18. (a) $f'(x) = \dfrac{-2x}{(x^2 + 3)^2}$; $f(3.01) \doteq f'(3) + f'(3)(0.01) \doteq 0.082916$
    (b) $f(3.01) \doteq 0.082918$
19. (a) $h'(x) = 112 - 32x$ and $h'(0) = 112$ ft/sec is the initial velocity. (b) $h'(2) = 48$ ft/sec (c) $h(3.5) = 196$ ft
20. Since $f'(x) = 3x^2 - 12x$ and since $15x - y = 11$ has 15 as slope, we need to solve $3x^2 - 12x = 15$ for x. We find $x^2 - 4x - 5 = 0$, $(x - 5)(x + 1) = 0$, $x = 5$ or $x = -1$. Since $f(5) = -23$ and $f(-1) = -5$, we find the two lines with slope 15 containing the points $(5, -23)$ and $(-1, -5)$; they are $15x - y = 98$ and $15x - y = -10$.

CHAPTER 3. FURTHER APPLICATIONS OF THE DERIVATIVE

3.2 INCREASING AND DECREASING FUNCTIONS

1. (a) $x \geq 0$ (b) $x \leq 0$ 3. (a) All x (b) None
5. (a) $-8 \leq x \leq 8$ (b) None 7. (a) $x \geq 0$ (b) $x \leq 0$
9. (a) $x \leq 3$ (b) $x \geq 3$ 11. (a) $x < 0$ (b) $x > 0$
13. (a) $-\tfrac{1}{2} \leq x \leq 0$, $2 \leq x \leq 4$ (b) $0 \leq x \leq 2$
15. (a) $x \leq -3$ or $x \geq \tfrac{1}{3}$ (b) $-3 \leq x \leq \tfrac{1}{3}$
17. (a) $-\tfrac{3}{2} \leq x \leq -1$, $1 \leq x \leq 4$ (b) $-1 \leq x \leq 1$
19. (a) $x \leq -2$, $x \geq \tfrac{4}{3}$ (b) $-2 \leq x \leq \tfrac{4}{3}$
21. (a) $x < -5$, $x > -5$ (b) None
23. (a) $x \geq 1$ (b) $x < 0$, $0 < x \leq 1$
25. $V'(r) = 4\pi r^2$ 27. (a) 1.598% (b) 3.142%
29. (a) 68 units (b) 80 units (c) 68 units
    (d) $f'(t) = 60 + 18t - 3t^2 = -3(t^2 - 6t + 9) + 60 + 27$
    $= -3(t - 3)^2 + 87 > 0$ for $0 \leq t \leq 8$.
    (e) $f''(t) = 18 - 6t$. $f''(t) > 0$ if $t < 3$ and $f''(t) < 0$ if $t > 3$. Rate of increase in production increases during the first 3 hours and then decreases.
31. $f'(x) = \dfrac{12{,}960x}{(x^2 + 108)^2} > 0$ if $x > 0$
33. (a) $f''(x) = \dfrac{[38{,}880(36 - x^2)]}{(x^2 + 108)^3}$. $f''(x) > 0$ if $0 \leq x \leq 6$ and $f''(x) < 0$ if $x > 6$ (b) $f'(6) = 3.75$ is approximately the increase in the number of words per minute during the sixth week. This is the greatest weekly increase.
35. (a) $f'(6) = 3.75$ and $f'(20) = 1.004$ (b) 10.71% and 1.49%

3.3 RELATIVE MAXIMA AND RELATIVE MINIMA

1. (a) None (b) (0, 0) (c) None (d) $f(0) = 0$ (e) None (f) Always concave upward

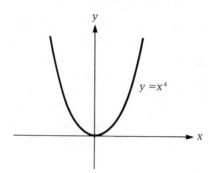

Figure for answer to Exercise 1

3. (a) None (b) None (c) None (d) None (e) $(\frac{1}{2}, 0)$ (f) Concave downward for $x < \frac{1}{2}$ and concave upward for $x > \frac{1}{2}$

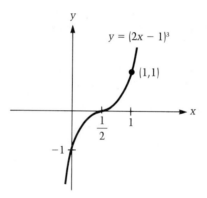

Figure for answer to Exercise 3

5. (a) None (b) None (c) $f(8) = 32$ (d) $f(-8) = -32$ (e) (0, 0) (f) Concave downward for $x < 0$ and concave upward for $x > 0$

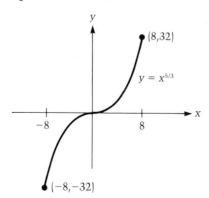

Figure for answer to Exercise 5

7. (a) None (b) None (c) None
   (d) None (e) None
   (f) Concave upward for $x < 0$ and for $x > 0$

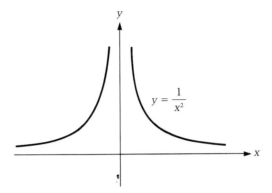

Figure for answer to Exercise 7

9. (a) $(0, 0)$ (b) $(2, -4)$ (c) $f(4) = 16$
   (d) $f(2) = -4$ (e) $(1, -2)$
   (f) Concave downward for $-\frac{1}{2} < x < 1$ and concave upward for $1 < x < 4$.

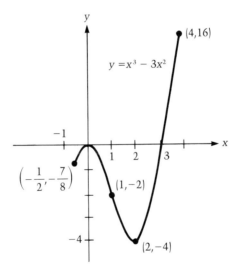

Figure for answer to Exercise 9

11. (a) $(-3, 19)$ (b) $(\frac{1}{3}, \frac{13}{27})$ (c) None
    (d) None (e) $(-\frac{4}{3}, \frac{263}{27})$
    (f) Concave downward for $x < -\frac{4}{3}$ and concave upward for $x > -\frac{4}{3}$

C18 APPENDIX C ANSWERS TO ODD-NUMBERED EXERCISES

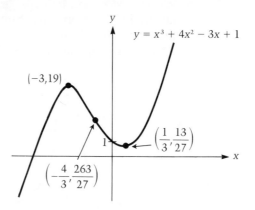

Figure for answer to Exercise 11

13. (a) $(-1, 4)$ (b) $(1, -4)$ (c) $f(4) = 104$ (d) $f(1) = -4$ (e) $(0, 0)$
(f) Concave downward for $x < 0$ and concave upward for $x > 0$

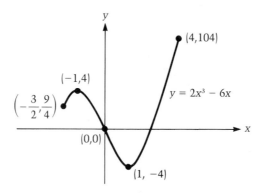

Figure for answer to Exercise 13

15. (a) $(-2, 17)$ (b) $(\frac{4}{3}, -\frac{41}{27})$ (c) None (d) None (e) $(-\frac{1}{3}, \frac{209}{27})$
(f) Concave downward for $x < -\frac{1}{3}$ and concave upward for $x > -\frac{1}{3}$

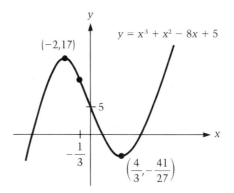

Figure for answer to Exercise 15

CHAPTER 3. FURTHER APPLICATIONS OF THE DERIVATIVE C19

17. (a) None (b) (3, −26) (c) None
(d) $f(3) = -26$ (e) (0, 0) and (2, −15)
(f) Concave downward where $0 < x < 2$ and concave upward for $x < 0$ or $x > 2$

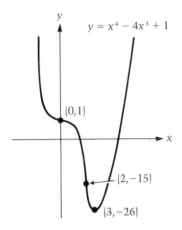

Figure for answer to Exercise 17

19. (a) None (b) (1, 3) (c) None
(d) None (e) $(-\sqrt[3]{2}, 0)$
(f) Concave downward where $-\sqrt[3]{2} < x < 0$; concave upward for $x > 0$ and $x < -\sqrt[3]{2}$

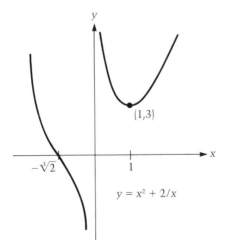

Figure for answer to Exercise 19

21. (a) $p'(t) = \dfrac{2400t}{(t^2 + 12)^2} > 0$. Yes (b) $p''(t) = \dfrac{7200(4 - t^2)}{(t^2 - 12)^3}$.
Since $p''(t) > 0$ where $0 < t < 2$ and $p''(t) < 0$ for $t > 2$, $p'(2) = 18.75$ is approximately the greatest increase.

23. $P(x) = R(x) - C(x)$ and $P'(x) = R'(x) - C'(x)$. Since $P'(t) = 0$ when $R'(t) = C'(t)$, $P(t)$ can be a maximum or minimum at such a t. Assuming $P'(x) > 0$ for $x < t$ and $P'(x) < 0$ for $x > t$ makes $P(t)$ a maximum value for the profit function.

25. (a) $\overline{C}(x) = \dfrac{C(x)}{x} = x^2 + 6 + \dfrac{54}{x}$.

   (b) $\overline{C}'(x) = 2x - \dfrac{54}{x^2} = 0$ at $x = 3$. Since $\overline{C}''(x) = 2 + \dfrac{108}{x^3} > 0$ for $x > 0$, $x = 3$ thousand pencils minimize average cost.

   (c) The minimum average cost is $\overline{C}(3) = \$33$ per thousand or 3.3 cents apiece.

3.4 MORE APPLIED MAXIMUM AND MINIMUM PROBLEMS

1. $\overline{C}'(q) = 2 - \dfrac{50}{q^2}$ and $\overline{C}''(q) = \dfrac{100}{q^3}$. Since $\overline{C}'(5) = 0$ and $\overline{C}''(5) > 0$, $\overline{C}(5) = 24$ is a relative minimum. In fact, since $\overline{C}''(x) > 0$ for $x > 0$, 24 is the absolute minimum.

3. Let x be the number of \$5 increases for each apartment. The rental income is $R(x) = (180 - x)(600 + 5x) = 108{,}000 + 300x - 5x^2$. $R'(x) = 300 - 10x$ and $R''(x) = -10$. $R'(x) = 0$ at $x = 30$. A rent of $600 + 5(30) = 750$ would maximize the rental income; it is $R(30) = (150)(750) = \$112{,}500$ per month.

5. Let x be the number of persons over 100 attending. $R(x) = (100 + x)(3 - 0.01x) = 300 + 2x - 0.01x^2$. $R'(x) = 2 - 0.02x$ and $R''(x) = -0.02$. $R'(x) = 0$ at $x = 100$. Thus, $R(100) = (200)(2) = \$400$ is the maximum income. The minimum income is an endpoint extremum. Since $R(0) = 300$ and $R(200) = 300$, the minimum income is \$300; it results with either 100 or 300 persons attending.

7. (a) $C(5) = 2.5 - 18.5 + 63 = \$47{,}000$.
   (b) $C'(x) = 0.2x - 3.7$ and $C''(x) = 0.2$. $C'(x) = 0$ at $x = 18.5$. Thus, 18,500 units will minimize cost.
   (c) $C(18.5) = 28.775$, or \$28,775.

9. (a) $R(x) = C(x)$ when
   $0.1x^2 - 3.7x + 63 = -0.1x^2 + 5.1x - 5$,
   $0.2x^2 - 8.8x + 68 = 0$, $x^2 - 44x + 340 = 0$,
   $(x - 10)(x - 34) = 0$, $x = 10$ or $x = 34$
   (b) $P(x) = -0.2x^2 + 8.8x - 68$ (c) $P'(x) = -0.4x + 8.8$ and $P''(x) = -0.4$. $P'(x) = 0$ at $x = 22$, or 22,000 units. The maximum profit is $P(22) = 33.8$, or \$33,800.

CHAPTER 3. FURTHER APPLICATIONS OF THE DERIVATIVE

11. Let x be the number of 10-cent decreases in the price per bulb. Then the revenue function is
$$R(x) = (8.60 - 0.10x)(2000 + 50x)$$
and the cost function is
$$C(x) = 2(2000 + 50x).$$
The profit function is
$$P(x) = R(x) - C(x) = 13{,}200 + 130x - 5x^2.$$
$P'(x) = 130 - 10x$ and $P''(x) = -10$. $P'(x) = 0$ at $x = 13$. The selling price should be $\$8.60 - 13(0.10) = \$7.30$ to obtain a maximum profit of $P(13) = \$14{,}045$.

13. $x^2 h = 200$ and $h = \dfrac{200}{x^2}$. $2.5xh$ is the cost of each of the four sides. $2(4x^2)$ is the total cost of top and bottom.
$$C(x) = 10x\left(\frac{200}{x^3}\right) + 8x^2 = \frac{2000}{x} + 8x^2.$$
$$C'(x) = -\frac{2000}{x^2} + 16x.$$
$C'(x) = 0$ when $16x^3 = 2000$, $x^3 = 125$, and $x = 5$. The dimensions of the top and bottom should be 5 inches by 5 inches and the height should be $h = \frac{200}{25} = 8$ inches.

15. Let x be the width of the *printed page* and let y be its (vertical) length. Then $xy = 48$ and $y = \dfrac{48}{x}$. The area of the page is
$$A(x) = (x + \tfrac{3}{2})\left(\frac{48}{x} + 2\right) = 51 + \frac{72}{x} + 2x.$$
$A'(x) = -\dfrac{72}{x^2} + 2$ and $A'(x) = 0$ when $2x^2 = 27$, $x^2 = 36$, and $x = 6$. Thus $y = 8$. The dimensions of the page should be 7.5 inches by 10 inches.

17. $V(x) = x(17 - 2x)(12 - 2x) = 4x^3 - 58x^2 + 204x$, $0 \le x \le 6$,
$V'(x) = 12x^2 - 116x + 204 = 4(3x^2 - 29x + 51)$,
$V''(x) = 24x - 116.$

$V'(x) = 0$ where $3x^2 - 29x + 51 = 0$, $x = \dfrac{29 \pm \sqrt{229}}{6}$, $x \doteq 2.31$ or $x \doteq 7.76$. The only possible answer is 2.31; furthermore, $V''(2.31) < 0$. Thus, approximately 2.31 inches for the lengths of the squares will maximize the volume.

19. (a) $x(10) = 640$, $x(30) = 320$, and $x(50) = 0$

    (b) $16p = 800 - x$ and $p(x) = 50 - \dfrac{x}{16}$

    (c) $P(x) = 50x - \dfrac{x^2}{16} - 18x - 2400 = 32x - \dfrac{x^2}{16} - 2400$.

    $P'(x) = 32 - \dfrac{x}{8}$ and $P''(x) = -\tfrac{1}{8}$. $P'(x) = 0$ at $x = 256$. Thus 256 units per week maximizes profit. (d) $p(256) = \$34$ is the selling price to maximize profit. (e) $P(256) = \$1696$ is the maximum weekly profit.

21. (a) Let x be the number over 150 taking the cruise. Then

    $$R(x) = (150 + x)(8 - 0.02x) = 1200 + 5x - 0.02x^2$$
    $$R'(x) = 5 - 0.04x \text{ and } R''(x) = -0.04.$$
    $$R'(x) = 0 \text{ at } x = 125.$$

    Thus, $150 + 125 = 275$ persons will maximize income.
    (b) $\$8.00 - 0.02(125) = \$5.50$ (c) $R(125) = \$1512.50$

23. Let y be the length of the rectangular part of the field and let x be the width, which is also the diameter of each semicircular end. Since the perimeter is $2y + \pi x = 440$, $y = \dfrac{440 - \pi x}{2}$. Thus, the area is $A(x) = \tfrac{1}{2}(440x - \pi x^2)$. $A'(x) = 220 - \pi x$ and $A''(x) = -\pi$. Since $A'(x) = 0$ at $x = \dfrac{220}{\pi} \doteq 70.03$ and since $A''(x)$ is negative, the dimensions should be $x = 70.03$ by $y = 110$ to maximize the area.

25. Let k be the cost per square unit of the side. Then $3k$ is the cost per square unit of the ends. If the volume is V, then $\pi r^2 h = V$ and $h = \dfrac{V}{\pi r^2}$. The cost of the two ends is $3k(2\pi r^2)$ and the cost of the side is $k(2\pi rh)$. The total cost is $C(r) = 6\pi kr^2 + 2kV/r$. $C'(r) = 12k\pi r - 2kV/r^2$ and $C''(r) = 12\pi k + 4kV/r^3$. Note: $C''(r) > 0$ for $r > 0$. $C'(r) = 0$

when $12\pi kr = 2kV/r^2$, $r^3 = V/6\pi$. This is the cube of the radius which minimizes cost. Since $h = V/\pi r^2$, $h/r = V/\pi r^3 = 6$ is the ratio which minimizes cost.

27. (a) $\dfrac{C(10,000)}{10,000} = \dfrac{175,000 + 100,000}{10,000} = \$27.50$.
$C(x + 1) - C(x) = \$10$.
(b) $x(10) = 10,000 + 60,000(\frac{2}{3}) = 50,000$ and $x(30) = 10,000$.
$1 - p/30 = \dfrac{x - 10,000}{60,000}$, $\dfrac{p}{30} = 1 - \dfrac{x - 10,000}{60,000}$, and
$p(x) = 35 - \dfrac{x}{2000}$. Note: $P(50,000) = \$10$.
Note: $P(50,000) = \$10$.
(c) $P(x) = 25x - \dfrac{x^2}{2000} - 175,000$. $P'(x) = 25 - \dfrac{x}{1000}$ and $P''(x) = -\frac{1}{1000}$. $P'(x) = 0$ at $x = 25,000$. This is the number which maximizes profit; the price which maximizes profit is $p(25,000) = \$22.50$

29. Since the length is fixed, we find x to maximize the area $A(x) = x(16 - 2x) = 16x - 2x^2$. $A'(x) = 16 - 4x$ and $A''(x) = 4$. $A'(x) = 0$ at $x = 4$. The maximum volume is $4 \times 8 \times 336 = 10,752$ in^3.

3.5 RELATED RATES (OPTIONAL)

1. -14.4 cm^3/min   3. $\frac{25}{72}$ cm/sec   5. 47 in^2/min
7. 8 cm^2/min   9. $\dfrac{52}{9\pi}$ ft/min   11. $\frac{8}{3}$ ft^2/min
13. $15 - \dfrac{9\pi}{5} \doteq 9.345$ ft^3/sec is draining out

3.6 SUMMARY AND REVIEW EXERCISES

1. (a) $f'(x) = 6x^2 - 4x - 32 = 2(3x^2 - 2x - 16) = 2(3x - 8)(x + 2)$. f is increasing where $x \leq -2$ and where $x \geq \frac{8}{3}$. (b) f is decreasing on $[-2, \frac{8}{3}]$.
2. (a) $f'(x) = \dfrac{(x^2 + 1) - x(2x)}{(x^2 + 1)^2} = \dfrac{1 - x^2}{(1 + x^2)^2}$. f is increasing on $[-1, 1]$.
(b) f is decreasing where $x \geq 1$ or $x \leq -1$.
3. $P(x) = \dfrac{f'(x)}{f(x)}(100) = \dfrac{200x}{x^2 + 1}$. $P'(x) = \dfrac{(x^2 + 1)200 - 400x^2}{(x^2 + 1)^2}$ and $P'(x) > 0$ where $200 - 200x^2 > 0$, $x^2 < 1$, $-1 < x < 1$. Hence, percentage rate of change increases on $-1 \leq x \leq 1$.

4. $f'(x) = \dfrac{(x^2 + 3) - x(2x)}{(x^2 + 3)^2} = \dfrac{3 - x^2}{(x^2 + 3)^2}$ is the slope function.

$f''(x) = \dfrac{(x^2 + 3)^2(-2x) - (3 - x^2)2(x^2 + 3)(2x)}{(x^2 + 3)^4}$

$= \dfrac{(-2x)(x^2 + 3) - 4x(3 - x^2)}{(x^2 + 3)^3} = \dfrac{2x(x^2 - 9)}{(x^2 + 3)^3}$

$= \dfrac{2x(x - 3)(x + 3)}{(x^2 + 3)^3}$

Since $f''(x) > 0$ for $-3 < x < 0$ or $x > 3$, the slope increases when $-3 \leq x \leq 0$ or $x \geq 3$.

5. (a) $(-\sqrt{2}, 4\sqrt{2})$ (b) $(\sqrt{2}, -4\sqrt{2})$ (c) None (d) None (e) $(0, 0)$ (f) Concave upward for $x > 0$. Concave downward for $x < 0$

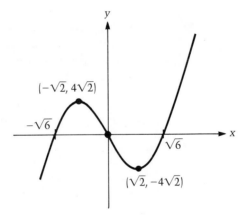

Figure for answer to Exercise 5

6. (a) $(\sqrt{3}, 6\sqrt{3})$ (b) $(-\sqrt{3}, -6\sqrt{3})$ (c) None (d) None (e) $(0, 0)$ (f) Concave downward for $x < 0$. Concave upward for $x > 0$.

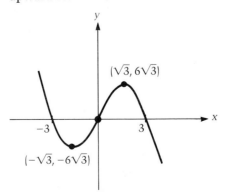

Figure for answer to Exercise 6

CHAPTER 3. FURTHER APPLICATIONS OF THE DERIVATIVE · C25

7. (a) (0, 0) (b) (−2, −4), (2, −4) (c) None
(d) $f(-2) = f(2) = -4$ (e) $(-\sqrt[4]{\frac{4}{3}}, -\frac{20}{9}), (\sqrt[4]{\frac{4}{3}}, -\frac{20}{9})$ (f) See figure.

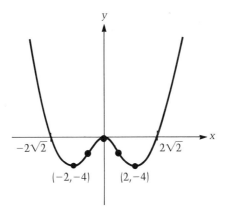

Figure for answer to Exercise 7

8. (a) None (b) $(\sqrt[3]{4}, 7.56)$ (c) None (d) None (e) (−2, 0)
(f) Concave downward: $-2 < x < 0$. Concave upward: $x < -2, x > 0$

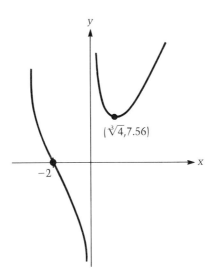

Figure for answer to Exercise 8

9. (a) (0, 0) (b) (2, −16), (−2, −16) (c) $f(-4) = 128$
(d) $f(-2) = f(2) = -16$ (e) $(\sqrt[4]{\frac{4}{3}}, -\frac{80}{9}), (-\sqrt[4]{\frac{4}{3}}, -\frac{80}{9})$ (f) See figure.

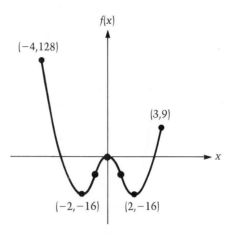

Figure for answer to Exercise 9

10. (a) $f'(x) = -3x^2 + 12x = -3x(x - 4)$. $f''(x) = -6x + 12$. Since $f'(4) = 0$ and $f''(4) = -12 < 0$, $(4, 60)$ is a relative maximum point. (b) Since $f'(0) = 0$ and $f''(0) = 12$, $(0, 0)$ is a relative minimum point. (c) $f(4) = 32$. (d) $f(0) = f(6) = 0$. (e) $(2, 16)$ is an inflection point. (f) Concave upward: $(-1, 2)$. Concave downward: $(2, 6)$.

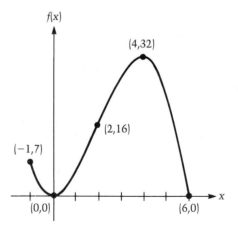

Figure for answer to Exercise 10

11. $V(x) = x(12 - 2x)(18 - 2x) = 216x - 60x^2 + 4x^3$, $0 < x < 6$

$V'(x) = 12x^2 - 120x + 216 = 12(x^2 - 10x + 18)$.

$V'(x) = 0$ where $x = \dfrac{10 \pm \sqrt{100 - 72}}{2} = 5 \pm \sqrt{7}$.

$x \doteq 2.35$ inches

12. Let x be the width and length of the base and let h be the altitude. Cost of the base is $3x^2$ and cost of the four sides

is $4(xh)(2) = 8xh$. Since $3x^2 + 8xh = 36$, $h = \dfrac{36 - 3x^2}{8x}$.

Hence, volume $V(x) = x^2 h = (\tfrac{1}{8})(36x - 3x^3)$.
$V'(x) = (\tfrac{1}{8})(36 - 9x^2)$ and $V'(x) = 0$ at $x = 2$. Thus, 2 meters by 2 meters by 1.5 meters should be the dimensions and 6 m³ is the maximum volume for such a box.

13. Let x be the width of the page. Then $x - 1$ is the width of the print, $\dfrac{96}{x-1}$ is the length of the print, and $\dfrac{96}{x-1} + \dfrac{3}{2}$ is the length of the page. Hence
$A(x) = 96x(x-1) + \dfrac{3x}{2}$ and $A'(x) = -\dfrac{96}{(x-1)^2} + \dfrac{3}{2}$.
$A'(x) = 0$ if $2(96) = 3(x-1)^2$, $(x-1)^2 = 64$, $x - 1 = \pm 8$. Since $A'(9) = 0$ and $A''(9) > 0$, a width of 9 inches and a length of 13.5 inches minimize the surface area of the paper and, thus, its cost.

14. Let x be the number of machines. Installation cost is $200x$; x machines produce $90x$ units per hour and would require $\dfrac{181{,}500}{90x} = \dfrac{6050}{3x}$ hours to do the job. Therefore
$C(x) = 200x + \dfrac{12(6050)}{3x} = 200x + \dfrac{24{,}200}{x}$ and
$C'(x) = 200 - \dfrac{24{,}200}{x^2}$. $C'(x) = 0$ at $x = 11$ and $C''(11) > 0$.
Therefore, $C(11) = \$4400$ is the minimum cost.

15. Let x be the number of 25-cent reductions. Therefore $P(x) = (14.75 - 0.25x)(1560 + 40x)$ is the profit function. Hence, $P'(x) = 180 - 20x$. $P'(x) = 0$ at $x = 9$ and $P''(9) < 0$. Therefore, $38.75 - 9(0.25) = \$36.50$ is the selling price that maximizes profits and $P(9) = \$23{,}040$ is the maximum monthly profit possible.

16. $\dfrac{PR}{x} = \dfrac{6}{(x^2 - 9)^{1/2}}$, $PR = \dfrac{6x}{(x^2 - 9)^{1/2}}$, and

$L(x) = x + \dfrac{6x}{(x^2 - 9)^{1/2}}$ and $L'(x) = 1 + \dfrac{-54}{(x^2 - 9)^{3/2}}$.

$L'(x) = 0$ where $(x^2 - 9)^{3/2} = 54$, $x^2 - 9 = 54^{2/3}$, $x^2 = 9 + 9(2)^{2/3}$, and $x = 3(x + 2^{2/3})^{1/2}$. The first derivative test shows that this x minimizes the length; the minimum length is approximately 12.5 feet.

17. Let x be the number of \$5 increases.
$R(x) = (600 + 5x)(200 - x)$, $R'(x) = 400 - 10x$, and $R''(x) = -10$. Since $R'(40) = 0$ and $R''(40) < 0$, it follows that $600 + 40(5) = \$800$ should be the rent to maximize revenues.

18. Let P be the nearest point on shore, let D be his destination point, and let x be the distance in miles from P to the point T on shore to which he should row in order to minimize the time required. Using the Pythagorean theorem, we find that the distance from the island to point T is $(x^2 + 9)^{1/2}$. The distance from T to D is $4 - x$ miles. Since distance $\div$ rate = time, the total time required to make the trip is given by

$$f(x) = \tfrac{1}{2}(x^2 - 9)^{1/2} + \tfrac{1}{4}(4 - x). \text{ Thus } f'(x) = \frac{x}{2}(x^2 + 9)^{-1/2} - \tfrac{1}{4}.$$

$f'(x) = 0$ where $2x = (x^2 + 9)^{1/2}$, $4x^2 = x^2 + 9$, $3x^2 = 9$, $x^2 = 3$, and $x = \sqrt{3}$. Since $f''(x) > 0$ for each x in the interval, $f(\sqrt{3})$ is not only a relative minimum but the absolute minimum. Thus, $\sqrt{3}$ miles is the distance from P on the shore toward which he should row in order to minimize the time.

19. Since $x = \left(10 - \dfrac{p}{300}\right)^2$, $x^{1/2} = 10 - \dfrac{p}{300}$,
$p(x) = 3000 - 300x^{1/2}$, and since $P(x) = xp(x) - C(x)$, $P(x) = 2700x - 300x^{3/2} - 1400$. Therefore, $P'(x) = 2700 - 450x^{1/2}$. Since $P''(x) < 0$ and $P'(x) = 0$ at $x = 36$, $p(36) = \$1200$ is the selling price to maximize profits and $P(36) = \$31{,}000$ is the maximum profit possible.

20. If $PR = x$, then $PQ = 500 - x$. By the Pythagorean theorem, $PS = ((500 - x)^2 + 800^2)^{1/2}$ and $PT = (x^2 + 300^2)^{1/2}$. Hence the total pumping distance is

$$f(x) = (x^2 + 300^2)^{1/2} + ((500 - x)^2 + 800^2)^{1/2}.$$

Differentiating, we obtain

$$f'(x) = \frac{x}{(x^2 + 300^2)^{1/2}} + \frac{(500 - x)(-1)}{((500 - x)^2 + 800^2)^{1/2}}.$$

Now $f'(x) = 0$ for $0 < x < 500$ if and only if

$$\frac{x}{(x^2 + 300^2)^{1/2}} = \frac{500 - x}{((500 - x)^2 + 800^2)^{1/2}},$$

$$\frac{x^2}{x^2 + 300^2} = \frac{(500 - x)^2}{(500 - x)^2 + 800^2},$$

$$x^2(500 - x)^2 + 800^2 x^2 = x^2(500 - x)^2 + 300^2(500 - x)^2,$$

$$800^2 x^2 = 300^2(500 - x)^2,$$

$$800x = 300(500 - x),$$

$$8x = 1500 - 3x,$$

$$11x = 1500,$$

$$x = \frac{1500}{11} \doteq 136.36.$$

The physical aspects of the problem allow us to conclude that the pumping station should be 136.36 m from R.

CHAPTER 4. EXPONENTIAL AND LOGARITHMIC FUNCTIONS

4.2 THE CHAIN RULE AND INVERSE FUNCTIONS

1. (a) 27 (b) 56   3. (a) 4 (b) 49   5. (a) 2 (b) 27
7. (a) $F(x) = 6x - 13$ (b) $H(x) = 6x - 1$
9. (a) $F(x) = 9x^2 - 6x + 1$ (b) $H(x) = 3x^2 - 1$
11. (a) $F(x) = x^{3/2}$ (b) $H(x) = x^{3/2}$
13. (a) $F(x) = 2x^3 + 1$ (b) $H(x) = 8x^3 - 12x^2 + 6x$
15. (a) $F(x) = (x^3 + 4)^{1/2}$ (b) $H(x) = x^{3/2} + 4$
17. (a) $f'(x) = 2x$, $f'(g(x)) = 6x - 2$, and $g'(x) = 3$
    (b) $F'(x) = 18x - 6$ (c) $F'(x) = f'(g(x))g'(x)$
19. (a) $g'(x) = 3x^2$, $g'(f(x)) = 3(2x - 1)^2$, and $f'(x) = 2$
    (b) $H'(x) = 24x^2 - 24x + 6$
21. (a) $3x + 6 = 5$ at $x = -\frac{1}{3}$ and $f'(x) = 3$. Thus
    $g'(5) = \dfrac{1}{f'(-\frac{1}{3})} = \frac{1}{3}$. (b) $g(x) = \dfrac{(x - 6)}{3}$, $g'(x) = \frac{1}{3}$, and $g'(5) = \frac{1}{3}$

23. (a) $x^2 + 1 = 5$ at $x = 2$ and $f'(x) = 2x$. Thus $g'(5) = \dfrac{1}{f'(2)} = \dfrac{1}{4}$.
(b) $g(x) = (x + 1)^{1/2}$, $g'(x) = (\tfrac{1}{2})(x - 1)^{-1/2}$, and $g'(5) = \tfrac{1}{4}$
25. (a) $\sqrt{4x - 3} = 5$ at $x = 7$ and $f'(x) = 2(4x - 3)^{-1/2}$. Thus $g'(5) = \dfrac{1}{f'(7)} = \dfrac{1}{(\tfrac{2}{5})} = \dfrac{5}{2}$ (b) $g(x) = \dfrac{x^2 + 3}{4}$, $g'(x) = \dfrac{x}{2}$, and $g'(5) = \tfrac{5}{2}$

4.3 EXPONENTIAL AND LOGARITHMIC FUNCTIONS

1. 3   3. $\tfrac{1}{1000}$ 5. 2   7. 7
9. -2 11. 5   13. 2   15. e^3
17. e^2 19. 275.42   21. 300   23. 0.09
25. 0.1   27. 101.75   29. 346

4.4 DIFFERENTIATION OF THE EXPONENTIAL AND LOGARITHMIC FUNCTIONS

1. $f'(x) = 5e^{5x}$ and $f''(x) = 25e^{5x}$
3. $f'(x) = 3x^2 e^{x^3}$ and $f''(x) = e^{x^3}(9x^4 + 6x)$
5. $f'(x) = -3xe^{-3x} + e^{-3x}$ and $f''(x) = e^{-3x}(9x - 6)$
7. $f'(x) = \dfrac{x}{x + 1} + \ln(x + 1)$ and $f''(x) = \dfrac{x + 2}{(x + 1)^2}$
9. $f'(x) = 3^x(\ln 3)$ and $f''(x) = 3^x(\ln 3)^2$
11. $f'(x) = \dfrac{1 + e^x}{x + e^x}$ and $f''(x) = \dfrac{xe^x - 2e^x - 1}{(x + e^x)^2}$
13. $f'(x) = 1 + \ln x$ and $f''(x) = \dfrac{1}{x}$
15. $f'(x) = 1 - x - 2x(\ln x)$ and $f''(x) = -3 - 2(\ln x)$
17. $f'(x) = \dfrac{2(\ln x)}{x}$ and $f''(x) = \dfrac{2(1 - \ln x)}{x^2}$
19. $f'(x) = \dfrac{\ln x - 1}{(\ln x)^2}$ and $f''(x) = \dfrac{2 - \ln x}{x(\ln x)^3}$
21. $f'(x) = 1 + \ln x$ and $f''(x) = \dfrac{1}{x}$, $f'(x) = 0$ at $\ln x = -1$, $x = e^{-1}$. No relative or absolute maximum. $f\left(\dfrac{1}{e}\right) = -\dfrac{1}{e}$ is a relative and absolute minimum. The curve is always concave upward and has no inflection points. f is

decreasing where $0 < x \leq \dfrac{1}{e}$ and f is increasing where $x \geq \dfrac{1}{e}$. (See graph.)

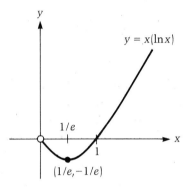

Figure for answer to Exercise 21

23. $f'(x) = xe^x + e^x = e^x(x + 1)$, $f''(x) = xe^x + 2e^x = e^x(x + 2)$. $f(-1) = -\dfrac{1}{e}$ is a relative and absolute minimum. There is no absolute or relative maximum. $\left(-2, -\dfrac{2}{e^2}\right)$ is an inflection point. f increases where $x \geq -1$ and f decreases where $x \leq -1$. (See graph.)

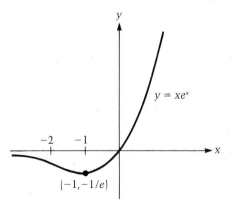

Figure for answer to Exercise 23

25. $f'(x) = 3e^{3x}$ and $f'(0) = 3$. Tangent line is $y = 3x$.

4.5 EXPONENTIAL GROWTH AND CONTINUOUS INTEREST

1. (a) 10% (b) 0.10986 (c) 30 days (d) See Exercise 17 for a general proof.

3. (a) $f(x) = 132{,}165{,}000\left(\dfrac{151{,}326}{132{,}165}\right)^{x/10} = 132{,}165{,}000(1.145)^{x/10}$
   (b) $f(40) = 227{,}163{,}000$ (d) $f(60) = 297{,}817{,}000$

5. $A = 10{,}000e^{0.3} = \$13{,}498.58$ 7. $P = \$2744.06$

9. $4717.57 = 4000e^{2r}$, $e^{2r} = 1.17939$, $r = 0.0825$, or 8.25%

11. $g(x) = 1000e^{kx}$ and $5000 = 1000e^{5x}$. Thus $(e^k)^5 = 5$ and $e^k = 5^{1/5}$. $g(\tfrac{1}{2}) = 1000e^{k/2} = 1000(5^{0.1}) \doteq 1175$

13. $(0.5)^{x/1600} = \tfrac{3}{4}$, $(0.5)^x = (0.75)^{1600}$, $x \doteq 664$ years

15. (a) $f(x) = 126 - Ae^{-kx}$. Since $f(0) = 0$, $A = 126$. Since $f(3) = 42$, $42 = 126 - 126e^{-3k}$ and $e^{-k} = \left(\tfrac{84}{126}\right)^{1/3}$. Now solving $84 = 126 - 126(\tfrac{2}{3})^{x/3}$ for x yields $x \doteq 8.12$ hours.
    (b) $120 = 126 - 126(\tfrac{2}{3})^{x/3}$ and $x \doteq 22.53$ hours

17. Since $f'(x) = \dfrac{ca^2 ke^{-akx}}{(1 + ce^{-akx})^2}$, it follows that

$$kf(x)[a - f(x)] = \frac{ka}{1 + ce^{-akx}}\left[a - \frac{a}{1 + ce^{-akx}}\right]$$
$$= \frac{ka^2[1 + ce^{-akx} - 1]}{(1 + ce^{-akx})^2} = f'(x)$$

19. We find that $f''(x) = ca^3 k^2 e^{-akx} \dfrac{(ce^{-akx} - 1)}{(1 + ce^{-akx})^3}$. Now $f(x) < \dfrac{a}{2}$ implies

$$\frac{a}{1 + ce^{-akx}} < \frac{a}{2}, \quad 2 < 1 + ce^{-akx},$$

and $ce^{-akx} - 1 > 0$. Thus $f''(x) > 0$ if $f(x) < \dfrac{a}{2}$; hence, $f'(x)$ is increasing for these values. Similarly, $f''(x) < 0$ if $f(x) > \dfrac{a}{2}$.

4.6 IMPLICIT AND LOGARITHMIC DIFFERENTIATION (OPTIONAL)

1. (a) $2yy' = 4$, $yy' = 2$, $y' = \dfrac{2}{y}$ (b) $yy'' + (y')^2 = 0$, $y'' = \dfrac{-(y')^2}{y}$

3. (a) $3x^2 - 2yy' = 0$, $y' = \dfrac{3x^2}{2y}$ (b) $6x - 2yy'' - 2(y')^2 = 0$,
   $y'' = \dfrac{3x - (y')^2}{y}$

5. (a) $xy' + y + 2yy' = -1$, $y' = -\dfrac{y+1}{x+2y}$
   (b) $xy'' + y' + y' + 2y(y'') + 2(y')^2 = 0$, $y'' = -\dfrac{2y' + 2(y')^2}{x+2y}$

7. (a) $2xy' + 2y + 2yy' = 0$, $y' = -\dfrac{y}{x+y}$
   (b) $xy'' + y' + y' + yy'' + (y')^2 = 0$, $y'' = -\dfrac{2y' + (y')^2}{x+y}$

9. (a) $3x^2 + 3y^2 y' = 0$, $y' = -\dfrac{x^2}{y^2}$
   (b) $2x + y^2 y'' + 2y(y')^2 = 0$, $y'' = -\dfrac{2x + 2y(y')^2}{y^2}$

11. (a) $xy' + y + 3 + y' = 0$, $y' = -\dfrac{y+3}{x+1}$
    (b) $xy'' + y' + y' + y'' = 0$, $y'' = -\dfrac{2y'}{x+1}$

13. $y' = -\dfrac{2x^3 e^y + 2xye^y + 2x}{x^4 e^y + x^2 y e^y + 1}$

15. $y' = \dfrac{3x^2 y - y(\ell n\, y) - y^2 e^{xy}}{xye^{xy} + x}$

17. $\ell n\, f(x) = \ell n\, x + (\tfrac{3}{2})\ell n(2x+1)$. Differentiating yields
    $$\dfrac{f'(x)}{f(x)} = \dfrac{1}{x} + \dfrac{3}{2x+1} = \dfrac{5x+1}{x(2x+1)}.$$
    Therefore, $f'(x) = (2x+)^{1/2}(5x+1)$.

19. $\ell n\, f(x) = \ell n(3x+7) - \ell n(5x-1)$. Differentiating yields
    $$\dfrac{f'(x)}{f(x)} = \dfrac{3}{3x+7} - \dfrac{5}{5x-1} = \dfrac{-38}{(3x+7)(5x-1)}.$$
    Thus, $f'(x) = \dfrac{-38}{(5x-1)^2}$.

21. $\ln f(x) = 2\ln x + (\tfrac{7}{5})\ln(5x - 1)$. Differentiating yields
$$\frac{f'(x)}{f(x)} = \frac{2}{x} + \frac{7}{5x - 1} = \frac{17x - 2}{x(5x - 1)}.$$
Thus, $f'(x) = x(17x - 2)(5x - 1)^{2/5}$.

23. $\ln f(x) = x[\ln(\ln x)]$. Differentiating yields
$$\frac{f'(x)}{f(x)} = \frac{x}{\ln x}\left(\frac{1}{x}\right) + \ln(\ln x).$$
Thus, $f'(x) = (\ln x)^x\left[\dfrac{1}{\ln x} + \ln(\ln x)\right]$.

25. $2x + 2xy' + 2y + 2yy' = 0$, $y' = 1$. Since $x^2 + 2xy + y^2 = (x + y)^2$, $(x + y)^2 = 4$ and $x + y = 4$ or $x + y = -4$. The graph is two parallel lines having 1 as slope.

27. $2x - xy' + y - 2yy' + 2 = 0$. At $(1, 1)$, $y' = -\tfrac{5}{3}$. An equation of the tangent is $5x + 3y = 8$.

29. $2 - xy'' - y' - 2yy'' - 2(y')^2 = 0$. At $(1, 1)$, we find that $2 - y'' + \tfrac{5}{3} - 2y'' - 2(\tfrac{5}{3})^2 = 0$ and $y'' = -\tfrac{17}{27}$.

4.7 SUMMARY AND REVIEW EXERCISES

1. (a) $b^{-5} = 32$, $b^5 = \tfrac{1}{32}$, $b = \tfrac{1}{2}$. (b) $e^{-1/2}$ (c) 8
2. $200 = 500 - A$ implies $A = 300$. Then $300 = 500 - 300e^{-3k}$ implies $(e^{-k})^3 = \tfrac{2}{3}$, $e^{-k} = (\tfrac{2}{3})^{1/3}$. Therefore, $f(5) = 500 - 300(e^{-k})^5 = 500 - 300(\tfrac{2}{3})^{5/3} \doteq 347.37$.
3. $f'(x) = \dfrac{2x}{x^2 + 1}$ and $f''(x) = \dfrac{2 - 2x^2}{(x^2 + 1)^2}$
4. $f'(x) = -xe^{1/x} + 3x^2 e^{1/x}$ and $f''(x) = e^{1/x}\left(\dfrac{1}{x} - 4 + 6x\right)$
5. $f'(x) = x^{-1}e^x - x^{-2}e^x$ and $f''(x) = e^x(x^{-1} - 2x^{-2} + 2x^{-3})$
6. $f'(x) = x^2 + 3x^2(\ln x)$ and $f''(x) = 5x + 6x(\ln x)$
7. $f'(x) = \dfrac{e^{2x}}{x} + 2e^{2x}(\ln x)$ and
$$f''(x) = e^{2x}\left[-\dfrac{1}{x^2} + \dfrac{4}{x} + 4(\ln x)\right]$$
8. $f'(x) = x^{-1/2} + (\tfrac{1}{2})x^{-1/2}(\ln x)$ and $f''(x) = (-\tfrac{1}{4})x^{-3/2}(\ln x)$
9. (a) $3x + 15 = 8$ at $x = -\tfrac{7}{3}$; $f'(x) = 3$ and $g'(8) = \dfrac{1}{f'(-\tfrac{7}{3})} = \tfrac{1}{3}$

CHAPTER 5. THE ANTIDERIVATIVE AND ITS APPLICATIONS

(b) $3g(x) + 15 = x$, $g(x) = \dfrac{x - 15}{3}$, $g'(x) = \frac{1}{3}$, and $g'(8) = \frac{1}{3}$

10. (a) $g'(8) = \dfrac{1}{f'(-\frac{1}{2})} = -\dfrac{1}{2}$ (b) $g(x) = \dfrac{x - 9}{-2}$ and $g'(8) = -\frac{1}{2}$

11. (a) $g'(8) = \dfrac{1}{f'(1)} = \dfrac{1}{3}$ (b) $g(x) = (x - 7)^{1/3}$ and $g'(8) = \frac{1}{3}$

12. (a) $g'(8) = \dfrac{1}{f'(2)} = \dfrac{1}{6}$ (b) $g(x) = -1 + (1 + x)^{1/2}$ and $g'(8) = \frac{1}{6}$

13. $f'(x) = 3xe^{3x} + e^{3x} = e^{3x}(3x + 1)$ and $f''(x) = e^{3x}(9x + 6)$. Since $f'(-\frac{1}{3}) = 0$ and $f''(-\frac{1}{3}) > 0$, $f(-\frac{1}{3}) = -\dfrac{1}{3e}$ is a relative minimum. No relative maximum and no absolute maximum or minimum. $(-\frac{2}{3}, f(-\frac{2}{3}))$ is an inflection point. The function decreases where $x \leq -\frac{1}{3}$ and increases where $x \geq -\frac{1}{3}$.

14. $f'(x) = x^2 + 3x^2(\ln x)$, $m = f'(1) = 1$ and $f(1) = 2$. The line containing $(1, 2)$ with slope 1 is $x - y = -1$.

15. $f'(x) = 3x^2e^{3x} + 2xe^{3x} + 2$, $m = f'(0) = 2$, and $f(0) = 1$. The line containing $(0, 1)$ with slope 2 is $2x - y = -1$.

16. (a) $A = 5000e^{0.33} = \$6954.84$ (b) 8.4 years

17. $4647.33 = 4000e^{2r}$, $e^{2r} = 1.161\,832\,5$, $r = 0.075$, or 7.5%

18. $f(x) = 100{,}000e^{kx}$ and $300{,}000 = 100{,}000e^{2k}$. Thus, $e^{2k} = 3$ and $e^k = (3)^{1/2}$. Hence, $f(\frac{7}{2}) = 100{,}000(3)^{7/4} \doteq 683{,}852$.

19. $f(x) = 1 - e^{-kx}$. Since $f(15) = 0.60$, $0.60 = 1 - 3^{-15k}$, $e^{-15k} = 0.4$ and $e^{-k} = (0.40)^{1/15}$. Now $0.90 = 1 - (e^{-k})^x$, $(0.40)^{x/15} = 0.10$, and $x \doteq 37.69$ days.

20. 12 years

CHAPTER 5. THE ANTIDERIVATIVE AND ITS APPLICATIONS

5.2 THE ANTIDERIVATIVE OF A FUNCTION

1. $(\frac{1}{6})x^6 + C$
3. $-x^{-1} + C$
5. $(\frac{3}{5})x^{5/3} + C$
7. $-2x^{-1/2} + C$
9. $(\frac{6}{5})x^{10/3} + C$
11. $x^3 - 3x^2 + 7x + C$
13. $2x^4 - 2x^3 + 21x + C$
15. $(\frac{3}{5})x^{5/3} + 4x^{-1/2} + C$
17. $(\frac{1}{4})(3x + 5)^4 + C$
19. $x^3 + x^2 - x + C$
21. $(\frac{1}{30})(5x - 1)^6 + C$
23. $(\frac{2}{5})(x^2 + 4)^{5/2} + C$

25. $(\frac{1}{5})(x^{3/2} - 1)^{10/3} + C$ 27. $(\frac{5}{162})(6x^3 + 11)^{9/5} + C$
29. $-(\frac{1}{4})(x^2 + 1)^{-2} + C$

5.3 ANTIDERIVATIVES INVOLVING EXPONENTIAL AND LOGARITHMIC FUNCTIONS

1. $(\frac{1}{5})e^{5x} + C$
3. $-(\frac{1}{2})e^{-2x} + C$
5. $(\frac{1}{3})e^{3x} + (\frac{4}{7})x^{7/4} + C$
7. $(\frac{1}{3})e^{x^3} + C$
9. $2e^{\sqrt{x}} + C$
11. $\ell n(4x + 5) + C$
13. $(\frac{1}{2})\ell n(2x - 5) + C$
15. $(\frac{1}{4})\ell n(2x^2 + 1) + C$
17. $(\frac{1}{3})\ell n(x^3 + 1) + C$
19. $\dfrac{3^x}{\ell n\,3} + C$
21. $(\frac{1}{2})[\ell n(x - 1) - \ell n(x + 1)] + C$

5.4 APPLICATIONS OF THE ANTIDERIVATIVE

1. $f(x) = x^3 + x^2 - 13$ 3. $f(x) = -x^2 + 4x$
5. $R(x) = 18x^2 - 0.01x^3$
7. $P(x) = 84x - x^2 - 400$; \$80 loss; 42 units; \$1364 profit.
9. (a) $C(x) = 100x - 0.1x^2 + 210$ (b) $C(50) = \$4960$
   (c) \$99.20   (d) $C'(50) = \$90$
11. (a) $V(t) = 5000e^{-7/5} + 7000$ (b) $V(2) \doteq \$10,351.60$
    (c) \$7676.68
13. $h(t) = -16t^2 + 800t$
15. $P(t) = (\frac{100}{3})e^{0.03t} + \frac{80}{3}$ and $P(20) \doteq 87.4$ million
17. $f(t) = 500t + 20t^{3/2} + 100,000$ and $f(9) = 105,040$

5.5 INTEGRATION BY SUBSTITUTION

1. Let $u = x^2 + 1$. Then $du = (2x)\,dx$ and $x\,dx = \frac{1}{2}\,du$. Substituting, we have

$$\int u^{4/3}(\tfrac{1}{2})\,du = \tfrac{3}{14}u^{7/3} + C$$

Hence

$$\int x(x^2 + 1)^{4/3}\,dx = \tfrac{3}{14}(x^2 + 1)^{7/3} + C$$

3. Let $u = x^4 + 1$. Then

$$du = 4x^3\,dx \text{ and } x^3\,dx = \tfrac{1}{4}du.$$

Substituting yields

$$\frac{1}{4}\int u^{1/2}\,du = \frac{1}{6}u^{3/2} + C.$$

Thus

$$\int x(x^4 + 1)^{1/2}\,dx = \frac{1}{6}(x^4 + 1)^{3/2} + C$$

CHAPTER 5. THE ANTIDERIVATIVE AND ITS APPLICATIONS

5. Let $u = 9x + 2$. Then $x = (\frac{1}{9})(u - 2)$ and $dx = (\frac{1}{9})\, du$. Substituting yields

$$\frac{1}{81} \int (u^{3/2} - 2u^{1/2})\, du = \frac{2}{405} u^{5/2} - \frac{4}{243} u^{3/2} + C.$$

Therefore,

$$\int x(9x + 2)^{1/2}\, dx = \frac{2}{405}(9x + 2)^{5/2} - \frac{4}{243}(9x + 2)^{3/2} + C$$

7. $\frac{1}{9}[\frac{1}{14}(3x + 1)^{14} - \frac{1}{13}(3x + 1)^{13}] + C = \frac{1}{1638}(3x + 1)^{13}(39x - 1) + C$
9. $\frac{1}{27}[\frac{2}{7}u^{7/2} + \frac{24}{5} u^{5/2} + 24u^{3/2}] + C$ where $u = 3x - 6$
11. $\frac{1}{3}x^3 + \frac{1}{15}(2x + 1)^{3/2}(3x - 1) + C$
13. $\frac{2}{135}(3x + 4)^{3/2}(9x - 8) + 200 + \frac{128}{135}$

5.6 INTEGRATION BY PARTS

1. Let $f(x) = x$ and let $g'(x) = (3x + 1)^{1/2}$. Then $f'(x) = 1$ and $g(x) = (\frac{2}{9})(3x + 1)^{3/2}$. Therefore,

$$\int x(3x + 1)^{1/2}\, dx = \frac{2}{9} x(3x + 1)^{3/2} - \frac{2}{9} \int (3x + 1)^{3/2}\, dx$$

$$= \frac{2}{9} x(3x + 1)^{3/2} - \frac{4}{135}(3x + 1)^{5/2} + C$$

3. Let $f(x) = x$ and let $g'(x) = (x - 4)^5$. Then $f'(x) = 1$ and $g(x) = (\frac{1}{6})(x - 4)^6$. Therefore,

$$\int x(x - 4)^5\, dx = \frac{x}{6}(x - 4)^6 - \frac{1}{6} \int (x - 4)^6\, dx$$

$$= \frac{x}{6}(x - 4)^6 - \frac{1}{42}(x - 4)^7 + C$$

5. Let $f(x) = x$ and $g'(x) = e^{-x}$. Then $f'(x) = 1$ and $g(x) = -e^{-x}$.

$$\int xe^{-x}\, dx = -xe^{-x} + \int e^{-x}\, dx = -xe^{-x} - e^{-x} + C$$

7. Let $f(x) = x$ and let $g'(x) = (2x + 1)^{1/3}$. Then $f'(x) = 1$ and $g(x) = (\frac{3}{8})(2x + 1)^{4/3}$. Therefore,

$$\int x(2x + 1)^{1/3}\, dx = \frac{3x}{8}(2x + 1)^{4/3} - \frac{3}{8} \int (2x + 1)^{4/3}\, dx$$

$$= \frac{3x}{8}(2x + 1)^{4/3} - \frac{9}{112}(2x + 1)^{7/3} + C$$

9. Let $f(x) = x^3$ and $g'(x) = e^x$. Then $f'(x) = 3x^2$ and $g(x) = e^x$. Hence,

$$\int x^3 e^x \, dx = x^3 e^x - 3\int x^2 e^x \, dx$$

Using results of Example 3, we get

$$\int x^3 e^x \, dx = x^3 e^x - 3(x^2 e^x - 2xe^x + 2e^x) + C$$

11. Let $f(x) = x$ and let $g'(x) = (3x + 8)^{2/3}$. Then $f'(x) = 1$ and $g(x) = (\frac{1}{5})(3x + 8)^{5/3}$. Therefore,

$$\int x(3x + 8)^{2/3} \, dx = \frac{x}{5}(3x + 8)^{5/3} - \frac{1}{5}\int (3x + 8)^{5/3} \, dx$$

$$= \frac{x}{5}(3x + 8)^{5/3} - \tfrac{1}{40}(3x + 8)^{8/3} + C$$

Since $y(0) = \tfrac{3}{5}$, $\tfrac{3}{5} = -(\tfrac{1}{40})(2)^8 + C$ and $C = 7$. Consequently,

$$y = \frac{x}{5}(3x + 8)^{5/3} - \frac{1}{40}(3x + 8)^{8/3} + 7$$

13. Let $f(t) = t$ and $g'(t) = e^{t/6}$. Then $f'(t) = 1$ and $g(t) = 6e^{t/6}$. Therefore,

$$\int te^{t/6} \, dt = 6te^{t/6} - 6\int e^{t/6} \, dt = 6te^{t/6} - 36e^{t/6} + C$$

If $s(t) = 6te^{t/6} - 36e^{t/6} + C$, then $s(0) = 0$ and $C = 36$. Thus, $s(t) = 6te^{t/6} - 36e^{t/6} + 36$ and $s(18) \doteq 1482.16$ feet.

5.7 DIFFERENTIAL EQUATIONS (OPTIONAL)

1. $y^3 = Cx$
3. $xe^{-x} + e^{-x} = \ln y + C$
5. $y^2 = 6[\ell n(x + 1)] + C$
7. $x^2 = y^2 - 2y + C$
9. $\ell n\, y = \dfrac{x^2}{2} + e^x + C$

11. (a) $y(0) = A$ is initial amount (b) $2A = Ae^k$, $e^k = 2$, and $y(x) = A(2^x)$. Thus, $y(2) = 4A$ and $k = \ell n\, 2$.

5.8 SUMMARY AND REVIEW EXERCISES

1. $2x^{5/2} + \dfrac{1}{x} + C$
2. $\dfrac{x^5}{5} + \dfrac{2x^3}{3} + C$
3. $\tfrac{5}{14}(x^2 + 4)^{7/5} + C$
4. $\tfrac{1}{5}(x^3 + 1)^{5/3} + C$

CHAPTER 6. THE DEFINITE INTEGRAL AND ITS APPLICATIONS

5. $-e^{-x} - (\frac{1}{2})\ln(2x+1) + C$
6. $\frac{1}{2}[e^{2x} + \ln(x^2+1)] + C$
7. $(\frac{1}{60})(4x+1)^{3/2}(6x-1) + C$
8. $(\frac{2}{315})(3x+1)^{5/2}(15x-4) + C$
9. $(\frac{1}{2})(\ln x)^2 + C$
10. $\ln(\ln x) + C$
11. $f(x) = 3x^{4/3} + x^2 + 1$
12. $f(x) = 3x^2 + 3^x + 1$
13. $C(x) = 300x - \dfrac{x^2}{4} + 1525$
14. $P(x) = 84x - x^2 - 240$ and $P(4) = \$80$ profit. Since $P'(x) = 0$ at $x = 42$ and since $P''(42) < 0$, $P(42) = \$1524$ is maximum profit.
15. $f''(x) = -6$ and $f'(x) = 6x + C_1$. Since $f'(2) = 0$, $0 = -12 + C_1$ and $C_1 = 12$. Since $f'(x) = -6x + 12$, $f(x) = -3x^2 + 12x + C_2$. Since $f(2) = 4$, $4 = -12 + 24 + C_2$ and $C_2 = -8$. Hence, $f(x) = -3x^2 + 12x - 8$.
16. $h''(t) = -32$ and $h'(t) = -32t + C_1$. Since $h'(0) = 1200$, $C_1 = 1200$ and $h'(t) = -32t + 1200$. Therefore, $h(t) = -16t^2 + 1200t + C_2$. Since $h(0) = 0$, $C_2 = 0$ and $h(t) = -16t^2 + 1200t$.
17. $s''(t) = 4t$ and $s'(t) = 2t^2 + C_1$. Since $s'(0) = 0$, $C_1 = 0$ and $s'(t) = 2t^2$. Hence, $s'(6) = 72$ ft/sec is its velocity at the end of 6 seconds. Since $s'(t) = 2t^2$, $s(t) = (\frac{2}{3})t^3 + C_2$. Since $s(0) = 0$, $C_2 = 0$ and $s(t) = (\frac{2}{3})t^3$. Thus, $s(6) = 144$ feet.
18. $P'(t) = e^{0.04t}$ and $P(t) = 25e^{0.04t} + C$. Since $P(0) = 40$, $40 = 25 + C$ and $C = 15$. Hence, $P(t) = 25e^{0.04t} + 15$ and $P(20) = 25e^{0.8} + 15 \doteq 70.6385$ million.
19. $B'(t) = 120 - 4t$ and $B(t) = 120t - 2t^2 + C$. Since $B(0) = 0$, $C = 0$ and $B(t) = 120t - 2t^2$. Hence, $B(15) = 1350$.
20. $s''(t) = 3t$ and $s'(t) = (\frac{3}{2})t^2 + C_1$. Since $s'(0) = 0$, we have $s'(t) = (\frac{3}{2})t^2$ and $s'(8) = 96$ ft/sec. Since $s'(t) = (\frac{3}{2})t^2$, $s(t) = (\frac{1}{2})t^3 + C_2$. Since $s(0) = 0$, $s(t) = (\frac{1}{2})t^3$ and $s(8) = 256$ ft.

CHAPTER 6. THE DEFINITE INTEGRAL AND ITS APPLICATIONS

6.2 THE DEFINITE INTEGRAL

1. $\frac{1023}{5}$ 3. 16 5. -2 7. $\frac{27}{2}$ 9. $\frac{14}{3}$
11. $-\frac{49}{10}$ 13. $\frac{243}{5}$ 15. $\frac{259}{8}$ 17. 0 19. $\frac{61}{3}$
21. 20 23. $\dfrac{(e^3 - 1)}{3}$ 25. $\frac{1}{2}(\ln \frac{11}{3})$ 27. $\dfrac{148{,}672}{315}$
29. (a) $(\frac{1}{2})(11+3)(4) = 28$ (b) $(x^2 + x)\big|_1^5 = 28$ (see figure)

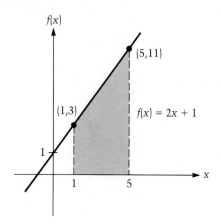

Figure for answer to Exercise 29b

31. 20 (see figure)

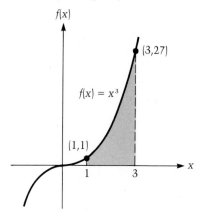

33. 96.8 (see figure)

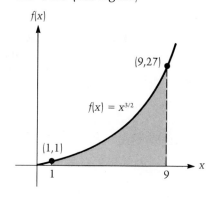

35. $\frac{47}{3}$ (see figure)

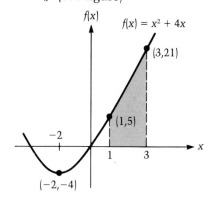

37. $\ln 3$ (see figure)

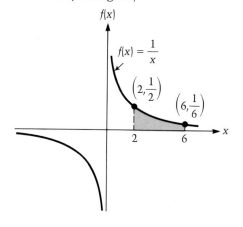

CHAPTER 7. MULTIVARIABLE CALCULUS AND ITS APPLICATIONS

6.3 MORE ON AREA

1. $\frac{1}{6}$ 3. 36 5. $\frac{37}{12}$ 7. $\frac{1}{2}$ 9. $\frac{1}{12}$
11. $\frac{32}{3}$ 13. $\frac{1}{3}$ 15. $\frac{9}{2}$ 17. $\frac{3}{14}$ 19. $32\sqrt{3}/9$

6.4 OTHER APPLICATIONS OF THE DEFINITE INTEGRAL

3. (a) $\frac{31}{3}$ (b) $\sqrt{\frac{31}{3}} \doteq 3.214$
5. (a) 66 mph (b) At 5 o'clock and 8 o'clock
   (c) $v(5) = 70$ mph (d) $v(3) = 50$ mph (e) $\frac{1325}{20} = 66.25$ mph
7. $CS = 144$, $PS = 36$ 11. $\left(\frac{3}{38}\right)\int_2^4 (x^2 - x)\,dx = 1$
13. $\left(\frac{1}{39}\right)\int_2^5 x^2\,dx = 1$ 15. $\left(\frac{1}{117}\right)(x^3 - 8)$
17. $\left(\frac{\pi}{16}\right)\int_2^4 x^4\,dx = \frac{3093\pi}{80}$ 19. $\pi\int_1^9 x^3\,dx = 1640\pi$

6.5 SUMMARY AND REVIEW EXERCISES

1. 1002 2. 0 3. $\frac{176}{3}$ 4. 64
5. $\frac{1}{3}(e^3 - 2)$ 6. $(\frac{1}{2})(e^2 + 1)$ 7. $\frac{149}{30}$ 8. $\dfrac{148{,}672}{315}$
9. $e^{-1} - e^{-3} + \ln 3$ 10. $(\frac{1}{2})[(\ln 4)^2 - 1]$
11. $\frac{26}{3}$ 12. $6 - (\frac{3}{2})(4)^{1/3}$
13. $(\frac{1}{4})(e^2 + 1)$ 14. $\int_4^2 (8 - x^2 - 2x)\,dx = 36$
15. $\int_{-2}^2 (4 - x^2)\,dx = \frac{32}{3}$ 16. (a) $72,296 (c) $71,200

CHAPTER 7. MULTIVARIABLE CALCULUS AND ITS APPLICATIONS

7.2 RECTANGULAR COORDINATE SYSTEM IN THREE DIMENSIONS

1.

3.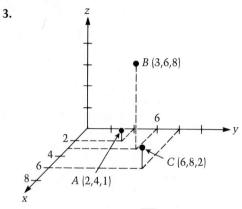

5. $|AB| = 6$, $|BC| = \sqrt{33}$, $|AC| = \sqrt{73}$

7. $|AB| = \sqrt{54}$, $|BC| = 7$, $|AC| = \sqrt{33}$

9. AB: $(2, 4, 3)$, BC: $(\frac{7}{2}, 4, 7)$, AC: $(\frac{5}{2}, 2, 5)$

11. AB: $(\frac{5}{2}, 5, \frac{9}{2})$, BC: $(\frac{9}{2}, 7, 5)$, AC: $(4, 6, \frac{3}{2})$

13. $|AB| = \sqrt{41}$, $|BC| = \sqrt{59}$, and $|AC| = \sqrt{18}$. $|AB|^2 + |AC|^2 = |BC|^2$

15., **17.**, **19.**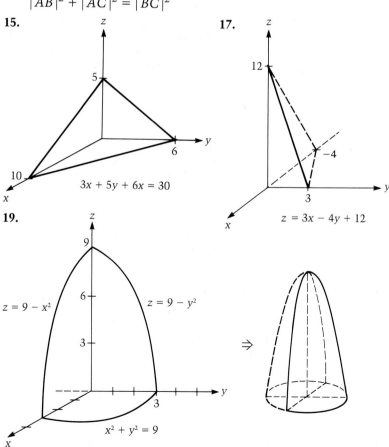

7.3 FUNCTIONS OF TWO VARIABLES

1. $f(3, 4) = -2$ 3. $f(3, 4) = 30$ 5. $f(3, 4) = 165$
7. $f(3, 4) = \frac{21}{11}$ 9. $f(3, 4) = 8 + \ln 12$
11. (a) $V(x, y) = x^2 y$ (b) $V(4, 7) = 112$
13. (a) $V(4.02, 7) = 113.1228$ (b) $V(4.02, 7) - V(4, 7) = 1.1228$
15. $xyh = 300$ and $h = \dfrac{300}{xy}$

$$C(x, y) = 4xy + 8xy + 5(2xh) + 5(2yh),$$

$$C(x, y) = 12xy + \frac{3000}{y} + \frac{3000}{x}$$

17. $xyh = 400$ and $h = \dfrac{400}{xy}$. $S(x, y) = xy + \dfrac{800}{x} + \dfrac{800}{y}$
19. (a) $F(500, 500) = 500$ (b) $F(1000, 500) = 350$
21. (a) $T(x, y) = S(x, y) + F(x, y) = 1400 - 0.2x - 0.1y$

$$T(1000, 1000) - T(500, 500) = 200 - 350 = -150,$$

a decrease of 150 cars per year

(b) $\left(\dfrac{-150}{1250}\right)(100) = -12$, a decrease of 12%.

7.4 PARTIAL DERIVATIVES

1. $f_x = 3y + 1$; $f_y = 3x$; $f_{xx} = 0$; $f_{yy} = 0$; $f_{xy} = f_{yx} = 3$
3. $f_x = 2x + 7$; $f_y = 6y$; $f_{xx} = 2$; $f_{yy} = 6$; $f_{xy} = f_{yx} = 0$
5. $f_x = 2xy^3 + 6$; $f_y = 3x^2y^2 + 5$; $f_{xx} = 2y^3$; $f_{yy} = 6x^2y$; $f_{xy} = f_{yx} = 6xy^2$
7. $f_x = 2ye^{2x}$; $f_y = e^{2x}$; $f_{xx} = 4ye^{2x}$; $f_{yy} = 0$; $f_{xy} = f_{yx} = 2e^{2x}$
9. $f_x = (\tfrac{5}{2})(x + 2y)^{3/2}$; $f_y = 5(x + 2y)^{3/2}$; $f_{xx} = (\tfrac{15}{4})(x + 2y)^{1/2}$; $f_{yy} = 15(x + 2y)^{1/2}$; $f_{xy} = f_{yx} = (\tfrac{15}{2})(x + 2y)^{1/2}$
11. $f_x = 4(x + y)^3$; $f_y = 4(x + y)^3 + 3(y - 2)^2$; $f_{xx} = 12(x + y)^2$; $f_{yy} = 12(x + y)^2 + 6(y - 2)$; $f_{xy} = f_{yx} = 12(x + y)^2$
13. $f_x = \dfrac{y^2 - x^2}{(x^2 + y^2)^2}$; $f_y = \dfrac{-2xy}{(x^2 + y^2)^2}$; $f_{xx} = \dfrac{2x^3 - 6xy^2}{(x^2 + y^2)^3}$; $f_{yy} = \dfrac{6xy^2 - 2x^3}{(x^2 + y^2)^3}$; $f_{xy} = f_{yx} = \dfrac{6x^2y - 2y^3}{(x^2 + y^2)^3}$
15. $f_x = 5x(x^2 + y^2)^{3/2}$; $f_y = 5y(x^2 + y^2)^{3/2}$; $f_{xx} = 5(4x^2 + y^2)(x^2 + y^2)^{1/2}$; $f_{yy} = 5(x^2 + 2y^2)(x^2 + y^2)^{1/2}$; $f_{xy} = f_{yx} = 15xy(x^2 + y^2)^{1/2}$
17. $f(x, y) = \dfrac{x}{y} + \dfrac{1}{x}$; $f_x = \dfrac{1}{y} - \dfrac{1}{x^2}$; $f_y = -\dfrac{x}{y^2}$; $f_{xx} = \dfrac{2}{x^3}$; $f_{yy} = \dfrac{2x}{y^3}$; $f_{xy} = f_{yx} = -\dfrac{1}{y^2}$

19. $f_x = e^{xy}(xy + 1)$; $f_y = x^2 e^{xy}$; $f_{xx} = e^{xy}(xy^2 + 2y)$; $f_{yy} = x^3 e^{xy}$; $f_{xy} = f_{yx} = xe^{xy}(xy + 2)$

21. $C_x = 16y - \dfrac{2000}{x^2} = 0$ and $C_y = 16x - \dfrac{2000}{y^2} = 0$ where $16x^2y = 16xy^2$, $x = y$. Note: $x \neq 0$ and $y \neq 0$. Since $16x - \dfrac{2000}{x^2} = 0$ at $x = 5$, the point is $(5, 5)$.

23. (a) $C_{xx} = \dfrac{4000}{x^3} > 0$ (b) $C_{yy} = \dfrac{4000}{y^3} > 0$
    (c) $C_{xy} = C_{yx} = 16 > 0$

7.5 TANGENT PLANES AND THE TOTAL DIFFERENTIAL

1. (a) 111.24 (b) 0.01448   3. (a) 16.25526 (b) 16.256
5. (a) $f(4, 3) + f_x(4, 3)\, dx + f_y(4, 3)\, dy = 24.1175$
   (b) $f(4.01, 3.02) \doteq 24.117545$
7. $z = 6 + 5(x - 2) - 2(y + 1)$; $5x - 2y - z = 6$
9. $z = -6 - 5(x - 2) + 8(y + 1)$; $5x - 8y + z = 12$
11. $z = \tfrac{3}{5} - (\tfrac{7}{25})(x - 2) - (\tfrac{7}{25})(y + 1)$; $7x + 7y + 25z = 22$
13. (a) 12.9783 (b) 12.96128.   15. (a) -0.00188 (b) 0
17. (a) $P(800, 525) - P(800, 500) = \$11{,}375$ is the total change in profit with an \$800 profit for each Fastback model but with a change from \$500 to \$525 profit on each Slowback model. (b) $P_y = 0.2x + 500 - 0.4y$ and $P_y(800, 500)(25) = \$11{,}500$
19. (a) $P_x\, dx + P_y\, dy = (300 - 0.6x + 0.2y)\, dx + (0.2x + 500 - 0.4y)\, dy = (-20)(50) + (440)(25) = \$10{,}000$.
    (b) $P(700, 525) - P(700, 500) = 342{,}375 - 333{,}000 = \$9375$

7.6 APPLIED MAXIMA AND MINIMA PROBLEMS

1. $f_x = 2x + 2y$, $f_y = 2x + 2y$, and $f_x = f_y = 0$ where $y = -x$. Since $f(x, y) = (x + y)^2 + 4$, we see that f has a relative minimum at any point where $y = -x$; the relative (and absolute) minimum is 4.
3. $f_x = 12y - 12xy - 3y^2 = 3y(4 - 4x - y)$ and $f_y = 12x - 6x^2 - 6xy = 6x(2 - x - y)$. $f_x = f_y = 0$ at $(0, 0)$, $(0, 4)$, $(2, 0)$, and $(\tfrac{2}{3}, \tfrac{4}{3})$. $f_{xx} = -12y$, $f_{yy} = -6x$, and $f_{xy} = 12 - 12x - 6y$. At $(0, 0)$, $(0, 4)$, and $(2, 0)$, $f_{xx}f_{yy} - [f_{xy}]^2 < 0$; thus, there is a saddle point at each of these points. At $(\tfrac{2}{3}, \tfrac{4}{3})$, $AC - B^2 > 0$ and $f_{xx} < 0$; therefore, $f(\tfrac{2}{3}, \tfrac{4}{3})$ is a relative maximum.
5. $f_x = 25 - \dfrac{49}{x^2}$, $f_y = 36 - \dfrac{64}{y^2}$, and $f_x = f_y = 0$ at $(\tfrac{7}{5}, \tfrac{4}{3})$,

$(-\frac{7}{5}, \frac{4}{3})$, $(\frac{7}{5}, -\frac{4}{3})$, and $(-\frac{7}{5}, -\frac{4}{3})$. $f_{xx} = \frac{98}{x^3}$, $f_{yy} = \frac{128}{y^3}$, and $f_{xy} = 0$. Since $f_{xx}f_{yy} - [f_{xy}]^2 < 0$ at $(-\frac{7}{5}, \frac{4}{3})$ and $(\frac{7}{5}, -\frac{4}{3})$, each are the x- and y-coordinates of a saddle point. At $(\frac{7}{5}, \frac{4}{3})$ and at $(-\frac{7}{5}, -\frac{4}{3})$, $f_{xx}f_{yy} - [f_{xy}]^2 > 0$. Since $f_{xx}(\frac{7}{5}, \frac{4}{3}) > 0$, $f(\frac{7}{5}, \frac{4}{3})$ is a local minimum. Since $f_{xx}(-\frac{7}{5}, -\frac{4}{3}) < 0$, $f(-\frac{7}{5}, -\frac{4}{3})$ is a local maximum.

7. $f_x = 6 - y - 2xy^2$ and $f_y = -x - 2x^2y$. $f_x = f_y = 0$ at $(0, 6)$. $f_{xx} = -2y^2$, $f_{yy} = -2x^2$, and $f_{xy} = -1 - 4xy$. Since $AC - B^2 < 0$, there is a saddle point at $(0, 6)$.

9. $f_x = 3x^2 - 6x$, $f_y = 4y$, and $f_x = f_y = 0$ at $(0, 0)$ and at $(2, 0)$. $f_{xx} = 6x - 6$, $f_{yy} = 4$, $f_{xy} = 0$. Since $AC - B^2 < 0$ at $(0, 0)$, $(0, 0, 0)$ is a saddle point. At $(2, 0)$, $AC - B^2 > 0$ and $f_{xx} > 0$; consequently, $f(2, 0) = -4$ is a relative minimum.

11. Both $f_x = f_y = 0$ and $g_x = g_y = 0$ at $(0, 0)$. It is obvious that $f(0, 0) = 0$ is the minimum value of f; however, $g(0, y) > 0$ for $y > 0$ and $g(0, y) < 0$ for $y < 0$, so $g(0, 0) = 0$ is neither a relative maximum nor a relative minimum for g. In both cases, $AC - B^2 = 0$.

13. No relative maximum inside the domain. The maximum, which is on the boundary, is $P(1166.67, 2000) = \$608,333.33$.

15. $xyz = 500$ where z is the height. Therefore $z = \dfrac{500}{xy}$ and the surface area is $S(x, y) = 2xz + 2yz + xy = \dfrac{1000}{y} + \dfrac{1000}{x} + xy$. $S_x = -\dfrac{1000}{x^2} + y$, $S_y = -\dfrac{1000}{y^2} + x$, and $S_x = S_y = 0$ at $(10, 10)$. The height should be 5 inches with a square bottom 10 inches on a side.

17. 18 inches wide, 18 inches high, and 36 inches long

19. $P_{xx} = -50(1 - y)$, $P_{yy} = -5000\left(2 + \dfrac{x}{4}\right)$,

$P_{xy} = 50x - 1250y + 525$. At $(7, 0.4)$, $P_{xx}P_{yy} - (P_{xy})^2 = (-30)(-18,750) - (375)^2 > 0$ and $P_{xx} = -30 < 0$.

7.7 MAXIMA AND MINIMA WITH CONSTRAINTS

1. $f(4, 4) = 64$
3. $f(\frac{6}{49}, \frac{72}{49}) = \frac{5292}{2401}$
5. $f(36, 36, 36) = 46,656$
7. $f(\frac{2}{3}, \frac{2}{3}, \frac{4}{3}) = \frac{24}{9}$

7.8 METHOD OF LEAST SQUARES (OPTIONAL)

3. (a) $f(x) = \dfrac{9x}{5} + \dfrac{3}{5}$ (c) $\dfrac{140}{25}$

5. (a) $f(x) = x^2 + \dfrac{6x}{6} - \dfrac{3}{5}$ (c) $\dfrac{4}{5}$

7.9 SUMMARY AND REVIEW EXERCISES

1. (a) $(-1, 5, \tfrac{3}{2})$ (b) 3

2. (a) $(5, 1, 7)$ (b) $2\sqrt{6}$

3.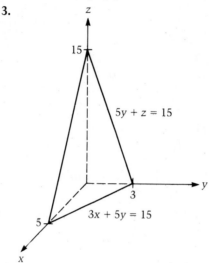

4.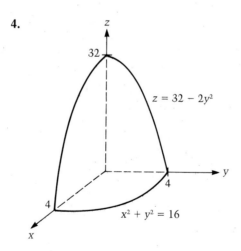

5. $f_x = 2xy^3 + \dfrac{1}{x}$; $f_y = 3x^2y^2 + \dfrac{1}{y} + 1$; $f_{xx} = 2y^3 - \dfrac{1}{x^2}$; $f_{yy} = 6x^2y - \dfrac{1}{y^2}$; $f_{xy} = f_{yx} = 6xy^2$

6. $f_x = y^4 + ye^{xy} + 3$; $f_y = 4xy^3 + xe^{xy}$; $f_{xx} = y^2 e^{xy}$; $f_{yy} = 12xy^2 + x^2 e^{xy}$; $f_{xy} = f_{yx} = 4y^3 + xye^{xy} + e^{xy}$

7. Since $f_x = 2y^2 - y + 2x$ and $f_y = 4xy - x$, we have $f_x(1, 2) = 8$ and $f_y(1, 2) = 7$. $z = 7 + 8(x - 1) + 7(y - 2)$.

8. Since $f_x = 6xy + 5y$ and $f_y = 3x^2 + 5x - 6y^2$, we have $f_x(2, 1) = 17$ and $f_y(2, 1) = -7$. $z = 4 + 17(x - 2) - 7(y - 1)$.

9. $f(4.01, 7.98) \doteq f(4, 8) + f_x(4, 8)(0.01) + f_y(4, 8)(-0.02)$
$= 104 + 0.25 - 0.25\overline{33}... = 103.99\overline{66}...$

10. $f(1.01, 3.98) \doteq f(1, 4) + f_x(1, 4)(0.01) + f_y(1, 4)(-0.02)$
$= 2(-\frac{46}{7})(0.01) + (\frac{1}{4})(-0.02) \doteq 1.9883$

11. $f_x = f_y = 0$ only at $(0, 0)$. But $AC - B^2 < 0$ so there are no relative maxima or minima.

12. $f_x = f_y = 0$ only at $(0, 0)$. But $AC - B^2 < 0$ so there are no relative maxima or minima.

13. $f_x = f_y = 0$ at the points $(5, 4)$, $(-5, -4)$, $(5, -4)$, and $(-5, 4)$. $f(5, 4)$ is a relative minimum, $f(-5, -4)$ is a relative maximum, and there are saddle points at $(5, -4)$ and $(-5, 4)$.

14. $f_x = f_y = 0$ at $(0, 0)$ and $(-2, 0)$. $f(-2, 0)$ is a relative maximum and there is a saddle point at $(0, 0)$.

15. $\frac{50}{3}$ inches wide, $\frac{50}{3}$ inches high, and $\frac{100}{3}$ inches long.

16. $P(x, y) = 40x + 30y - \dfrac{x^2}{200} - \dfrac{y^2}{100} - 1000$. The maximum profit is attained when $x = 4000$ and $y = 1500$. The selling price of Product A should be $30 and that of Product B should be $35.

17. $x = y = (\frac{1}{2})^{1/3}$ and $z = 2(\frac{1}{2})^{1/3}$ ft.   19. 4

CHAPTER 8. SEQUENCES AND MATHEMATICS OF FINANCE

8.2 ARITHMETIC SEQUENCES

1. (a) 3 (b) 17, 20, 23 (c) 798
3. (a) 4 (b) 13, 17, 21 (c) 861
5. (a) 1.1 (b) 5.6, 6.7, 7.8 (c) 279.3
7. (a) -5 (b) $-12, -17, -22$ (c) -987
9. (a) $\frac{1}{6}$ (b) $\frac{7}{6}, \frac{4}{3}, \frac{3}{2}$ (c) 49
11. (a) 249 (b) 6325   13. 3.1
15. (a) 3 (b) 21 (c) 735   17. 1395

19. (a) $26,200 (b) $309,000
21. (a) (i) $411,000 (ii) $414,000
23. $3339.75   (b) Yes. $\frac{n(n+1)}{2} = 300$; $n = 24$
27. $289.38   29. $987.75

8.3 GEOMETRIC SEQUENCES

1. -1 and 3 3. $-\frac{1}{2}$ and $\frac{1}{16}$ 5. $-\frac{1}{3}$ and $\frac{1}{9}$
7. $\frac{1}{2}$ and $\frac{1}{96}$ 9. $1.2 = \frac{6}{5}$ and 8.957952
11. $\frac{819}{128}$ 13. (a) $3\sqrt{2}$ (b) 24 and 36
15. (a) $13{,}200(1.05)^9 = \$20{,}477.53$ (b) $\$166{,}028.18$
17. (a) $20{,}000(1.06)^9 = \$33{,}789.58$ (b) $\$263{,}615.90$
19. (a) $19,500 from (1) and $20,273.75 from (2). (b) $172,500 from (1) and $158,169.54 from (2). (c) $395,000 from (1) and $441,427.09 from (2).
21. $29,428.47

8.4 INTEREST: SIMPLE AND COMPOUND

1. $j = 0.005$ and $m = 36$ 3. $j = 0.025$ and $m = 8$
5. $j = 0.06$ and $m = 10$ 7. $1050.94
9. $3804.72   11. 5.0945%   13. 12.6825%
15. $618.31   17. 47 quarters   19. 70 months
21. $3642.23   23. $6231.67   25. $12,405.21

8.5 CONTINUOUS INTEREST

1. 5.127%   3. 10.517%   5. 16.183%
7. $6615.65   9. $40.50   11. $5538.70
13. $7261.49   15. 9%   17. 5%

8.6 INTEREST ON PERIODIC SAVINGS

1. $7011.88   3. $9241.50   5. $15,197.96
7. $2444.31   9. $31.30   11. $99.93
13. (a) $14,023.77 (b) 94 months (c) $5200 approximately.
15. $283.30   17. (i) $9012.22 (ii) $8733.20
19. (i) $25,155.78 (ii) $26,413.57

CHAPTER 8. SEQUENCES AND MATHEMATICS OF FINANCE

8.7 INSTALLMENT PAYMENTS AND RETIREMENT INCOME

1. $8206.64
3. (a) $2037.04 (b) $40,740.80
5. $166.07
7. $514.31
9. $438.79
11. $141.11
13. 16.24%

15. $A = \dfrac{200}{0.005}[(1.005)^{241} - 1.005] = \$92{,}870.22$. Now,

$$92{,}870.22 = \dfrac{1000}{0.005}[1 - (1.005)^{-m}]$$

$(1.005)^{-m} = 0.5356489$, $(1.005)^m = 1.8668945$,

$m = 125^+$ months $\doteq$ 10 years 5 months.

17. $V = \left(\dfrac{3000}{0.015}\right)[1 - (1.015)^{-40}] = \$89{,}747.53$ is the amount they will need in 30 years to buy the 10-year annuity. Thus, we now need to find D where

$$89{,}747.53 = \dfrac{D}{0.015}[(1.015)^{121} - 1.015]; \quad D = \$266.90.$$

8.8 SUMMARY AND REVIEW EXERCISES

1. (a) -8 (b) 16, 8, 0 (c) -2880
3. (a) $\tfrac{1}{2}$ (b) $\tfrac{13}{6}, \tfrac{8}{3}, \tfrac{19}{6}$ (c) $\tfrac{475}{2}$
5. (a) $-\tfrac{1}{4}$ (b) $-1, \tfrac{1}{4}, -\tfrac{1}{16}$
7. (a) 1.4 (b) $x = 3.5$ and $y = 4.9$
9. (a) -41 (b) 5   11. (a) $28,800 (b) $234,000
13. $181.41   15. $7284.46
17. 7.5%   19. $148.62

INDEX

A

Absolute maximum value, 45
  of a function of two variables, 360, 389
Absolute minimum value, 45
  of a function of two variables, 360, 389
Acceleration, 116 (Ex. 33)
Add-on interest rate, 436
Algebraic expression, A20
Antiderivative
  of a constant times a function, 246, 279
  definition of, 244, 278
  of exponential functions, 250, 253, 279
  notation for, 244
  of the sum of two functions, 246, 279
  of x^r, 245, 279
Antidifferentiation, 244–280
Annual percentage rate, 438
Annuity
  due, 429
  ordinary, 429–430
  simple, 429
Applications
  of linear inequalities, 7–12
  of linear functions, 50–64
  of linear systems, 30–39
  of a rectangular coordinate system, 20–21
APR, 438
Area
  of a circle, 43
  of a region in a plane, 290, 301–308, 324
Arithmetic sequence, 397
Asymptote, 48
Average
  acceleration, 95 (Ex. 34), 129
  cost, 47, 66
  marginal cost, 83, 129
  marginal profit, 95 (Ex. 33), 129

Average—Continued
  marginal revenue, 94 (Ex. 32), 129
  value of a function, 309, 325
  velocity, 89
Axes, coordinate, 12, 332

B

Boundary conditions, 272
Break-even point, 54, 66

C

Calculus, Fundamental Theorem of, 291–293
Cartesian coordinate system, 13, 332
Chain rule, 199, 237
Change of variable, 261, 298
Coefficient of inequality, 311
Common difference, 397, 441
Common ratio, 406, 441
Composite function, 196
  derivative of, 198–199, 237
Compound interest, 412–420, 441
Concave downward, 157, 187
Concave upward, 157, 187
Concavity, 157, 187
Constant function
  derivative of, 101
Constant of integration, 244
Constant of proportionality, 51, 66
Constraints, 372
Consumer's surplus, 316, 325
Continuous compounding of interest, 219–220, 236, 421–426
Continuous function, 79, 345, 388
Coordinate axes, 12, 332
Cost function, 47, 52, 54, 66
Cumulative distribution function, 320, 325

D

Decreasing function, 140, 187
Definite integral, 289, 324
Demand function, 58
Dependent variable, 42, 65
Derivative
   of a composite function, 198–199, 237
   of a constant function, 101
   of a constant times a function, 103
   definition of, 98, 112, 129
   of the difference of two functions, 104
   of exponential functions, 212, 236
   of higher order, 124
   of inverse function, 202, 237
   of logarithmic functions, 214–215, 236
   notation for, 98, 123–124
   partial, 347–353
   of the product of two functions, 109
   of the quotient of two functions, 111
   as a rate of change, 144, 188
   second order, 105
   of the sum of two functions, 104
   symbols for, 98, 123–124
   of x^r, 103
Descartes, René, 69
Difference quotient, 83
Differential, 117
Differential equations, 270
Differentiable, 98, 130
Differentiation,
   implicit, 228
   logarithmic, 232
Differentiation theorems, 101–111
Diffusion curve, 222
Directly proportional, 51, 66
Discontinuous function, 80
Distance, 13, 333, 387
Domain of a function, 65
Domain of a relation, 39

E

e (base of natural logarithm), 78
Effective interest rate, 418, 422, 442
Elasticity of cost, 150
Elimination technique, 29
Endpoint extrema, 152

Equation
   of a line, 21–27
   differential, 270
   of perpendicular bisector of line segment, 26
   of a plane, 335
   of a tangent line, 98
   of a tangent plane, 357, 388
Equilibrium, market, 58, 66
   point, 58, 66
   price, 58, 66
   quantity, 58, 66
Error formula for differential, 118, 150
Exponent, A15
Exponential functions, 204, 235, 250, 279
Extrema
   absolute, 45
   first derivative test for, 152, 187
   of a function of one variable, 45, 151
   of a function of two variables, 360–362
   relative, 151, 187, 360–362
   second derivative test for, 155, 187

F

Factoring, A23
First derivative, 98
First derivative test for relative extrema, 152, 187
First partial derivatives, 348
Fixed costs, 10
Function
   average cost, 47, 66
   composite, 196
   continuous, 79, 345
   cost, 47, 52, 54, 66
   decreasing, 140, 187
   definition of, 40, 65
   demand, 58
   differentiable, 98
   discontinuous, 80
   domain of, 39, 65
   exponential, 204, 235, 250, 253
   extrema of, 45, 151, 360–362
   graph of, 39, 340
   identity, 200
   increasing, 140, 187
   inverse, 200
   limit of, 74, 345
   limited growth, 221, 236

Function—Continued
  linear, 41, 65
  logarithmic, 205, 235
  marginal cost, 120
  marginal revenue, 120
  marginal profit, 120
  maximum value of, 45, 65, 360
  minimum value of, 45, 65, 360
  monotonic, 141
  of more than one variable, 339–347
  natural logarithmic, 207, 236
  of one variable, 40
  polynomial, 76
  profit, 54, 66
  quadratic, 143
  range of, 39, 65
  revenue, 54, 66
  square root, 200
  supply, 58
  of two variables, 339–347
  unlimited growth/decay, 221, 236
Fundamental Theorem of Calculus, 291–293

G

Gauss, Carl Friedrich, 239
Geometric sequence, 406
General power rule for antidifferentiation, 247, 279
General power rule for differentiation, 108
Graph
  of a function, 39, 340, 387
  of a line, 24–25
  of a relation, 39
Greater than, 5
Growth
  exponential, 217
  limited, 221, 236, 275
  unlimited, 221, 236, 274

H

Half-life, 218
Higher-order derivatives, 124

I

Identity function, 200
Implicit differentiation, 228

Increasing function, 140, 187
Indefinite integral, 244, 278
Independent variable, 42, 65
Inequality, 5
  properties of, 5–6, 64
  solution set of, 6
Inflection point, 157, 187
Initial conditions, 272
Instantaneous rate of change, 218
Integral
  definite, 289
  indefinite, 244
  notation for, 244
  Riemann, 289
Integrand, 244
Integration
  constant of, 244
  indefinite, 244
  by parts, 265, 279
  by substitution, 261
Interest
  add-on, 437
  compound, 412–420, 441
  compounded continuously, 219–220, 236, 421–426
  effective rate of, 418, 422, 442
  on periodic savings, 426–433
  rate, 412
  rate per period, 416
  simple, 412
Intersection of lines, 28
Inverse function, 200
Inverse relation, 200
Irrational numbers, 4

K

Kovalevsky, Sonya, 281

L

Lagrange, Joseph Louis, 392
Lagrange multipliers, 372, 374, 389
Laws of exponents, A15
Learning curve, 222
Least squares approximation, 380–385
Left-hand coordinate system, 332–333
Leibniz, Gottfried Wilhelm, 191
Leibniz notation, 123

Less than, 4
Limit of a function, 74, 345
Limited growth functions, 221, 236, 275
Line
  equation of, 22, 24, 65
  point-slope form for equation of, 22, 65
  slope of, 64–65
  slope-intercept form for equation of, 22, 65
  two-point form for equation of, 22
Linear
  approximation, 118, 357
  depreciation, 51
  equation in three variables, 335
  equation in two variables, 22, 65
  function, 41, 65
  least squares approximation, 381
  system, 28–29
Local
  extrema, 151, 360
  maximum, 151, 187, 360, 362, 389
  minimum, 151, 187, 360, 362, 389
Logarithmic differentiation, 232
Logarithmic functions, 205, 235
Logarithmic properties, 206, 236
Logistic equation, 223, 236
Lorentz curve, 311
Lower sum, 289

M

Marginal
  analysis, 120
  cost, 120, 129
  profit, 120, 129
  revenue, 120, 129
Market equilibrium point, 58, 66
Maximum value of a function, 45, 65
  with constraints, 371–380
Method of least squares, 380–385
Midpoint, 14, 64, 334, 387
Minimum value of a function, 45, 65
  with contraints, 371–380
Montonic function, 141

N

Natural logarithmic function, 207, 236
Neighborhood of (a, b), 360, 389

Newton, Isaac, 133
Newton quotient, 85, 128
Noether, Emmy, 281
Normal equations, 384
Notation for
  antiderivative, 244
  definite integral, 289
  derivative, 98, 123–124
  indefinite integral, 244
  limit of a function, 74, 345
  partial derivatives, 349, 351, 388
  Riemann integral, 289
Numbers
  irrational, 4
  rational, 4
  real, 4

O

Order of a differential equation, 272
Ordinary annuity, 429
Origin, 4, 12

P

Parabola, 143
Paraboloid of revolution, 337
Parallel lines, 17, 65
Partial derivative, 347–353, 388
Particular solution of a differential equation, 272
Partition of an interval, 287
Percentage rate of change, 145, 188
Periodic savings, 426–433, 442
Perpendicular lines, 17, 65
Point
  in three dimensions, 332
  in two dimensions, 12
  of the number line, 4
Point of inflection, 157, 187
Point-slope form for equation of a line, 22
Polynomial function, 76, A20
Power rule, 103, 108
Present value, 433–434, 442
Principal nth root, A17
Probability density function, 320, 325
Producer's surplus, 316, 325
Profit function, 54

Q

Quadratic equation, A24
Quadratic formula, A25
Quadratic function, 143
Quadratic least squares, 385

R

Radicals, A18
Range of a function, 65
Range of a relation, 39
Rate of change, 144, 188
Rational exponents, A18
Rational numbers, 4
Real numbers, 4
Rectangular coordinate system, 13, 332
Regular partition, 287
Related rates, 183
Relation, 39
Relative extrema, 151
  first-derivative test for, 152, 187
  second-derivative test for, 155, 187
Relative maximum value, 151, 187
  of a function of two variables, 360, 362, 389
Relative minimum value, 151, 187
  of a function of two variables, 360, 362, 389
Revenue function, 54
Riemann, George F. B., 327
Riemann integral, 289, 324
Right-hand coordinate system, 332–333

S

Saddle point, 361
Second derivative, 105
Second derivative test, 155, 187
Second partial derivative, 351
Separable differential equation, 273
Sequence
  arithmetic, 397
  geometric, 406
Simple annuity, 429
Simple interest, 412
Slope
  of a line, 16, 64–65
  of a tangent line, 98, 129

Slope-intercept form for equation of a line, 22
Solution of a differential equation, 270
Solution set of an inequality, 6
Solution set of a linear system, 28
Special algebraic products, A22
Substitution technique
  for integration, 261
  to solve linear systems, 30
Sum
  of an arithmetic sequence, 399, 400, 441
  of a geometric sequence, 407, 441
Supply function, 58
Symbol
  for antiderivative, 244
  for definite integral, 289
  for derivative, 98, 123–124
  for indefinite integral, 244
  for limit of a function, 74
  for partial derivative, 349, 351, 388
  for Riemann integral, 289

T

Tangent line, 98
Tangent plane, 357, 388
Tests for relative extrema, 152, 155, 187
Three dimensions, 332
Total differential, 355, 388
Trace of a surface in a plane, 335
Two-point form for an equation of a line, 22
Two dimensions, 13

U

Unlimited growth functions, 221, 236, 274

V

Variable
  dependent, 42, 65
  independent, 42, 65
Velocity, 114
Volume
  of a cone, 184
  by disk method, 321–324, 325
  of a sphere, 43
von Neumann, John, 445

X

x-axis, 12, 332
x-coordinate, 12, 332
x-intercept, 25 (Ex. 3)

Y

y-axis, 12, 332
y-coordinate, 12, 332
y-intercept, 25 (Ex. 3)

Z

z-axis, 332
z-coordinate, 332

DIFFERENTIATION THEOREMS AND APPLICATIONS

1. If $f(x) = c$, c a constant, then $f'(x) = 0$.
2. If $f(x) = cg(x)$, c a constant, then $f'(x) = cg'(x)$.
3. If $F(x) = f(x) + g(x)$, then $F'(x) = f'(x) + g'(x)$.
4. If $F(x) = f(g)g(x)$, then $F'(x) = f(x)g'(x) + f'(x)g(x)$.
5. If $F(x) = \dfrac{f(x)}{g(x)}$, then
$$F'(x) = \frac{g(x)f'(x) - f(x)g'(x)}{[g(x)]^2}.$$
6. If $F(x) = f(g(x))$, then $F'(x) = f'(g(x)) \cdot g'(x)$.
7. For inverse functions f and g, $g'(x) = \dfrac{1}{f'(g(x))}$.
8. If $F(x) = [g(x)]^r$, r a constant, then
$$F'(x) = r[g(x)]^{r-1}g'(x).$$
   In particular, if $f(x) = x^r$, then $f'(x) = rx^{r-1}$.
9. If $F(x) = e^{g(x)}$, then
$$F'(x) = e^{g(x)}g'(x).$$
   In particular, if $F(x) = e^x$, then $F'(x) = e^x$.
10. If $F(x) = \ell n\, g(x)$, then
$$F'(x) = \frac{g'(x)}{g(x)}.$$
    In particular, if $F(x) = \ell n\, x$, then $F'(x) = \dfrac{1}{x}$.
11. If $f(x) = b^x$, then $f'(x) = b^x(\ell n\, a)$.
12. If $f(x) = \log_b x$, then
$$f'(x) = \frac{1}{x}(\log_b e) = \frac{1}{x(\ell n\, b)}.$$
13. If $f'(x) > 0$ for each x in some interval, then f is an *increasing function* in the interval. Similarly, if $f'(x) < 0$, then f is a *decreasing function* in the interval.
14. If $f''(x) > 0$ for each x in some interval, then the graph of f is *concave upward* (bends up) in the interval. Similarly, if $f''(x) < 0$, then the graph of f is *concave downward* in the interval.
15. If $s(x)$ is the distance an object travels in time x, then $v(x) = s'(x)$ is the *velocity* and $v'(x) = s''(x)$ is the *acceleration* at time x.